McGraw-Hill Ryerson
SCIENCE LINKS 9

Authors

Beth Lisser
B.Sc. (Hons.), B.Ed., M.Ed.
Peel District School Board
Brampton, Ontario

Barbara Nixon-Ewing
B.Sc. (Hons.), B.Ed.
Ontario Institute for Studies in Education
Seconded from Toronto District
 School Board
Toronto, Ontario

Sandy Searle
B.Sc. (Hons.), MBA
Calgary Board of Education
Calgary, Alberta

Contributing Authors

Jonathan Bocknek
B.A.
Science Writer

Lois Edwards
Ph.D.
Science Writer

Glen Hutton
B.Sc., B.Ed., M.Ed.
Science Writer

Michael Lattner
B.Sc., M.Sc.F., B.Ed.
Algonquin & Lakeshore Catholic District
 School Board
Napanee, Ontario

Natasha Marko
B.Sc., M.Sc., M.A.
Science Writer

Rob Smythe (ret.)
B.Sc., M.Sc., B.Ed.
Halton District School Board

Christine Weber
B.Sc.
Science Writer

Program Consultant

Beth Lisser
B.Sc. (Hons.), B.Ed., M.Ed.
Peel District School Board
Brampton, Ontario

Curriculum and Pedagogical Consultants

Tigist Amdemichael
B.Sc., B.A., B.Ed., M.Ed.
Toronto District School Board
Toronto, Ontario

Stephanie Grant
B.Sc., B.Ed.
Dufferin-Peel Catholic District School Board
Mississauga, Ontario

Assessment Consultant

Katy Farrow
B.Sc., B.Ed., M.Ed.
Thames Valley District School Board
London, Ontario

Literacy Consultants

Steve Bibla
B.Sc., B.Ed.
Toronto District School Board
Toronto, Ontario

Kay Moreland Stephen
B.Sc., B.Ed.
Ottawa Carleton Catholic School Board
Ottawa, Ontario

Differentiated Instruction Consultant

Nadine Morrison
B.Sc., B.A., B.Ed., French & I.B. Cert.
Hamilton Wentworth District School Board
Hamilton, Ontario

ELL Consultants

Wendy Campbell
B.A., B.Ed.
Waterloo Region District School Board

Maureen Innes
B.A.
Nipissing University

Al Tordjman
B.A., B.Ed.
Waterloo Region District School Board

Advisors

Dave Erb (ret.)
B.Sc., B.Ed.
Keewatin-Patricia District School Board
Red Lake, Ontario

Katy Farrow
B.Sc., B.Ed., M.Ed.
Thames Valley District School Board
London, Ontario

Patricia Gaspar
B.Sc. (Hons.), B.Ed.
York Region District School Board
King City, Ontario

Keith Gibbons
B.Sc., B.Ed.
London District Catholic School Board
London, Ontario

John Hallett
B.Sc., B.Ed.
Peel District School Board
Caledon, Ontario

Michael Lattner
B.Sc., M.Sc.F., B.Ed.
Algonquin & Lakeshore Catholic District
 School Board
Napanee, Ontario

Mirella Sanwalka
B.Sc., B.Ed.
York Region District School Board
Newmarket, Ontario

Toronto Montréal Boston Burr Ridge, IL Dubuque, IA Madison, WI New York San Francisco
St. Louis Bangkok Bogotá Caracas Kuala Lumpur Lisbon London Madrid Mexico City
Milan New Delhi Santiago Seoul Singapore Sydney Taipei

McGraw-Hill Ryerson

Copies of this book may be obtained by contacting:
McGraw-Hill Ryerson Limited

e-mail:
orders@mcgrawhill.ca

Toll-free fax:
1-800-463-5885

Toll-free call:
1-800-565-5758

or by mailing your order to:
McGraw-Hill Ryerson Limited
Order Department
300 Water Street
Whitby, ON L1N 9B6

Please quote the ISBN and title when placing your order.

Science Links 9

Copyright © 2009, McGraw-Hill Ryerson Limited, a Subsidiary of The McGraw-Hill Companies. All rights reserved. No part of this publication may be reproduced or transmitted in any form or by any means, or stored in a data base or retrieval system, without the prior written permission of McGraw-Hill Ryerson Limited, or, in the case of photocopying or other reprographic copying, a licence from The Canadian Copyright Licensing Agency (Access Copyright). For an Access Copyright licence, call toll free to 1-800-893-5777.

The information and activities in this textbook have been carefully developed and reviewed by professionals to ensure safety and accuracy. However, the publisher shall not be liable for any damages resulting, in whole or in part, from the reader's use of the material. Although appropriate safety procedures are discussed and highlighted throughout the textbook, the safety of students remains the responsibility of the classroom teacher, the principal, and the school board district.

ISBN-13: 978-0-07-072690-1
ISBN-10: 0-07-072690-6

1 2 3 4 5 6 7 8 9 0 TCP 8 7 6 5 4 3 2 1 0 9

Printed and bound in Canada

Care has been taken to trace ownership of copyright material contained in this text. The publishers will gladly accept any information that will enable them to rectify any reference or credit in subsequent printings.

PUBLISHER: Diane Wyman
PROJECT MANAGER: Susan Girvan
DEVELOPMENTAL EDITORS: Jon Bocknek, Lois Edwards, Julie Karner, Christine Weber
MANAGING EDITOR: Crystal Shortt
SUPERVISING EDITOR: Janie Deneau
COPY EDITORS: Sheila Harris, Paula Pettitt-Townsend
PHOTO RESEARCH/PERMISSIONS: Linda Tanaka
ART BUYING: Pronk&Associates
REVIEW COORDINATOR: Jennifer Keay, Alexandra Savage-Ferr
EDITORIAL ASSISTANT: Michelle Malda
MANAGER, PRODUCTION SERVICES: Yolanda Pigden
PRODUCTION COORDINATOR: Sheryl MacAdam
INSTRUCTIONAL DESIGN CONCEPT: Jonathan Bocknek
INTERIOR DESIGN: Liz Harasymczuk
COVER DESIGN: Pronk & Associates
ELECTRONIC PAGE MAKE-UP: Pronk&Associates
COVER IMAGES:
Bottom right image ©NASA/JPL-Calech/University of Arizona
Bottom left image ©Peter Burian/Corbis
Top right image ©Larry Lilac/Alamy
Top middle image ©Oleg Shpak/Alamy
Top left image ©Walter Geiersperger/Corbis

The pages within this book were printed on paper containing 30% Post-Consumer Fiber.

Acknowledgements

Pedagogical Reviewers

Rosa Bellissimo
Toronto Catholic District School Board
North York, Ontario

Dave Black
Upper Canada District School Board
Brockville, Ontario

Monica Franciosa
York Catholic District School Board
Richmond Hill, Ontario

Elizabeth Frawley
Durham Catholic District School Board
Whitby, Ontario

Mari Hakim
Greater Essex County District
 School Board
Windsor, Ontario

John Hallett
Peel District School Board
Caledon, Ontario

Stephen Jacobs
Dufferin-Peel Catholic District
 School Board
Mississauga, Ontario

Benjamin Law
Unionville, Ontario
York Region District School Board

Dimitrios Melegos
Durham District School Board
Pickering, Ontario

Christopher Meyer
Toronto District School Board
North York, Ontario

David Mitchinson
Simcoe County District School Board
Barrie, Ontario

Robert Noble
Toronto Catholic District School Board
Scarborough, Ontario

Stephen Park
Durham District School Board
Oshawa, Ontario

Jason Pavelich
Peel District School Board
Mississauga, Ontario

Sharon Ramlochan
Toronto District School Board
Scarborough, Ontario

Jim Raso
Niagara Catholic District School Board
Welland, Ontario

Mary-Ann Rupcich
Dufferin-Peel Catholic District
 School Board
Mississauga, Ontario

Harvey Shear, Ph.D.
Department of Geography
University of Toronto Mississauga
Mississauga, Ontario

Philip Snider
Lambton-Kent District School Board
Chatham, Ontario

Brian Snow
Grand Erie District School Board
Simcoe, Ontario

Patricia Thomas
Ottawa-Carleton District School Board
Ottawa, Ontario

Frank Villella
Hamilton Wentworth Catholic District
 School Board
Hamilton, Ontario

Accuracy Reviewers

Jenna Dunlop (Unit 1)
Ph.D., M.B.A.
Science Writer
Toronto, Ontario

R. Tom Baker (Unit 2)
B.Sc., (Hons.), Ph.D.
Canada Research Chair in Catalysis
 Science for Energy Applications
Director, Centre for Catalysis Research
 and Innovation
University of Ottawa
Ottawa, Ontario

Paul Delaney (Unit 3)
B.Sc. (Hons.), M.Sc.
Senior Lecturer, Dept. of Physics
 and Astronomy
Director, Division of Natural Science
York University
Toronto, Ontario

T.J. Elgin Wolfe (Unit 4)
M.Ed.
Professor, Ontario Institute for Studies
 in Education
University of Toronto
Toronto, Ontario

Bias and Equity Reviewer

Nancy Christoffer
Scarborough, Ontario

Lab Testers

Clyde Ramlochan
Toronto District School Board
North York, Ontario

Julie Sylvestri
Hamilton Wentworth Catholic District School Board
Hamilton, Ontario

Safety Reviewer

Jim Agban
Past Chair, STAO Safety Committee
Mississauga, Ontario

Aboriginal Content Reviewer

Francis McDermott
Shabot Obaadjiwan Algonquin
 First Nation
Shawville, Quebec

Contents

Exploring *Science Links 9* x
Safety in Your Science Classroom xii

Unit 1 Sustainable Ecosystems 2

Unit 1 At A Glance 4
Get Ready for Unit 1 6

Topic 1.1 What are ecosystems, and why do we care about them? 8
Topic 1.2 How do interactions supply energy to ecosystems? 18
          Investigation 1A: Plot the Pathway 26
Topic 1.3 How do interactions in ecosystems cycle matter? 28
          Strange Tales of Science: Journey of an Immortal Carbon Atom .. 36
Topic 1.4: What natural factors limit the growth of ecosystems? 40
          Investigation 1B: Investigating Limiting Factors
                         for Algae Growth 46
          Strange Tales of Science: Bacteria Take Over the World 48
Topic 1.5 How do human activities affect ecosystems? 50
          Making a Difference: Dayna Corelli and Rebekah Parker 58
          Investigation 1C: Human Activity in a Local Ecosystem 59
Topic 1.6 How can our actions promote sustainable ecosystems? 62
          Making a Difference: Yvonne Su and Chaminade College 72
          Case Study Investigation: Securing a Bright Future
                         for Songbirds 74
          Investigation 1D: Investigating a Local Environmental Project ... 76

Science at Work: Fisheries Technician 78

Unit 1 Summary 80

Unit 1 Projects   **Inquiry Investigation:** Investigating Compost 82
                   An Issue to Analyze: Going Greener 83

Unit 1 Review 84

Unit 2 Exploring Matter 88

Unit 2 At A Glance 90
Get Ready for Unit 2 92

Topic 2.1 In what ways do chemicals affect your life? 94
  Strange Tales of Science: Minding Scientific Inquiry 102

Topic 2.2 How do we use properties to help us describe matter? 104
  Investigation 2A: Physical and Chemical Properties
  of Substances in the Home. 110

Topic 2.3 What are pure substances and how are they classified? 112
  Investigation 2B: Comparing the Physical Properties
  of Metals with Non-Metals. 118

Topic 2.4 How are properties of atoms used to organize elements
  into the periodic table? 120

Topic 2.5: In what ways do scientists communicate about
  elements and compounds? 130
  Case Study Investigation: Salt of the Earth 136

Topic 2.6: What are some of the characteristics
  and consequences of chemical reactions? 140
  Investigation 2C: Identifying an Unknown Gas 148
  Making a Difference: Adrienne Dulmering and
  Sarah Mediouni 150

Science At Work: Baker 152

Unit 2 Summary 154

Unit 2 Projects **Inquiry Investigation:** Rust Formation 156
  An Issue to Analyze: Evaluating the Use of Road Salt 157

Unit 2 Review 158

Unit 3 Space Exploration 162
Unit 3 At A Glance 164
Get Ready for Unit 3 166

Topic 3.1: What do we see when we look at the sky? 168
          Case Study Investigations: A Tale of the Bear 178
          Investigation 1A: Make a Star-Finder Wheel 180

Topic 3.2: What are the Sun and the Moon, and how are they linked to Earth? 182
          Case Study Investigation: Solar Storms 192

Topic 3.3: What has space exploration taught us about our solar system? ... 196

Topic 3.4: What role does Canada play in space exploration? 208
          Strange Tales of Science: The Phantom Torso 214
          Investigation 3B: You, Robot 216

Science at Work: Digital Compositor 218

Topic 3.5: How do we benefit from space exploration? 220
          Making a Difference: Shelby Mielhausen and Nishant Balakrishnan 228

Unit 3 Summary 230

Unit 3 Projects   **Inquiry Investigation:** Space Thirst 232
          An Issue to Analyze: The Cost and Benefits of Space Travel 233

Unit 3 Review 234

Unit 4 Electrical Applications 238

Unit 4 At A Glance 240
Get Ready for Unit 4 242

Topic 4.1: How do the sources used to generate
electrical energy compare? 244
Investigation 4A: Leapin' 'Lectricity 250

Topic 4.2: What are charges, and how do they behave? 252
Investigation 4B: Charging Materials 262

Topic 4.3: How can objects become charged and discharged? 264
Investigation 4C: Materials for Lightning Rods 272

Topic 4.4: How can people control and use the movement of charges? 274
Making a Difference: Vishvek Babbar and Ghufran Siddiqui ... 284
Investigation 4D: Using Ammeters and Voltmeters 286
Investigation 4E: Observing the Effects
of Resistance on Current 288
Investigation 4F: Potential Difference and Current 290

Topic 4.5: What are series and parallel circuits,
and how are they different? 292
Investigation 4G: Observing Characteristics of Series Circuits .. 298
Investigation 4H: Observing Characteristics of Parallel Circuits .. 300
Strange Tales of Science: Sparks of Genius 302

Topic 4.6: What features make an electrical circuit practical and safe? 304

Topic 4.7: How can we conserve our use of electrical energy at home? 314
Case Study Investigation: People Power 320
Making a Difference: Katie Pietzakowski and Patrick Bowman .. 322

Science at Work: Electronic Instrumentation Technician 324

Unit 4 Summary 326

Unit 4 Projects   **Inquiry Investigation:** Energy Savings 328
An Issue to Analyze: Choosing Energy Sources
in Ontario 329

Unit 4 Review 330

Guide to the Toolkits and Appendix 334

Science Skills Toolkit 1: Analyzing Issues—STSE 335
Science Skills Toolkit 2: Scientific Inquiry 339
Science Skills Toolkit 3: Technological Problem Solving 344
Science Skills Toolkit 4: Estimating and Measuring 346
Science Skills Toolkit 5: Precision and Accuracy 352
Science Skills Toolkit 6: Scientific Drawing 354
Science Skills Toolkit 7: Using Models and Analogies in Science 356
Science Skills Toolkit 8: How to Do a Research-Based Project 358
Science Skills Toolkit 9: Using Electric Meters 362
Science Skills Toolkit 10: Creating Data Tables 365

Numeracy Skills Toolkit 1: Exponents of Scientific Notation 366
Numeracy Skills Toolkit 2: Significant Digits and Rounding 367
Numeracy Skills Toolkit 3: The Metric System 368
Numeracy Skills Toolkit 4: Organizing and Communicating Scientific Results with Graphs 370
Numeracy Skills Toolkit 5: The GRASP Problem-Solving Method 376

Literacy Skills Toolkit 1: Preparing for Reading 377
Literacy Skills Toolkit 2: Reading Effectively 381
Literacy Skills Toolkit 3: Reading Graphic Text 385
Literacy Skills Toolkit 4: Word Study 388
Literacy Skills Toolkit 5: Organizing Your Learning: Using Graphic Organizers 390

Appendix: Properties of Common Substances 396

Glossary 400

Index 422

Credits 427

Periodic Table 434

Exploring *Science Links 9*

Answer the Questions, Reveal a Quote

In your notebook, answer the questions on these two pages as you explore *Science Links 9*. Use the numbered letters in each answer to find the missing words in the quote.

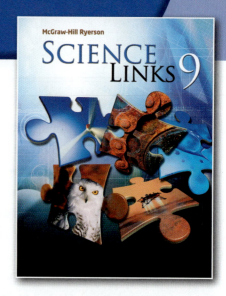

On September 10th, 2006, 48 Canadian youth organizations banded together to form the Canadian Youth Climate Coalition. The coalition members work in their communities and with students around the world to help change the way humans affect global climate. They use the Internet and gatherings both large and small to achieve their goals.

Our culture, health, security, environment, prosperity, and future are at stake. This is not a dress rehearsal. … We did not create this crisis, but …

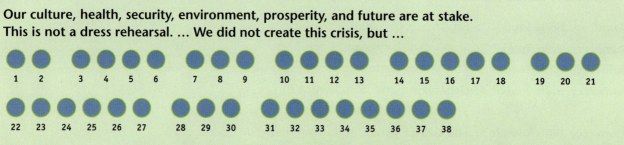

—Declaration, Canadian Youth Climate Coalition

The Canadian Youth Climate Coalition combines energy and creativity with scientific knowledge to help solve real world problems. How can you apply this approach to your own life?

Introducing *Science Links 9*

You will study four units in *Science Links 9*. What are they?

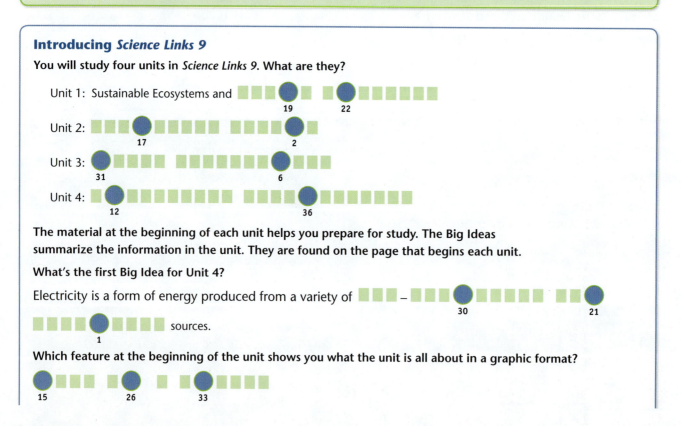

The material at the beginning of each unit helps you prepare for study. The Big Ideas summarize the information in the unit. They are found on the page that begins each unit.

What's the first Big Idea for Unit 4?

Electricity is a form of energy produced from a variety of ▯▯▯–▯▯▯▯▯▯▯▯ ▯▯▯▯▯▯▯ sources.

Which feature at the beginning of the unit shows you what the unit is all about in a graphic format?

The information in each unit is organized into topics. Each topic asks a question. What question does Topic 2.3 ask?

What are pure substances and  ?

Doing Science

What piece of safety equipment does this icon represent?

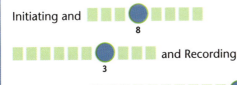

Each investigation features a Skill Check that tells you which science skills will be featured in the investigation. **What are these skills?**

Initiating and

Checking Your Learning

What feature gives you a chance to check your understanding as you read through a topic?

At the end of each topic, what are you asked to do?

Review the

It's important that you understand scientific terms and know how to use the words correctly. **Where would you look in the textbook if you wanted to review a definition?**

The

Finding Science in Unexpected Places

Each unit features a Canadian working in a career that uses scientific knowledge. **What is this feature called?**

Each unit features Canadians students who are using their scientific knowledge to do something in their community. **What is this feature called?**

Every unit tells you a "Strange Tale" about science. **What is the tale in Unit 3?**

The

Tools for Success

At the back of this textbook, there are three groups of Toolkits that explain and demonstrate skills that will help you in all your learning, both in and outside of school.

What are these toolkits?

Scientific Inquiry, Numeracy, and

Your teacher has asked you to create a data table. **Where can you find out how to do this?**

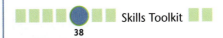 Skills Toolkit

Finally, you need to make a graph of your data. **Where can you find out how to do this?**

Safety in Your Science Classroom

Become familiar with the following safety rules and procedures. It is up to you to use them and your teacher's instructions to make your activities and investigations in *Science Links* 9 safe and enjoyable. Your teacher will give you specific information about any other special safety rules that need to be used in your school.

1. **Working with your teacher …**
   - Listen carefully to any instructions your teacher gives you.
   - Inform your teacher if you have any allergies, medical conditions, or other physical problems that could affect your work in the science classroom. Tell your teacher if you wear contact lenses or a hearing aid.
   - Obtain your teacher's approval before beginning any activity you have designed for yourself.
   - Know the location and proper use of the nearest fire extinguisher, fire blanket, first-aid kit, and fire alarm.

2. **Starting an activity or investigation …**
   - Before starting an activity or investigation, read all of it. If you do not understand how to do a step, ask your teacher for help.
   - Be sure you have checked the safety icons and have read and understood the safety precautions.
   - Begin an activity or investigation only after your teacher tells you to start.

3. **Wearing protective clothing …**
   - When you are directed to do so, wear protective clothing, such as a lab apron and safety goggles. Always wear protective clothing when you are using materials that could pose a safety problem, such as unidentified substances, or when you are heating anything.
   - Tie back long hair, and avoid wearing scarves, ties, or long necklaces.

4. **Acting responsibly …**
   - Work carefully with a partner and make sure your work area is clear.
   - Handle equipment and materials carefully.
   - Make sure stools and chairs are resting securely on the floor.
   - If other students are doing something that you consider dangerous, report it to your teacher.

5. **Handling edible substances …**
   - Do not chew gum, eat, or drink in your science classroom.
   - Do not taste any substances or draw any material into a tube with your mouth.

6. Working in a science classroom …
- Make sure you understand all safety labels on school materials or those you bring from home. Familiarize yourself, as well, with the WHMIS symbols and the special safety symbols used in this book, found on page xv.
- When carrying equipment for an activity or investigation, hold it carefully. Carry only one object or container at a time.
- Be aware of others during activities and investigations. Make room for students who may be carrying equipment to their work stations.

7. Working with sharp objects …
- Always cut away from yourself and others when using a knife or razor blade.
- Always keep the pointed end of scissors or any pointed object facing away from yourself and others if you have to walk with such objects.
- If you notice sharp or jagged edges on any equipment, take special care with it and report it to your teacher.
- Dispose of broken glass as your teacher directs.

8. Working with electrical equipment …
- Make sure your hands are dry when touching electrical cords, plugs, or sockets.
- Pull the plug, not the cord, when unplugging electrical equipment.
- Report damaged equipment or frayed cords to your teacher.
- Place electrical cords where people will not trip over them.

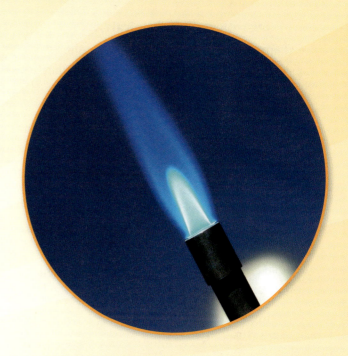

9. Working with heat …
- When heating an item, wear safety goggles and any other safety equipment that the text or your teacher advises.
- Always use heatproof containers.
- Point the open end of a container that is being heated away from yourself and others.
- Do not allow a container to boil dry.
- Handle hot objects carefully. Be especially careful with a hot plate that looks as though it has cooled down.
- If you use a Bunsen burner, make sure you understand fully how to light and use it safely.
- If you do receive a burn, inform you teacher, and apply cold water to the burned area immediately.

10. Working with various chemicals …

- If any part of your body comes in contact with a substance, wash the area immediately and thoroughly with water. If you get anything in your eyes, do not touch them. Wash them immediately and continuously for 15 minutes, and inform your teacher.
- Always handle substances carefully. If you are asked to smell a substance, never smell it directly. Hold the container slightly in front of and beneath your nose, and waft the fumes toward your nostrils.
- Hold containers away from your face when pouring liquids.

11. Working with living things …

On a field trip:

- Try not to disturb the area any more than is absolutely necessary.
- If you move something, do it carefully, and always replace it carefully.
- If you are asked to remove plant material, remove it gently, and take as little as possible.

In the classroom:

- Make sure that living creatures receive humane treatment while they are in your care.
- If possible, return living creatures to their natural environment when your work is complete.

12. Cleaning up in the science classroom …

- Clean up any spills, according to your teacher's instructions.
- Clean equipment before you put it away.
- Wash your hands thoroughly after doing an activity or an investigation.
- Dispose of materials as directed by your teacher. Never discard materials in the sink unless your teacher requests it.

13. Designing and building …

- Use tools safely to cut, join, and shape objects.
- Handle modelling clay correctly. Wash your hands after using modelling clay.
- Follow proper procedures when using mechanical systems and studying their operations.
- Use special care when observing and working with objects in motion.
- Do not use power equipment such as drills, sanders, saws, and lathes unless you have specialized training in handling such tools.

Safety Symbols

Science Links 9 Safety Symbols

The following safety symbols are used in *Science Links 9* to alert you to possible dangers. Be sure you understand each symbol used in an activity or investigation before you begin.

 Disposal Alert
This symbol appears when care must be taken to dispose of materials properly.

 Thermal Safety
This symbol appears as a reminder to use caution when handling hot objects.

 Sharp Object Safety
This symbol appears when a danger of cuts or punctures caused by the use of sharp objects exists.

 Electrical Safety
This symbol appears when care should be taken when using electrical equipment.

 Skin Protection Safety
This symbol appears when use of caustic chemicals might irritate the skin or when contact with micro-organisms might transmit infection.

 Clothing Protection Safety
A lab apron should be worn when this symbol appears.

 Fire Safety
This symbol appears when care should be taken around open flames.

 Eye Safety
This symbol appears when a danger to the eyes exists. Safety goggles should be worn when this symbol appears.

 Fume Safety
This symbol appears when chemicals or chemical reactions could cause dangerous fumes.

 Chemical Safety
This symbol appears when chemicals used can cause burns or are poisonous if absorbed through the skin.

Instant Practice—Safety Symbols

Find four of the *Science Links 9* safety symbols in activities or investigations in this textbook. For each symbol, identify the possible dangers in the activity or investigation that the symbol refers to.

WHMIS Symbols

Look carefully at the WHMIS (Workplace Hazardous Materials Information System) safety symbols shown here. The WHMIS symbols are used throughout Canada to identify dangerous materials. Make certain you understand what these symbols mean. When you see these symbols on containers, use safety precautions.

 Compressed Gas

 Flammable and Combustible Material

 Oxidizing Material

 Corrosive Material

 Poisonous and Infectious Material Causing Immediate and Serious Toxic Effects

 Poisonous and Infectious Material Causing Other Toxic Effects

 Biohazardous Infectious Material

 Dangerously Reactive Material

Instant Practice—Safety Symbols

Hydrogen gas is stored in containers under pressure. This gas is highly flammable.

1. What two symbols would you expect to see on a label for hydrogen gas?
2. Describe the following.
    a) the risks illustrated by the two symbols
    b) precautions someone would need to take when working with the gas
    c) where it could be safely stored
    d) first aid or emergency treatment
3. If you did not know the answer to part d., where would you find this information?

Unit 1: Sustainable Ecosystems and Human Activity

Big Ideas

- Ecosystems consist of a variety of components, including, in many cases, humans.
- The sustainability of ecosystems depends on balanced interactions between their components.
- Human activity can affect the sustainability of aquatic and terrestrial ecosystems.

From "Waiting On The World To Change" by John Mayer

Me and all my friends
We're all misunderstood
They say we stand for nothing and
There's no way we ever could

Now we see everything that's going wrong
With the world and those who lead it
We just feel like we don't have the means
To rise above and beat it

So we keep waiting
Waiting on the world to change
We keep on waiting
Waiting on the world to change

It's hard to beat the system
When we're standing at a distance
So we keep waiting
Waiting on the world to change

Instead of waiting for change that might never come, many people are choosing to be the change they are waiting for.

How is the person in the photo serving as an agent of change in the world?

Unit 1 At a Glance

In this unit you will learn about the characteristics of terrestrial and aquatic ecosystems. You will also learn about the interdependencies within and between these two types of ecosystems. You will analyze the impact of human activity on these ecosystems, and you will assess the effectiveness of selected initiatives on their sustainability.

Think about answers to each question as you work through the topic.

Topic 1.1: What are ecosystems, and why do we care about them?

Key Concepts
- Ecosystems are about connections.
- Ecosystems are made up of biotic (alive) and abiotic (not alive) parts that interact.
- Interactions between terrestrial (land) ecosystems and aquatic (water) ecosystems keep all ecosystems healthy.

Sustainable Ecosystems and Human Activity

Topic 1.6: How can our actions promote sustainable ecosystems?

Key Concepts
- We must understand and commit to sustainability.
- We must understand the link between biodiversity and sustainability.
- Our actions can maintain or rebuild sustainable ecosystems.
- You can choose actions that benefit ecosystems now and for the future.

Topic 1.5: How do human activities affect ecosystems?

Key Concepts
- We cannot always accurately predict the consequences of our actions.
- Introduced species can affect the health of ecosystems.
- Pollutants from human activities can travel within and between ecosystems.

Topic 1.2: How do interactions supply energy to ecosystems?

Key Concepts
- Photosynthesis stores energy, and cellular respiration releases energy.
- Producers transfer energy to consumers through food chains and food webs.
- Interactions are needed to provide a constant flow of energy for living things.

Topic 1.3: How do interactions in ecosystems cycle matter?

Key Concepts
- Abiotic and biotic interactions cycle matter in terrestrial and aquatic ecosystems.
- Photosynthesis and cellular respiration cycle carbon and oxygen in ecosystems.
- Human activities can affect ecosystems by affecting nutrient cycles.

Topic 1.4: What natural factors limit the growth of ecosystems?

Key Concepts
- Ecosystem growth is limited by the availability of resources.
- Abiotic and biotic factors limit populations in ecosystems.

Looking Ahead to the Unit 1 Project
At the end of this unit, you will do a project. The **Inquiry Investigation** examines the effects of compost on the growth of plants in a terrestrial ecosystem. The **Issue to Analyze** challenges you to make some lifestyle changes to reduce your environmental impact. Read pages 82–83. With tips from your teacher, start your project planning folder now.

Get Ready for Unit 1

Concept Check

1. Examine the forest in the picture below. A forest is an example of an ecosystem that contains living (biotic) parts and non-living (abiotic) parts that interact. Make a table like the one below with the headings "Biotic Parts" and "Abiotic Parts." List parts of the forest ecosystem under the correct heading in the table.

Biotic	Abiotic

2. In your notebook, copy and complete each sentence below, based on what you learned about ecosystems in earlier grades. (Do not write in this textbook.)
   a) A maple tree is a *producer* because...
   b. A chipmunk is a *consumer* because...

3. Name two other producers and two other consumers shown in the picture of the forest ecosystem.

4. Use the five words in the box to state one reason why plants are important to all life on Earth.

| oxygen | leaves | photosynthesis |
| food | plants | |

5. Choose one of the following events. Create a concept map to show which parts of a forest ecosystem would be affected by the event.
   a) A forest fire rages through the forest.
   b) A logging company clear-cuts the trees in the forest.
   c) A beaver builds a dam in a nearby pond.
   d) Hunters kill all of the wolves in the area.

6. One forest ecosystem food chain is shown below. Use some of the plants and animals in the forest ecosystem picture to draw a different forest ecosystem food chain.

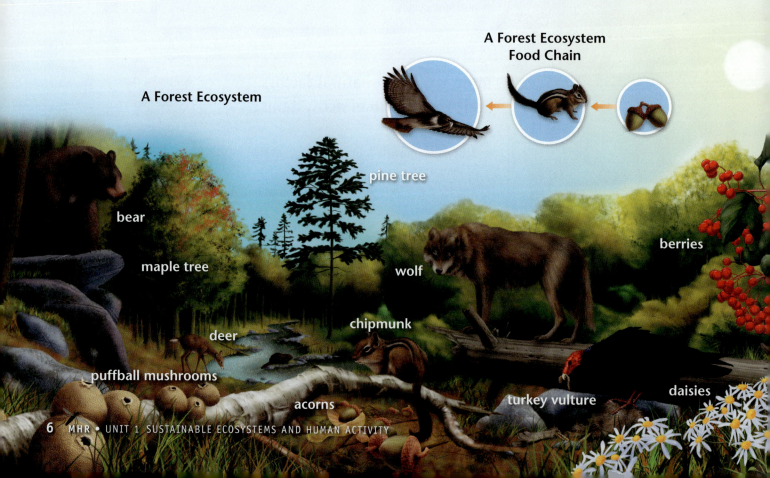

A Forest Ecosystem Food Chain

A Forest Ecosystem

Inquiry Check

Highway 60 runs through the middle of Algonquin Park, which is a forest ecosystem like the one in the picture. The Ontario Ministry of Transport has found out that most of the vehicle collisions with deer and bears occur on this highway in the months of May, June, October, November, and December.

7. **Analyze** In which two seasons do most collisions with deer and bears occur? Why do you think this is the case? Explain your answer.

8. **Predict** Which of the government strategies for reducing collisions listed below might be the most effective? Explain your answer.
   a) Installing fencing along major highways
   b) Draining salty ponds near highways
   c) Posting warning signs
   d) Adding highway lighting to improve night visibility
   e) Removing roadside brush so drivers can see the road better

9. **Plan** You are a scientist hired by the ministry to investigate its anti-collision strategies. Choose one of the five strategies above. Outline a procedure to test how well the strategy works.

Numeracy and Literacy Check

The five areas of Ontario with the highest number of reported vehicle collisions with wildlife are shown below. They are listed in alphabetical order.

Ontario's Highest Number of Reported Wildlife Collisions

Area	Human population	Number of incidents per year in 1997
Kenora	15 177	521
Lanark County	62 495	481
Ottawa	774 072	886
Simcoe County	266 100	656
Thunder Bay	109 140	463

10. **Ranking** List the areas in order from highest to lowest number of incidents.

11. **Graphing** Construct a bar graph to display the information shown in the table. Include a title and labels for your graph.

12. **Communicating a message** Create a poster to inform people about the dangers of collisions with wildlife on the roads and suggest ways to avoid them.

Topic 1.1

What are ecosystems, and why do we care about them?

Key Concepts

- Ecosystems are about connections.
- Ecosystems are made up of biotic (alive) and abiotic (not alive) parts that interact.
- Interactions between terrestrial (land) ecosystems and aquatic (water) ecosystems keep all ecosystems healthy.

Key Skills

Inquiry
Literacy

Key Terms

ecology
biotic
abiotic
ecosystem
terrestrial ecosystem
aquatic ecosystem

Everywhere in the world around you, living things are scurrying across the ground, burrowing through the soil, floating in the water, and soaring through the air. Each and every living thing makes its home somewhere. Even your armpit is an attractive home for many forms of life! In fact, dozens of kinds of microscopic organisms live on and inside your body.

eyelash mite

armpit bacteria

digestion-helping bacteria

athlete's foot fungus

Why do some organisms inhabit your body, while others make their home in spruce trees? It all comes down to basic needs. Living things make their homes in the places they do, because these places provide them with what they need to survive.

> ### Starting Point Activity
>
> 1. How many ecosystems can you see on these two pages? (Hint: First discuss what the word "ecosystem" means.)
> 2. Name at least three things that people, trees, and other living things need to survive.
> 3. State three reasons why a woodpecker can make its home on a spruce tree but not on a human being.

Ecosystems are about connections.

Imagine going back in time to the year 1966. Think about seeing the photo of Earth in **Figure 1.1**. This photo shocked the world because it showed for the first time what our home planet, our Earth, looks like from space.

When something is very large, as Earth is, it helps to take a step or two back so that we can see it and think about it in a new way. Photos of Earth from space helped people start to do just that. As we viewed Earth, as if with fresh eyes, a very different planet took shape in our minds. Saudi astronaut Sultan bin Salman bin Abdulaziz Al Saud put it this way, as the Space Shuttle *Discovery* carried him and six others away from Earth in 1985: "The first day or so we all pointed to our countries. The third or fourth day we were pointing to our continents. By the fifth day, we were aware of only one Earth."

Our journeys into space helped us start to see Earth in a way that First Nations and other Aboriginal people have seen it for a long time. We began to see an Earth where everything is connected.

Figure 1.1 In 1966, a satellite was sent to the Moon to search for landing sites for future missions. Looking back at Earth, the satellite took this photo. Two years later, astronaut William Anders was on board the first piloted mission to the Moon. Looking back at Earth, Anders said:

"We came all this way to explore the Moon, and the most important thing is that we discovered the Earth."

Studying the Connections

Ecology is a science that tries to explain the connections between everything on Earth. An ecologist is a scientist who studies these connections. So ecologists study how living things interact with each other and with everything else in their environment. Ecologists focus their attention on ecosystems in order to organize their studies.

An ecosystem can be very large and very small. A rotting log is an ecosystem. So is a pond, a forest, a desert, an ocean, a spruce tree, a human body, and even the whole Earth. The size and shape of an ecosystem depend on the types and numbers of connections an ecologist wants to study.

ecology: a branch of science that studies the relationships between living things and the environment

LEARNING CHECK

1. What is ecology?
2. What do ecologists study?
3. Name two large ecosystems and two small ecosystems.
4. What do you think this quotation means: "By the fifth day, we were aware of only one Earth." Explain your reasoning.

Literacy Focus

Activity 1.1

INSPIRING CONNECTIONS

"On Earth, we are all connected." What does this statement mean to you? In this activity, you will explore your ideas about connections in the world around you.

What To Do

1. Work together in small groups. Take turns reading each quote.
2. Share your ideas about each quote. Do you agree with it? Do you disagree? Why? Record your group's opinions.
3. Work together to write your own quote about the statement, "On Earth, we are all connected." Record your group's quote on a sheet of paper.

What Did You Find Out?

1. What does "On Earth, we are all connected" mean to you, personally?

"Connected means our planet is the ultimate recycler. Everything is reused around the planet."

Gary, age 13, Timmins, ON

"I think living things are our neighbours. Their lives affect us and we affect them."

Mira, age 14, Sarnia, ON

"To me this means that we need to stop polluting because what we do to the environment connects back to us."

Yan, age 14, Cornwall, ON

Ecosystems are made up of biotic and abiotic parts that interact.

> **Literacy Focus**
>
> ### Activity 1.2
> **PONDERING PONDS**
>
> Study **Figure 1.2**. With a partner, present your answers to these questions as a concept map.
>
> 1. In what parts of Ontario do you find ponds?
> 2. What kinds of plants live in and around a pond? Name at least three types that you can see or imagine.
> 3. What kinds of animals live in and around a pond? Name at least three types that you can see or imagine.
> 4. What other living things (but not plants or animals) live in and around a pond?
> 5. What non-living things would you find in and around a pond?

biotic: living

All the living things in an ecosystem such as the one in **Figure 1.2** are **biotic** parts of the ecosystem. The biotic parts of ecosystems might eat each other, defend themselves from each other, and compete with each other for living space, mates, and food. These are some of the ways that the biotic parts of ecosystems interact.

Within the Western tradition of modern science, some parts of the pond ecosystem are considered to be not alive. For instance, the water, the muddy pond soil, the air, and the sunlight shining on the pond are non-living. All the non-living things in a pond ecosystem, or any other ecosystem, are **abiotic** parts of the ecosystem. The temperature of the pond water is affected by how much sunlight it gets and by the weather patterns for the season. The shape and amount of shoreline is affected by the motion of the water. These are some ways that abiotic parts of ecosystems interact.

abiotic: not living (non-living)

Interacting Biotic and Abiotic Parts

To function and to stay healthy, the biotic parts of ecosystems interact with each other as well as with the abiotic parts of the ecosystem. For example, in a pond ecosystem, plants interact with soil and water. These abiotic parts help plants meet their basic needs for water and nutrients. Likewise, air provides plants with carbon dioxide and animals with oxygen they need to survive. The Sun supplies needed light and warmth.

A beaver is a mammal with large front teeth suited for gnawing wood and a flat, paddle-like tail suited for tamping mud. An average beaver is 1.5 m from nose to tail, with a mass of 20 kg.

▲ **Figure 1.2** Beaver dams are a common sight in many pond ecosystems of Ontario. Like all ecosystems, ponds have living (biotic) and non-living (abiotic) parts.

Defining the Term "Ecosystem"

To understand what an ecosystem is, you first have to know the words "biotic" and "abiotic." You also have to know that everything on Earth is connected. In other words, all the biotic and abiotic parts interact in ways that make the environment able to support and sustain life.

With these understandings, you can now understand how ecologists describe ecosystems. An **ecosystem** is all the biotic parts of a certain place, as well as all the ways that they interact with other biotic parts and with the abiotic parts of that place.

ecosystem: a system that is made up of all the interacting biotic and abiotic parts of a certain place

LEARNING CHECK

1. Use **Figure 1.2** to name two ways that biotic parts and two ways that abiotic parts of an ecosystem interact.
2. Explain how an ecosystem is like the Internet.
3. Use a graphic organizer to show the interactions you have with the abiotic and biotic parts of a neighbourhood ecosystem.

ACTIVITY LINK
Activity 1.4, on page 16

Interactions between terrestrial and aquatic ecosystems keep all ecosystems healthy.

terrestrial ecosystem: an ecosystem that is based mostly or totally on land

aquatic ecosystem: an ecosystem that is based mostly or totally in water

Some ecosystems are based on land. A forest, a desert, an ant colony, and a city are examples of land-based ecosystems. A land-based ecosystem is called a **terrestrial ecosystem**. (*Terra* means land or earth.)

Other ecosystems are based on water. A pond, a lake, a river, and an ocean are examples of water-based ecosystems. A water-based ecosystem is called an **aquatic ecosystem**. (*Aqua* means water.)

Activity 1.3

INTERACTION I.D.

Inquiry Focus

The picture shows an ecosystem that includes a land ecosystem and a water ecosystem. Identify (I.D.) three interactions in the land ecosystem. Then I.D. three interactions in the water ecosystem. Finally, I.D. three interactions that involve both the land and the water ecosystems.

Interactions within Ecosystems and between Ecosystems Sustain Life

Terrestrial and aquatic ecosystems are closely linked. **Figure 1.3** shows a few examples of ways that they are linked. Throughout this unit, you will observe and think about other ways that terrestrial and aquatic ecosystems are linked. These interactions keep biotic and abiotic parts of all ecosystems balanced. As a result, all ecosystems stay healthy.

In this unit, you also will look at ways that human activities can upset this balance. For instance, by cutting down a forest near a stream, tree roots that once trapped soil wither and die. Soil and nutrients wash into the stream, and this can harm or kill fish and other living things. This is just one way that the balance in ecosystems can be upset.

▲ **Figure 1.3A** Some organisms live in aquatic ecosystems for part of their lives and in terrestrial ecosystems for other parts of their lives. Frogs are an example.

◄ **Figure 1.3B** Animals that live in terrestrial ecosystems drink from ponds, rivers, and other bodies of water. A fox is one example.

▲ **Figure 1.3C** Many animals that live on land eat plants and animals that grow and live in aquatic ecosystems. *You* are an example.

LEARNING CHECK

1. Use a t-chart to brainstorm examples of five different aquatic ecosystems and five different terrestrial ecosystems. Compare your examples with a partner's.
2. What are some similarities and differences between an aquatic ecosystem and a terrestrial ecosystem?

Activity 1.4

ECOSYSTEMS WHERE YOU LIVE

Complete this mapping activity to assess your knowledge of terrestrial and aquatic ecosystems near your home.

What You Need
large sheet of paper
coloured markers

What To Do

1. Mark and label a dot at the centre of your paper. This represents your home.

2. Add details about local ecosystems to your map by completing the tasks that follow. Label each ecosystem you draw with a different colour. Details do not need to be drawn to scale. You may not be able to do all the tasks at this point.

   a) Indicate north, south, east, and west on your map.

   b) Show any prominent landforms near your home, such as hills, rivers, and lakes. Write the name beside each landform.

   c) Draw and label two terrestrial ecosystems found near your home. These may be natural areas (for example, a forest or field) or human-made areas (for example, a school field or park).

   d) Draw and label two aquatic ecosystems found near your home. These may be natural areas (for example, a bog or stream) or human-made areas (for example, a human-made pond or lake).

   e) Draw one very small ecosystem and one very large ecosystem near your home.

   f) At the side of your map, sketch three biotic parts of ecosystems that are naturally found near your home. Include at least one that is found in a terrestrial ecosystem and one found in an aquatic ecosystem.

   g) At the side of your map, name three abiotic parts of ecosystems near your home.

   h) Show an ecosystem that has been altered or is in the process of being altered by human activity.

   i) Show an ecosystem that is conserved or protected from development.

   j) Find the name(s) of the Aboriginal people who live in your area, or who lived there in the past. Determine if any local ecosystems are especially important to these Aboriginal people. Mark these ecosystems on your map.

3. Share your map with other students in your class. Discuss tasks a) to j) in step 2. If your classmates were able to complete any tasks that you had trouble with, you can use their information to complete your map. If you are still missing some information, use library or computer resources to complete further research. Add your research findings to your map.

What Did You Find Out?

1. One of the ecosystems you included on your map is being or has recently been changed by human activity. Describe how and why this ecosystem has changed.

2. Ecosystems consist of both biotic and abiotic parts. For an ecosystem you identified on your map, describe an interaction that might occur between

   a) two biotic parts
   b) two abiotic parts
   c) a biotic part and an abiotic part

3. Ecosystems are about connections. Describe one way that your daily activities are connected to the ecosystems around your home.

Topic 1.1 Review

Key Concepts Summary

- Ecosystems are about connections.
- Ecosystems are made up of biotic (alive) and abiotic (not alive) parts that interact.
- Interactions between terrestrial ecosystems and aquatic ecosystems help keep all ecosystems healthy.

Review the Key Concepts

1. **K/U** Answer the question that is the title of this topic. Copy and complete the graphic organizer below in your notebook. Fill in four examples from the topic using key terms as well as your own words.

   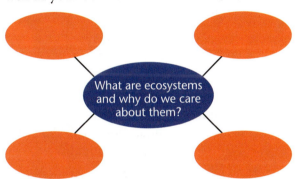

2. **A** Ecosystems are about connections. Thinking about the ecosystem in which you live, what are some of those essential connections?

3. **K/U** Refer to **Figure 1.2** to answer these questions.
   a) List three interactions that might occur between the biotic parts of this ecosystem.
   b) List three interactions that might occur between the biotic and abiotic parts of the ecosystem.

4. **C** Choose an ecosystem that is familiar to you (or one that is shown on the previous pages). Create a concept map showing how the abiotic parts are connected to the biotic parts of the ecosystem.

5. **K/U** As you read the paragraph below, complete a t-chart to list all the biotic and abiotic parts of ecosystems that are mentioned.

   > In a stream along the coast of British Columbia, a female salmon hatches. It eats microscopic living things in the stream and, as it grows, it swims out into the Pacific Ocean. Here, as it matures, it stores nutrients such as calcium, nitrogen, and phosphorus in its body tissues. After a few years, it starts the long swim back to the coastal stream where it began its life. Along the way, it dodges seals, sea lions, and other predators. Finally, it arrives at its stream. After it lays its eggs, it is caught and eaten by a grizzly bear. Later, the bear defecates. Decomposers release nutrients from the bear's feces (droppings) into the soil. Decomposers also release nutrients from the remains of the salmon carcass into the forest soil. Trees and other forest plants absorb these nutrients. In turn, the plants provide food for the forest-dwelling animals.

6. **A** Reread the paragraph in question 5. Explain how the interactions between terrestrial and aquatic ecosystems keep the forest ecosystem healthy.

Topic 1.2 How do interactions supply energy to ecosystems?

No matter what you eat each day, your food choices always have one main thing in common. Each and every food item carried with it some of the Sun—the Sun's energy, to be more precise. The energy that you depend on to sustain your life is stored in the food you eat. And that energy came originally from the Sun.

Key Concepts

- Photosynthesis stores energy, and cellular respiration releases energy.
- Producers transfer energy to consumers through food chains and food webs.
- Interactions are needed to provide a constant flow of energy for living things.

Key Skills

Numeracy

Key Terms

photosynthesis
cellular respiration
producer
consumer
food chain
food web

Starting Point Activity

1. How does the adult cow in the smaller photograph depend on energy that comes originally from the Sun?

2. How does the young calf in the smaller photograph depend on energy that comes originally from the Sun?

3. Think about the foods you have eaten during the past week. Write down three of these foods. Use words, pictures, or both to show how you think each of the foods is linked to the Sun.

Photosynthesis stores energy, and cellular respiration releases energy.

photosynthesis: a process in the cells of plants, algae, and some bacteria that converts light energy from the Sun into stored chemical energy

cellular respiration: a process in the cells of most organisms that converts the energy stored in chemical compounds into usable energy

Green plants make their own food. To do so, they use a process called **photosynthesis** to capture light energy from the Sun and transform the light energy into chemical energy. The chemical energy is stored in energy-rich food compounds such as glucose, which is a type of sugar. During photosynthesis, plants also produce oxygen gas.

All living things need the chemical energy stored in glucose and other energy-rich compounds to live. Most living things use a process called **cellular respiration** to break apart these compounds to release their stored energy. Once the energy is released, it can be used for life functions. During cellular respiration, living things also produce carbon dioxide gas and water vapour.

Table 1.1 Comparing Photosynthesis and Cellular Respiration

	Photosynthesis
1. What is it?	A series of chemical changes in which green plants capture the Sun's light energy and transform it into chemical energy that is stored in energy-rich food compounds such as sugars
2. Which living things use it?	Only green plants and certain kinds of single-celled organisms
3. How is energy changed?	Light energy is changed to chemical energy.
4. What substances does it use?	• carbon dioxide • water
5. What substances does it produce?	• glucose and other sugars • oxygen
6. How can it be represented?	light energy + carbon dioxide + water ⟶ glucose + oxygen
7. Why is it important?	1. Photosynthesis transforms the Sun's energy into a form that living things can use to survive. 2. Photosynthesis produces the oxygen that most living things need to survive.

Photosynthesis and Cellular Respiration Balance Each Other

Photosynthesis and cellular respiration take place in most ecosystems. **Table 1.1** shows how the two processes balance each other. In summary:
- Photosynthesis stores energy. Cellular respiration releases energy.
- Photosynthesis uses carbon dioxide and water, and produces glucose and oxygen. Cellular respiration uses glucose and oxygen, and produces carbon dioxide and water.

So each process makes the raw materials that the other process needs to store energy or to release energy. In this way, each process sustains the other. Together, both processes sustain life.

Go to **scienceontario** to find out more

LEARNING CHECK

1. What forms of energy are transformed during photosynthesis and cellular respiration?
2. Which substances are used and produced by photosynthesis and by cellular respiration?
3. Use a diagram to show how photosynthesis and cellular respiration balance each other.
4. Summarize your diagram using one or two paragraphs.

Cellular Respiration	
A series of chemical changes that let living things release the energy stored in energy-rich food compounds such as sugars to fuel all life functions	1. What is it?
Nearly all living things on Earth	2. Which living things use it?
Chemical energy is changed to other forms of energy such as kinetic (motion) energy and heat.	3. How is energy changed?
• glucose and other sugars • oxygen	4. What substances does it use?
• carbon dioxide • water	5. What substances does it produce?
glucose + oxygen → carbon dioxide + water vapour + usable energy	6. How can it be represented?
1. Cellular respiration releases the energy that living things use to survive. 2. Cellular respiration produces the carbon dioxide that green plants need to carry out photosynthesis.	7. Why is it important?

Producers transfer energy to consumers through food chains and food webs.

Many interactions between living things in ecosystems involve food and feeding. Because of this, you can describe living things based on how they get energy from food.

Producers are living things that *produce* (make) their own food to get the energy they need to live. They use photosynthesis to do this. Only green plants and some kinds of single-celled living things can carry out photosynthesis. So only these kinds of organisms are producers.

Consumers are living things that *consume* (eat) producers or other consumers to get the energy they need to live. Animals and most other kinds of living things are consumers.

producer: any living thing that gets the energy it needs by making its own food

consumer: any living thing that gets the energy it needs by eating producers or other consumers

Food Chains Chart the Flow of Energy from Producers to Consumers

A **food chain** is a model that describes how the stored energy in food is passed on from one living thing to another. You can use a food chain to show how energy flows in any ecosystem. Examples of food chains in a lawn ecosystem and a pond ecosystem are shown in **Figure 1.4A** and **B**. Notice that the flow of energy always goes from a producer to a consumer, and then onto one or more other consumers. Notice also that the path of energy always follows the path of a straight line. It doesn't matter if the path is shown up-and-down or side-to-side.

food chain: a model that describes how the energy that is stored in food is transferred from one living thing to another

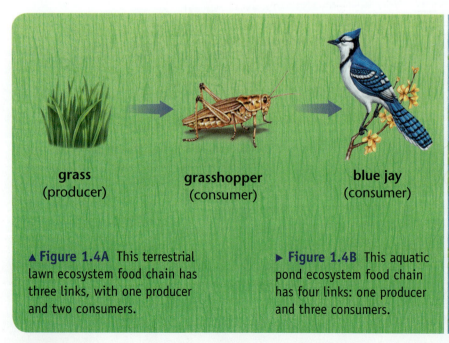

▲ **Figure 1.4A** This terrestrial lawn ecosystem food chain has three links, with one producer and two consumers.

grass (producer) → grasshopper (consumer) → blue jay (consumer)

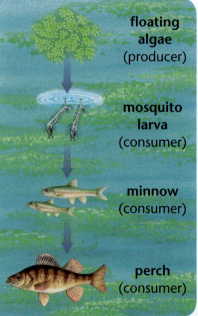

▶ **Figure 1.4B** This aquatic pond ecosystem food chain has four links: one producer and three consumers.

floating algae (producer) → mosquito larva (consumer) → minnow (consumer) → perch (consumer)

Food Webs Map Many Food Chains

You eat many different kinds of organisms that are producers and consumers. In other words, you are part of many different food chains. The same is true for other organisms. In any ecosystem, a more realistic model of feeding relationships shows a network of interacting and overlapping food chains. Such a model is called a food web. A **food web** weaves together two or more food chains within any given ecosystem. Refer to **Figure 1.5**.

All organisms in a food web are connected to each other through their feeding relationships. As a result, a change in the number of one organism could affect several food chains within the food web. In this sense, all organisms in an ecosystem are connected to and depend on each other for survival. Their interactions are key factors to sustaining life in aquatic and terrestrial ecosystems.

food web: a model that describes how energy in an ecosystem is transferred through two or more food chains

LEARNING CHECK

1. In **Figure 1.4**, grass and algae are labelled as producers. What makes them different from other organisms in the food chains?
2. A food web is a more realistic model for feeding relationships in an ecosystem than a food chain. Explain why.
3. Identify four food chains in the food web in **Figure 1.5**.

INVESTIGATION LINK
Investigation 1A, on page 26

◂ **Figure 1.5** A food web such as this one includes many different food chains.

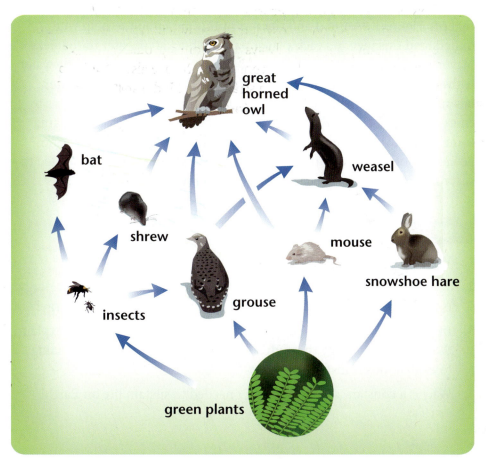

Interactions are needed for a constant flow of energy for living things.

Most food chains have three or four links. Some food chains have only two links. Some food chains might have as many as five or six links. Why are there limits to the length of a food chain?

There are limits, because just a small amount of energy is transferred from one living thing in a food chain to the next. Only about 10 percent of the food energy for a producer is available to a consumer that eats it. And only 10 percent of the food energy for that consumer is available to the next consumer. Here are some of the reasons why.

- Some of the original food energy has been used already to support life functions, such as growth and cellular respiration.
- Some energy is changed into heat that is given off into the environment. This energy cannot be used by other living things.
- Some energy is stored in wastes (urine and feces) that are excreted into the environment. Bacteria, fungi, and other decomposers extract some of this energy, but most is lost to the environment as heat.

The transfer of energy along a food chain is like a bucket-toss relay game. Refer to **Figure 1.6A** and **B** and the rest of this paragraph to help you understand this idea. As each player passes a bucket of water to the next player, some of the water spills from the bucket. There is only a little water left when the bucket reaches the last player, because some of the water has spilled each time the bucket was passed. This is like what happens to energy in a food chain. Each time energy is transferred in a food chain, some of it is lost as unusable heat. The energy that is lost as heat cannot be used by other living things. So a constant supply of energy is needed to sustain living things in terrestrial and aquatic ecosystems.

LEARNING CHECK

1. Why are there limits to the length of food chains?
2. When a mouse eats a plant, only about 10 percent of the plant's energy is transferred to the mouse. What happens to the rest of the energy?
3. In **Figure 1.6**, an analogy of a bucket-toss relay game is used to explain the transfer of energy through a food chain. Create your own analogy to explain this transfer of energy.
4. Why is a constant supply of energy needed to sustain life on Earth?

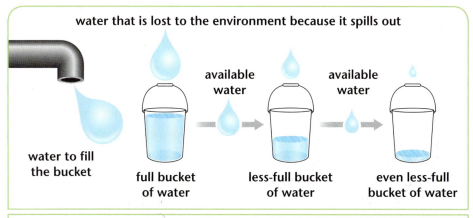

◀ **Figure 1.6A** Most of the water in the bucket that is transferred from one player to another in a bucket-toss relay game is lost to the environment. Less and less water is in the bucket for each player in the relay.

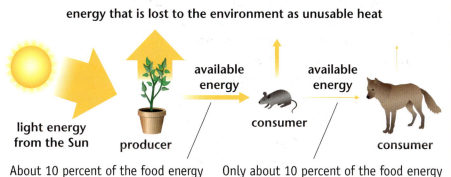

◀ **Figure 1.6B** Most of the energy that is transferred from one organism to another in a food chain is lost to the environment as unusable heat. Less and less energy is available to each organism in the food chain.

Numeracy Focus

Activity 1.5

PASS IT ON!

You will model how much energy is available to consumers in each link of a food chain.

What You Need

- 100 pennies (or plastic game chips)
- calculator

What To Do

1. Work in a group of three or four. Make a food chain that has one producer and two consumers.
2. Have one person get 100 pennies and bring them back to your group.
3. Food energy is measured with a unit called the kilojoule (kJ). Each penny represents 1 kJ of energy. So if you have 100 pennies, your producer has 100 kJ of stored food energy.
4. Only 10 percent of the energy stored in an organism is transferred to the organism that eats it. How many pennies is 10 percent of the pennies you started with? Use a calculator to find out. (Hint: Multiply the number of pennies by 0.1.) This is the amount of energy available to the next consumer in the chain.
5. Determine the amount of energy available to the last consumer in the chain.

What Did You Find Out?

1. How did the amount of energy stored in the producer compare with the amount of energy available to the last consumer in the chain?
2. Most food chains in any ecosystem are 5 or 6 links long at the most. Use your experience in this activity to help you explain why.

TOPIC 1.2 HOW DO INTERACTIONS SUPPLY ENERGY TO ECOSYSTEMS? • MHR 25

Investigation 1A

Skill Check

Initiating and Planning

Performing and Recording

✓ Analyzing and Interpreting

✓ Communicating

What You Need
- writing materials
- 12 small pieces of paper

Plot the Pathway

Interactions between biotic and abiotic parts of an ecosystem provide a constant flow of energy through it. You will investigate a possible pathway of energy flow through a common Ontario forest ecosystem to better understand how these interactions affect energy transfer.

What To Do

1. Work with a group to complete this activity. On each piece of paper, write the name of one of the ecosystem components below.

   grasshopper
   willow tree
   Sun
   weasel
   red fox
   grass
   caterpillar
   great grey owl
   hermit thrush (a small ground-nesting bird)
   eastern fox snake
   bunch berry (a low-growing flowering plant)
   deer mouse

2. Beginning with the Sun, arrange the pieces of paper into the longest food chain you can. You will not use all the pieces of paper.

3. Draw your food chain in your science notebook.

What Did You Find Out?

1. Identify the following ecosystem components in your food chain:
   a) abiotic parts
   b) biotic parts
   c) producers
   d) consumers

2. a) Which of the organisms in your food chain stores energy through photosynthesis?
   b) Which of the organisms in your food chain releases energy through cellular respiration?
   c) What would happen to your food chain if the photosynthetic organisms died out? Explain your answer.

3. How many organisms are in your food chain? Why are all food chains limited in length?

4. The eastern fox snake is a threatened species in Ontario. This means that the numbers of eastern fox snakes are low enough that all could die out at some point in the near future. How might a food chain that includes this snake be affected if the number of eastern fox snakes continues to decline? Why would this be the case?

Topic 1.2 Review

Key Concept Summary

- Photosynthesis stores energy and cellular respiration releases energy.
- Producers transfer energy to consumers through food chains and food webs.
- Interactions are needed to provide a constant flow of energy for living things.

Review the Key Concepts

1. **K/U** Answer the question that is the title of this topic. Copy and complete the graphic organizer below in your notebook. Fill in four examples from the topic using key terms as well as your own words.

   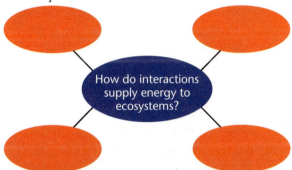

2. **C** Think about everything that you ate yesterday. Draw a food web that includes you.
   a) Look at your food web and determine how many links are in the longest food chain within it.
   b) What process did your body use to release the energy from the food you ate?

3. **A** Some people who adopt a vegetarian diet claim to feel more energetic than they felt when they were eating meat. Using your knowledge of energy and food chains, explain why they might feel this way.

4. **K/U** When two things are complementary, they balance each other. Explain why photosynthesis and cellular respiration are considered complementary processes.

5. **C** Refer to **Table 1.1**. Should photosynthesis win the Most Important Process on Earth Award? Write a supported opinion paragraph explaining your position.

6. **K/U** Why do food webs require a continual input of energy from the Sun?

7. **K/U** Copy this table into your notebook. Complete the table to compare photosynthesis and cellular respiration. (Do not write in this textbook.)

Comparing Photosynthesis and Cellular Respiration

Process	Photosynthesis	Cellular Respiration
Organisms in which the process occurs (Give three examples.)		
Substances used by the process		
Substances produced by the process		

Topic 1.3 How do interactions in ecosystems cycle matter?

Key Concepts

- Abiotic and biotic interactions cycle matter in terrestrial ecosystems and aquatic ecosystems.
- Photosynthesis and cellular respiration cycle carbon and oxygen in ecosystems.
- Human activities can affect ecosystems by affecting nutrient cycles.

Key Skills

Inquiry
Literacy

Key Terms

decomposer
nutrient
nutrient cycle

Day changes to night, which changes to day. Spring leads to summer, then to autumn, to winter, and back to spring. These examples of changes whose ending leads back to where they begin are known as cycles. Another cycle that plays a major role in your life is the yearly celebration of the date of your birth. In fact, all calendar systems—Julian, Gregorian, Islamic, Jewish, Chinese, Indian, Mayan, Baháʼí, Aboriginal, or any other—involve cycles.

You know other cycles from science class, too. There are life cycles of frogs, moths, and other animals. There are product life cycles, which depend on you and your commitment to recycling. There are also cycles that involve substances in nature. Water is one example. A simple diagram of the water cycle is shown here.

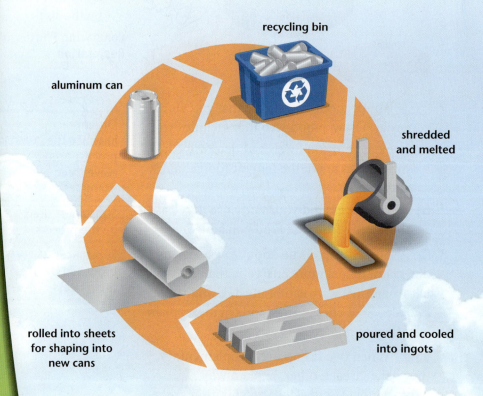

The product life cycle of aluminum is based on the fact that 100 percent of an aluminum can is able to be recycled.

Starting Point Activity

1. A cycle is a pattern of change that repeats itself forever. In what way does the water cycle demonstrate the features of a cycle?

2. On a map of Canada or Ontario, locate Toronto and Thunder Bay. The distance between these two cities is nearly 1400 km. Now picture a ball with a diameter of 1400 km. (In other words, the ball is 1400 km across.) All the water on Earth could fit into that ball. Do you think that is a lot or not? Explain.

3. During a drought (lack of rain for a long time), the amount of water in a body of water can drop a great deal. What happens to this missing water? Is it really missing? Explain.

4. The water cycle ensures that Earth will never run out of water. In fact, the total amount of water on Earth (the amount in that 1400 km-diameter ball) always stays the same. So why are we concerned about conserving water resources? (Hint: What is the difference between the total amount of water on Earth and the amount of water that is available in any one place at any one time?)

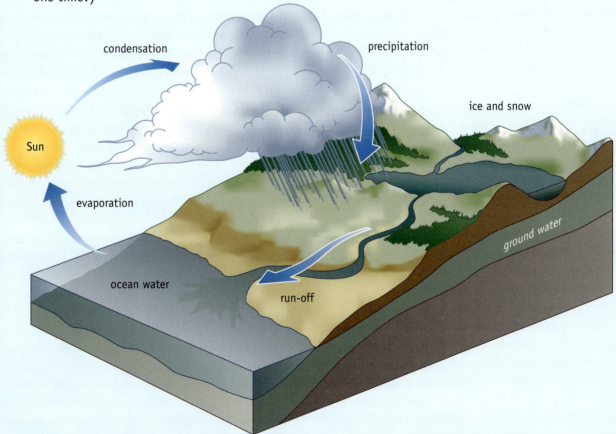

There is water on Earth's surface in the form of ponds, lakes, rivers, and the ocean. There is water under Earth's surface in the form of ground water. And there is water in the air in the form of water vapour. All this water continuously cycles through ecosystems by means of the interaction of three processes: evaporation, condensation, and precipitation.

Abiotic and biotic interactions cycle matter in terrestrial and aquatic ecosystems.

decomposer: organism that obtains energy by consuming dead plant and animal matter

nutrient: any substance that a living thing needs to sustain its life

nutrient cycle: the pattern of continual use and re-use of a nutrient

Some people love searching the soil for bugs and worms and all types of "creepy-crawlies." For other people, these organisms really do give them "the creeps." But you and all other living things owe your lives to these organisms. They are a group of consumers called **decomposers**.

In addition to soil insects and earthworms, decomposers include moulds, mushrooms, and certain kinds of bacteria. They get their food energy by digesting wastes such as urine, feces, and the bodies of dead organisms. As decomposers digest these wastes, some of the chemical substances that make up the wastes enter the soil, water, and air. These substances include carbon, nitrogen, iron, and other chemicals that living things need and use as nutrients. A **nutrient** is any substance that a living thing needs to sustain its life.

All producers and consumers use nutrients to grow and build their bodies and to help them carry out their life functions. When organisms die, decomposers return the nutrients to the environment. Then the nutrients are available to be used once again by living things. This pattern of use and re-use of nutrients has been taking place for millions of years in all ecosystems, all over the Earth. The pattern of continual use and re-use of the nutrients that living things need is called a **nutrient cycle**.

LEARNING CHECK

1. Use pictures, words, or a graphic organizer to explain the following terms: nutrient, nutrient cycle.
2. What role do decomposers play in nutrient cycles?
3. Explain what is meant by the statement "You and all other living things owe your lives to decomposers."
4. The living things that decompose dead organisms link the biotic and abiotic parts of ecosystems. How do they do this?

Activity 1.6

INTERACTIONS AND NUTRIENT CYCLES

What To Do

The diagram below is a general nutrient cycle. It shows the path of energy and matter in an ecosystem. Use the questions to interpret what the diagram shows.

1. What path (arrow colour) does energy follow in the diagram?
2. What path (arrow colour) does matter follow in the diagram?
3. Which part of the diagram shows where photosynthesis takes place?
4. Which part of the diagram shows where cellular respiration takes place?
5. How does the diagram show that a constant flow of energy is needed for living things?

What Did You Find Out?

1. Use the information communicated by the diagram to show how you know that these two statements are true.
   - Statement 1: The biotic parts and the abiotic parts of an ecosystem work together in a nutrient cycle.
   - Statement 2: Decomposers link the biotic parts of an ecosystem with the abiotic parts of an ecosystem.

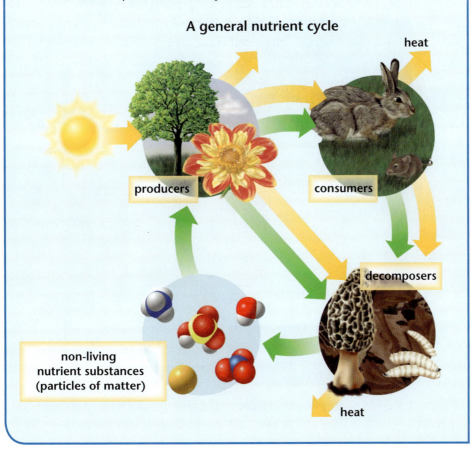

A general nutrient cycle

Photosynthesis and cellular respiration cycle carbon and oxygen in ecosystems.

In Topic 1.2, on pages 20 to 21, you learned the role that photosynthesis and cellular respiration play in the transfer of energy through ecosystems. These two processes also play a key role in the cycling of matter such as carbon and oxygen in ecosystems. Scan the large cycle picture shown in **Figure 1.7**. The labels will help you see how photosynthesis and cellular respiration are complementary processes for the cycling of carbon and oxygen. Because this is a cycle picture, you can start reading it anywhere and follow the arrows.

LEARNING CHECK

1. What substances do plants require to carry out photosynthesis?
2. What substances are released by all organisms—including both plants and animals—during cellular respiration?
3. Photosynthesis and cellular respiration are complementary processes. Use specific examples from **Figure 1.7** to support this statement.

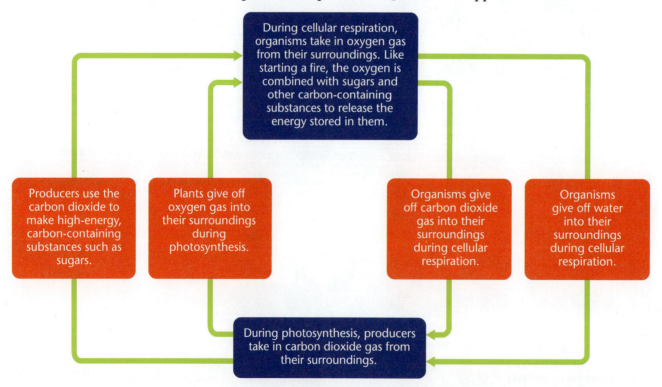

▲ **Figure 1.7** Photosynthesis and cellular respiration interact with each other as part of a cycle that uses and re-uses carbon and oxygen. This interaction takes place in both terrestrial and aquatic ecosystems.

Activity 1.7

CYCLE IT

What To Do

1. Work in a small group. Arrange yourselves so you are sitting in a circle.

2. With your group, examine the picture of interactions that are involved in the cycling of carbon and oxygen. Follow the coloured arrows for oxygen to see how it is cycled. Then do the same with carbon.

3. Help each other understand what happens to oxygen and to carbon as they cycle in an ecosystem. To do this, choose an ecosystem. Then choose one group member to start the story of how oxygen or carbon cycles through the ecosystem. This person will describe how the nutrient leaves the environment and enters a producer.

4. Each person in turn will describe what happens to the nutrient next.

5. End the story when the nutrient has returned to the person who started the story.

6. On your own, make a cycle diagram with labelled sketches to record the whole story that your group created together.

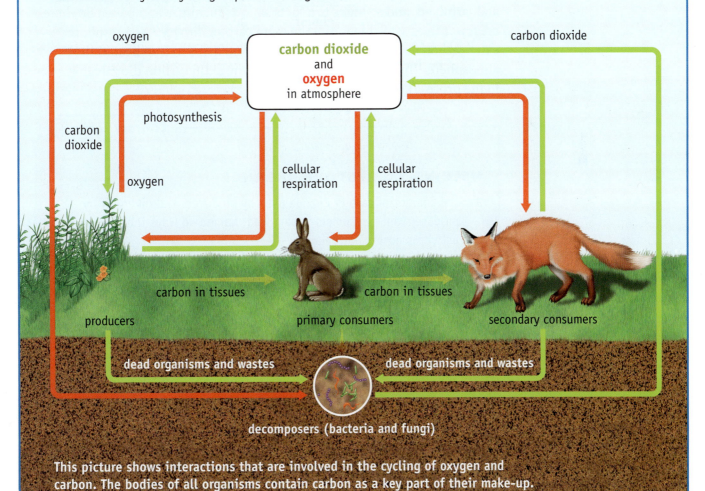

This picture shows interactions that are involved in the cycling of oxygen and carbon. The bodies of all organisms contain carbon as a key part of their make-up.

Human activities can affect ecosystems by affecting nutrient cycles.

Not all the carbon involved in the carbon cycle is used immediately by living things. Some is stored in the woody tissues of long-living trees. Some is stored in the slowly decomposing remains of organisms, which become buried deeply in the ground. With the passage of time, some of this stored carbon will eventually be transformed into the carbon-rich fuels that we know as coal, oil, and natural gas. This is what happened about 300 million years ago to form the coal, oil, and natural gas that we use today as fuel.

The amount of carbon dioxide that is used by photosynthesis and given off by cellular respiration is nearly the same. In other words, the amount of carbon dioxide is balanced. When we burn trees, coal, oil, and natural gas for fuel, the carbon stored long ago is released into the air in the form of carbon dioxide. So we upset that balance. As well, human activities have removed huge numbers of trees to make space for homes, buildings, and farmland, and to make products such as furniture and paper. So there are fewer trees available to use the extra carbon dioxide. As a result, the extra carbon dioxide builds up in the air and helps to trap heat in the atmosphere. This is one of the sources of the extra carbon dioxide that adds to the process known as global warming.

Literacy Focus

Activity 1.8
HELPING TO RESTORE BALANCE

1. The amount of carbon dioxide used by producers and given off by producers and consumers is balanced. Name two ways that human activities upset that balance.

2. Think about three human activities that could help to restore this balance. Use a cause-and-effect map to explain in each case.

ACTIVITY LINK
Activity 1.9, on page 38

Other Effects on Nutrient Cycles

Nitrogen is another nutrient that cycles in ecosystems. It is a major part of all cells and a key building block for proteins, which all cells need. Nitrogen makes up 78 percent of air, but most living things cannot use nitrogen from the air. Instead, they depend on certain kinds of bacteria in the soil and water to change the nitrogen into forms that plants can use.

Many human activities affect the nitrogen cycle. For instance, nitrogen is a key part of fertilizers. Farmers and gardeners use fertilizers to enhance the growth of their plants. Not all the nitrogen in the fertilizers is used by the plants, though. Some stays in the soil. When it rains, or when fields are watered, some of the nitrogen is carried into aquatic ecosystems. This excess nitrogen can cause an overgrowth of algae called an algal bloom. Figure 1.8 shows how an algal bloom can affect an aquatic ecosystem.

LEARNING CHECK

1. How does burning wood, coal, oil, and natural gas affect the amount of carbon dioxide in the atmosphere?
2. Why is nitrogen essential for all living things?
3. Use a graphic organizer to show how fertilizers, algal blooms, and the death of aquatic organisms are linked.

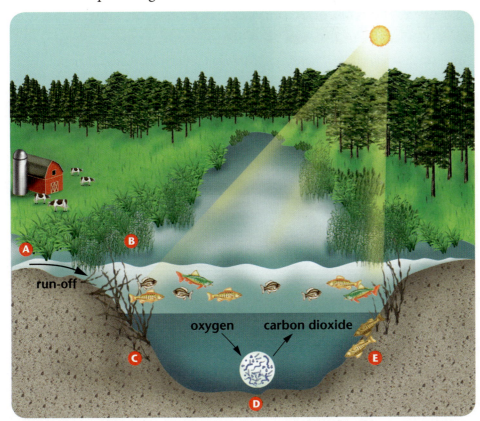

A: Rain carries nitrogen from farms, gardens, and lawns into aquatic ecosystems.

B: Algae and plants at the water's surface grow quickly. This blocks sunlight from reaching deeper water.

C: Deep-water plants get no sunlight. They cannot carry out photosynthesis, so they no longer give off oxygen, and they soon starve to death.

D: When the plants die, decomposers have lots of food. The number of decomposers increases quickly. They use up the oxygen in the water as they carry out cellular respiration.

E: As oxygen in the water is used up, aquatic organisms that need the oxygen suffocate and die.

Figure 1.8 An algal bloom is caused by too much of a nutrient, such as nitrogen, entering an aquatic ecosystem.

Strange Tales of Science

Journey of an Immortal Carbon Atom

Carbon, oxygen, and other nutrients have been cycling through ecosystems for eons, and they will continue to do so for eons more. But what actually happens to a nutrient as it travels though a nutrient cycle? Read the panels below to find out what adventures befall one lone carbon atom as it journeys through time and space in its quest for immortality.

65 million years ago...

A fern lives on the bank of an inland sea covering much of North America. On a sunny day, it takes up an atom of carbon to become part of a carbon dioxide molecule. The atom is one of many used to make carbon-rich compounds by photosynthesis. To do this, the fern also uses the Sun's energy and water from the soil.

Through cellular processes, the fern uses the carbon atom, along with some oxygen and hydrogen atoms, to build a fat molecule. It incorporates the fat in a membrane surrounding one of its cells.

MMMM ... BREAKFAST!

A plant-eating triceratops lumbers up to the fern and grabs a mouthful. The fat in the fern cell is digested in its stomach. The dinosaur uses oxygen to access the energy in the fat by cellular respiration.

MMMM ... LUNCH!

In the process, both carbon dioxide and water are released into the air again. But our carbon atom does not leave the triceratops. Instead, it ends up in a cell in the dinosaur's bony head frill—and, a short time later, in the stomach of a Tyrannosaurus rex (T.Rex).

MMMM... DINO DUNG!

The carbon in the triceratops' bone passes through the T. rex undigested and re-enters the environment in its dung. Decomposers return many nutrients in the dung to the soil. However, the bone with our carbon atom becomes fossilized and stays in the ground until...

1998
YES! DINO DUNG!

...1998. This is the year that Dr. Karen Chin, the world's leading expert on dinosaur dung, finds the bone fragment in a huge (44 cm) fossilized sample of T. rex dung in southern Saskatchewan.

The fossil is now stored at the Royal Saskatchewan Museum, along with the carbon atom we've been following for 65 million years!

AND THE JOURNEY CONTINUES... BUT HOW?

So... What do you think?

1. Continue this graphic novel to reveal the next installments in the journey of the immortal carbon atom.
2. Draw a diagram showing how photosynthesis and cellular respiration were involved in the cycling of this carbon atom.
3. The fossilized dung in this story is 65 million years old. Find out different methods paleontologists use to date fossils. (One of them involves carbon atoms.)
4. Find out more about Dr. Karen Chin and what dinosaur dung tells her about the prehistoric past.

Go to **scienceontario** to find out more

TOPIC 1.3 HOW DO INTERACTIONS IN ECOSYSTEMS CYCLE MATTER? • MHR 37

Activity 1.9
RECYCLING ON MARS

If humans ever colonize another planet, Mars is a good choice. Its conditions are more Earth-like than on any other planet. To live on Mars, we would have to create ecosystems that can sustain themselves for long periods of time, just as they do on Earth. For instance, a Mars colony would have to recycle and re-use all its materials. This includes nutrients such as water, carbon, oxygen, and nitrogen. In this activity, you will consider some of the factors that would be needed for a self-sustaining Mars colony.

What To Do

1. With your group, make a list of the things that a Mars colony would need and how you could maintain them over time. Use the following questions to help you:
   a) Mars has a maximum temperature of 20°C and a minimum temperature of −140°C. How could your colony maintain temperatures that are friendlier to life?
   b) How would your colony deal with food production and waste disposal?
   c) How would you generate energy for your colony?
   d) Because you can only bring supplies and materials to Mars once, all materials must be recycled. How would nutrients like water, oxygen, and carbon dioxide be continuously cycled within the colony?

What Did You Find Out?

1. How is the colony you created like an ecosystem? How is it different?
2. a) What cycles would have to be maintained to sustain a colony on Mars?
   b) How might the colony be affected if one of these cycles became disrupted?
   c) Which nutrient is the most difficult to cycle in your colony? Why?
3. You have learned that photosynthesis and cellular respiration are complementary processes. Did this knowledge help you cycle oxygen and carbon in your colony? Explain.
4. On Mars, large amounts of water are frozen in the polar icecaps and under the surface. How could this water be used to sustain your colony?
5. We don't know yet if the soil on Mars can be used to grow crops. One solution to this problem is to grow crops in greenhouses like the one below, which uses hydroponics. This technology uses nutrient-enriched water instead of soil to grow plants. Suggest a way that nitrogen, an important nutrient for plant growth, might be recycled within the greenhouse.

Canada's Devon Island, in Nunavut, is a barren landscape used by research scientists as a "stand-in" for the barren landscape of Mars. The Arthur C. Clarke Greenhouse experiment has been running since 2003.

Topic 1.3 Review

Key Concept Summary
- Abiotic and biotic interactions cycle matter in terrestrial and aquatic ecosystems.
- Photosynthesis and cellular respiration cycle carbon and oxygen in ecosystems.
- Human activities can affect ecosystems by affecting nutrient cycles.

Review the Key Concepts

1. Answer the question that is the title of this topic. Copy and complete the graphic organizer below in your notebook. Fill in four examples from the topic using key terms as well as your own words.

   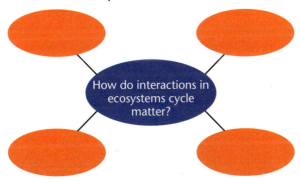

2. **A** What are some ways in which you and your family and friends affect nutrient cycles? Use a graphic organizer to demonstrate the cause-and-effect relationships involved in the examples you provide.

3. **C** In Activity 1.8, you identified three human activities that could restore the balance of carbon dioxide. Select one activity and write a letter to a classmate explaining why you would choose to do that activity.

4. **K/U** Water and chemical nutrients such as carbon and nitrogen are recycled through ecosystems. Explain why this recycling is necessary.

5. **K/U** Explain how biotic and abiotic interactions cycle matter in ecosystems. Use at least two examples to support your explanation.

6. **C** Refer to **Figure 1.7**. Create your own flowchart to show the cycling of carbon dioxide and oxygen through photosynthesis and cellular respiration.

7. **C** Use words or diagrams to illustrate how a carbon atom that was part of a dinosaur 70 million years ago could be part of you today.

8. **T/I** Methane is a substance that is made up of carbon and hydrogen. It is a greenhouse gas that is about 21 times more potent than carbon dioxide. The pie graph below shows sources of methane emissions around the world. Use the graph to answer the following questions.

   a) Which nutrient cycle is methane part of?
   b) What percentage of methane emissions come from human activities, as opposed to natural sources?

   Sources of Global Methane Emissions

   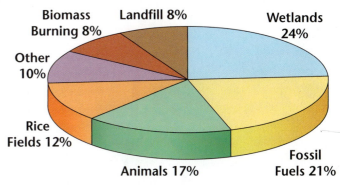

   Environment Canada, 2000

Topic 1.4

What natural factors limit the growth of ecosystems?

Key Concepts

- Ecosystem growth is limited by the availability of resources.
- Abiotic and biotic factors limit populations in ecosystems.

Key Skills

Inquiry

Key Terms

population
carrying capacity
limiting factor

In the year 1955, film star James Dean died in a car accident. Ford rolled out the Thunderbird. And Rosa Parks changed the course of history for African-Americans and African-Canadians. If you had been alive at that time, you would have shared the world with 2.8 billion people.

If that sounds like a lot, fast-forward to 1985. In that year, athlete Rick Hansen began his wheelchair trek around the world. The first Blockbuster video store opened. And 41 tornadoes tore a devastating path through Ohio, Pennsylvania, and Ontario. If you had been alive in 1955, you would have shared the world with 4.8 billion people.

Time-jump once more to today. Today, you are walking on Earth alongside about 7 billion other people. These numbers, and the graph on the right, show that Earth's human population has grown, and continues to grow, at a very fast rate.

Rosa Parks

Starting Point Activity

1. Name three resources that humans need to survive.
2. As our human population keeps growing in size, what could happen to our access to the resources we need?
3. As our population grows, what could happen to other living things that depend on many of the same resources that we do?
4. Do you think there is a limit to the size that our population can grow? Explain why.

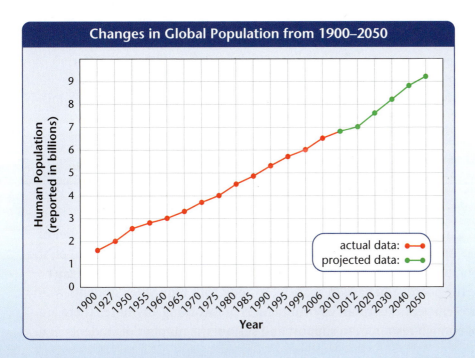

Rick Hansen

A fellow human

Ecosystem growth is limited by the availability of resources.

> **Activity 1.10**
>
> **UP FOR THE COUNT**
>
> It's noon, and a single-celled bacterium—a germ—has invaded your body. The warm, wet environment of your body provides this germ with lots of food and plenty of living space. And so the germ begins to divide (reproduce). In 20 min, the 1 germ divides to become 2 germs. After another 20 min, each of the 2 germs divides to become 4 germs. And 20 min after that, each of the 4 germs divides to become 8 germs. As this pattern continues, the germ population keeps growing in size.
>
> **Make a prediction:** Predict how long this germ population will be able to keep growing in size. Give reasons to support your prediction.

One day you walk into your science class and find that the number of students has doubled. Your classmates are sitting on desks and on the floor, because there are not enough chairs. You find a seat, but you can't see the board. You have to share a textbook with four other students. Your once-efficient classroom environment is not working anymore. There are not enough resources to support and sustain the number of students in it.

Carrying Capacity and Limiting Factors

Any ecosystem has a limited amount of resources. So it can only sustain a **population** of a certain size. The largest population size that an ecosystem can sustain is called its **carrying capacity**.

Carrying capacity is always limited by the resources that are available to a population. These resources are called **limiting factors**, because they limit the size to which the population can grow. In your classroom, limiting factors include the size of the room, the number of desks and chairs, and the number of textbooks. In a natural ecosystem, population growth is limited by factors such as the amount of living space, food, sunlight, and water.

In any ecosystem, a population can keep growing only if it has an endless supply of the resources that it needs. Without these resources, fewer new members of the population will be born, and more members of the population will die. So, limiting factors control the carrying capacity of an ecosystem and, therefore, the size of its populations.

population: all the individuals of a species that live in a certain place at a certain time

carrying capacity: the largest population size that an ecosystem can sustain

limiting factors: any resources that limit the size to which a population can grow

Limiting Factors Can Play Different Roles in Different Ecosystems

Most ecosystems are affected by the same limiting factors. However, a limiting factor might play a bigger role in one ecosystem than in another. For instance, look at the lake picture and the graph in **Figure 1.9**. In aquatic ecosystems such as this one, the amount of oxygen is a limiting factor. In terrestrial ecosystems, on the other hand, oxygen is always in the air. So it rarely affects carrying capacity. However, population growth in terrestrial ecosystems is often limited by something that is abundant in aquatic ecosystems—water!

LEARNING CHECK

1. Use pictures and words to explain "carrying capacity."
2. What are some examples of limiting factors in ecosystems?
3. Use **Figure 1.9** to explain how the limiting factors in an aquatic ecosystem can affect its carrying capacity.
4. Do you think that limiting factors also affect the human population? Explain why.

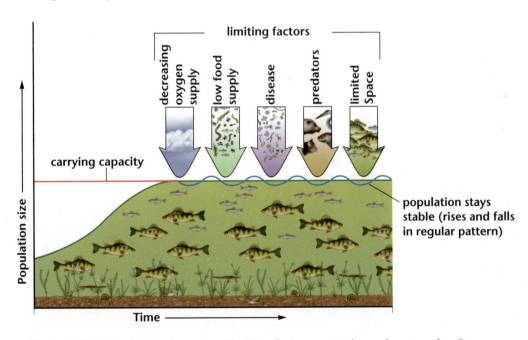

▲ **Figure 1.9** This picture shows how limiting factors control carrying capacity. Be sure you see the five limiting factors. Then look at the blue graph line. See how the limiting factors keep the size of a population from growing too big. Look also at the red line that is labelled "carrying capacity." This red line and the blue graph line show you the link between limiting factors and carrying capacity.

Abiotic and biotic factors limit populations in ecosystems.

INVESTIGATION LINK

Investigation 1B, on page 46

The factors that affect the carrying capacity of an ecosystem can be non-living and living. In other words, abiotic and biotic factors limit the size of populations in ecosystems. Abiotic factors that limit the size of populations include water, living space, nutrients, shelter, sunlight, and weather. Biotic factors that limit the size of populations include those described in the text boxes in **Figure 1.10**.

Figure 1.10 Biotic factors that limit the size of populations include those shown in this forest ecosystem. This is an ecosystem that might be found in Northwestern Ontario.

Parasites

Parasites are living things that live on or inside other living things and use them or their tissues for food. The living thing on which a parasite feeds is called the host. Most parasites weaken their hosts but rarely kill them.

ticks

Competition

Each member of a population has the same needs for the same resources. These resources include nutrients, shelter, light, water, and living space. Single members of the population are in competition with each other for these and other resources. Those members who are too young, too old, too weak, or who have injuries often will lose out to other members of the population.

white-tailed deer

lynx

Predators and Prey

A predator is an animal that hunts, kills, and eats other animals—its prey. The interaction between predators and prey is called predation. Predation affects the predator population as well as the prey population. Both populations benefit from this interaction. Predators benefit by getting the food they need. Some prey benefit because the predators often eat old, sick, or weak members of the prey population. The benefit is less competition among the prey population.

Activity 1.11

Inquiry Focus

WHAT'S THE LINK?

Share your ideas as you discuss these questions.

1. Why do palm trees grow in Florida but not in Ontario?
2. How could the number of foxes in a meadow affect the number of rabbits that also live there?
3. How could a severe drought affect the populations that live in and around a pond?
4. How is nitrogen a limiting factor in a lake ecosystem?

LEARNING CHECK

1. Refer to Figure 1.10. List three abiotic factors that limit the size of a population of deer.
2. Use pictures or words to explain the different ways in which competition can limit populations.
3. What kinds of resources might plant populations compete for?

yellow warbler

caterpillar

aspen

Different Populations Compete

Individual animals from different populations also compete for resources. For example, snowshoe hares eat many of the same foods that deer do. They may share some of the same predators. For instance, wolves eat deer and snowshoe hares. Bobcats and lynx prefer hares, but they will sometimes take a deer if it is too old, young, or sick.

Plant Competitors

Animals are not the only living things that compete. Plants also compete for the resources they need. Members from the same plant population compete with each other. They also compete with members of different plant populations.

snowshoe hare

Investigation 1B

Skill Check

✓ Initiating and Planning

✓ Performing and Recording

✓ Analyzing and Interpreting

✓ Communicating

Safety Precautions

What You Need

5 Erlenmeyer flasks or beakers

fertilizer solutions (five different concentrations)

dropper

well-lit space or grow light

algae

graph paper

Investigating Limiting Factors for Algae Growth

Algae are microscopic plant-like organisms commonly found in aquatic ecosystems. As is the case with all living things, the growth of an algae population is limited by abiotic and biotic factors. In this investigation, you will plan and conduct an experiment to explore how fertilizer affects the size of an algae population.

What To Do

1. Design a procedure to determine how different concentrations of fertilizer solutions affect the growth of algae. Use this checklist to help you plan your procedure.

   ☑ Because algae are producers, they need light for photosynthesis. Ensure that the algae have enough light.

   ☑ You will need to design a way to describe and compare the amount of algae growth in each test tube.

   ☑ Be sure to consider safety precautions and proper clean-up and disposal in your procedure. Why must you not pour the material in your flasks down the sink?

   ☑ Ensure that you design an experiment to test only one variable. The variable that you choose to test is called the independent variable. It is the variable that you make changes to. The variable that responds to the changes you make is the responding, or dependent, variable. All of the other variables that you are not testing are called controlled variables. You keep all the controlled variables the same. Turn to Science Skills Toolkit 2: Scientific Inquiry at the back of the book to review variables and how to conduct an experiment.

   ☑ Ask yourself:
   - What is the independent variable (the one you are changing) in this experiment?
   - What is the dependent variable (the one that changes as a result)?
   - What are the controlled variables (the ones that must be kept the same)? Hint: Consider any factors that might affect the outcome of the experiment. Examples include air temperature, water temperature, amount of light, and volume of pond water.)

2. Create a table to record your observations. Give your table a suitable title.

3. Ask your teacher to approve your procedure. Then carry it out.

4. When you have finished your observations, make a graph that compares the concentration of fertilizer solution to algae growth. Give your graph a suitable title. Turn to Numeracy Skills Toolkit 4: Organizing and Communicating Scientific Results with Graphs, to help you decide which of your variables goes on the *x*-axis and which goes on the *y*-axis.

What Did You Find Out?

1. What was the limiting factor that you investigated? Was it biotic or abiotic? Explain.

2. Explain how you controlled your experiment. As part of your answer, state your independent variable and your dependent variable. Also state the variables that you controlled.

3. If you were able to design your experiment again, what would you do differently? Why?

4. Using your graph, what can you conclude about the effect that different concentrations of fertilizer solution have on algae growth?

5. If algae have access to unlimited nutrients for growth, will an algae population keep growing forever? What other abiotic and biotic factors might limit the growth of the population?

6. Human activity can cause more nutrients than usual to enter aquatic ecosystems. For instance, farmers and gardeners often use nutrient-rich fertilizers to enhance plant growth. But not all the nutrients are used by the plants. Some stay behind in the soil. These excess nutrients are then carried into lakes, ponds, and other aquatic ecosystems by rain or run-off from watering. The excess nutrients can cause an overgrowth of algae called an algal bloom. How do you think an algal bloom might affect other living things in an aquatic ecosystem? How might it affect the ecosystem as a whole? Give reasons for your opinions.

Inquire Further

7. Many people believe that organic fertilizers such as manure and compost are better for the environment than synthetic (human-made) fertilizers. Is there less risk of an algal bloom if farmers and gardeners use organic fertilizers instead of synthetic fertilizers? Use print or electronic resources to find an answer.

Strange Tales of Science

Limiting factors limit the size to which a population can grow. Consider a population of bacteria, known as a colony. Bacteria grow by doubling: one bacterium becomes two, two become four, four become eight, and so on. If there were no limiting factors to keep its growth in check, a bacterial population could get very large, very quickly. How large? How quickly? *E. coli* bacteria divide once every 20 minutes. Without limiting factors, it would take a single *E. coli* bacterium (one cell) exactly 24 hours to create a super colony with the same mass as planet Earth!

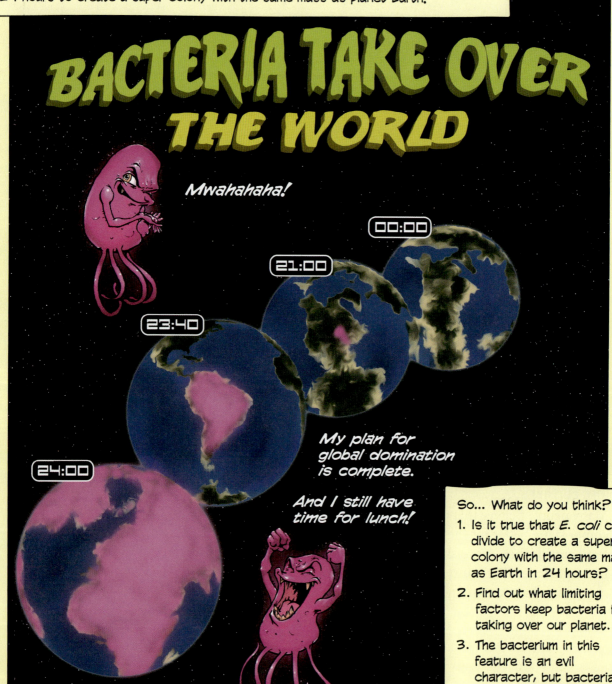

BACTERIA TAKE OVER THE WORLD

Mwahahaha!

My plan for global domination is complete.

And I still have time for lunch!

So... What do you think?

1. Is it true that *E. coli* could divide to create a super colony with the same mass as Earth in 24 hours?
2. Find out what limiting factors keep bacteria from taking over our planet.
3. The bacterium in this feature is an evil character, but bacteria also play beneficial roles in ecosystems. What would happen to ecosystems if there were no bacteria?

Topic 1.4 Review

Key Concepts Summary
- Ecosystem growth is limited by the availability of resources.
- Abiotic and biotic factors limit populations in ecosystems.

Review the Key Concepts

1. **K/U** Answer the question that is the title of this topic. Copy and complete the graphic organizer below in your notebook. Fill in four examples from the topic using key terms as well as your own words.

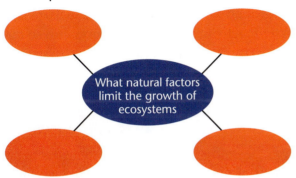

2. **K/U** Use a Venn diagram or other graphic organizer to compare the limiting factors in terrestrial and aquatic ecosystems.

3. **T/I** The two bacterial-population graphs below shows the growth patterns for two different populations of bacteria over a period of time.
   a) Describe, in words, what is happening to each of the bacteria populations.
   b) Has either of these populations reached its carrying capacity? Explain your answer.

Bacteria Population Growth over Time

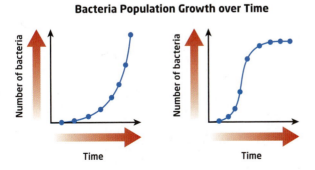

4. **C** List some examples of limiting factors on human populations. Then answer the questions below.
   a) Why would a government restrict the number of children that urban couples may have?
   b) Should governments be allowed to do this?
   c) Construct a list of pros and cons concerning government restrictions on the number of children that couples may have.
   d) Summarize your points in several paragraphs that support your opinion.

5. **A** How might the removal of dead timber from an area affect the carrying capacity of flying squirrels?

6. **K/U** Use a spider map to represent either the biotic factors or the abiotic factors that limit the size of populations in a forest ecosystem.

7. **T/I** The water-flea graph below shows how a population of water fleas changed during a laboratory experiment. Use the terms "carrying capacity" and "limiting factors" to explain how the population changed.

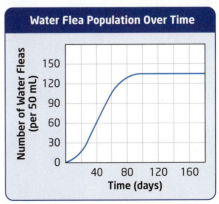

Topic 1.5 — How do human activities affect ecosystems?

Key Concepts

- We cannot always accurately predict the consequences of our actions.
- Introduced species can affect the health of ecosystems.
- Pollutants from human activities can travel within and between ecosystems.

Key Skills

Inquiry
Literacy

Key Terms

introduced species
species diversity
watershed

Human populations have a greater impact on ecosystems than populations of most other living things do. Some of us live in very large numbers in fairly small areas such as Cambridge, Sudbury, and Cornwall. Others of us sprawl out to fill large areas such as Toronto, Windsor, and Ottawa. Either way, our need for resources is very great, because there are so many of us.

We also use resources that other organisms don't. For instance, we dig deep into the ground for petroleum to make fuels, plastics, and other products. Our activities can change ecosystems so much that their health is often threatened. In some cases, ecosystem balance is so disrupted that living things that once thrived in a certain place can no longer get the resources they need to survive there.

Starting Point Activity

Around the world, vast numbers of trees are cut down. The timber from the harvested trees is used for a variety of purposes. In some cases, the cleared land is also used.

1. Name at least five different uses for timber.
2. Name at least two different uses for the land that is cleared of trees.
3. Name at least three effects of clearing trees on the organisms that make their homes in forested areas.
4. You use many products that come from trees or the land on which they grew. Is there a way to balance our need for resources with the needs of other organisms for those same resources? Share and discuss your ideas.

We cannot always accurately predict the consequences of our actions.

Consider this. A person cuts down a tree to let more sunlight shine on a small garden. Now think about some of the consequences of this action. For instance, before the tree was cut down, birds and insects used it for food, shelter, and nesting. Now the tree can no longer provide the birds and insects with these resources. As well, the insects, worms, and other organisms that lived in and among the tree's living roots can no longer get the resources they need to sustain them.

Human activities always cause changes to ecosystems. For example, in the situation in the first paragraph, many organisms have lost their living space and their source of food. As well, the area around the garden is now fully exposed to sunlight instead of being shaded from it. This will change the way that the soil holds rainwater. Plants that grew well in shaded conditions will also die.

The changes caused by human activities always have consequences for the biotic and abiotic parts of ecosystems. **Table 1.2** lists a few examples. Many of the problems that affect the health of ecosystems today come from the fact that we cannot always accurately predict the consequences of our actions. Sometimes, unfortunately, we do not even think about the consequences in the first place.

INVESTIGATION LINK
Investigation 1C, on page 59

Activity 1.12
PREDICT THE CONSEQUENCES

1. Work in groups. Choose one of the activities listed below.
   - logging
   - farming
   - outdoor recreation event

2. Brainstorm at least three consequences of the activity you chose. Be sure to think about things that could affect abiotic and biotic parts of the ecosystem. For instance, what are the effects on abiotic factors such as soil, sunlight, water, and air? What are the effects on biotic factors such as the size of populations, food webs, and nutrient cycles?

3. Research at least three of the consequences you have brainstormed to collect information about them.

4. Use a graphic organizer to help you organize your research and to record your findings.

5. Communicate your findings in a cause-and-effect map.

LEARNING CHECK

1. Use the text on page 52 to create a cause-and-effect map to summarize the consequences of cutting a tree in a garden.
2. Identify three human activities, different from those in **Table 1.2**, that could have consequences for the environment.
3. Create a t-chart to list the possible abiotic and biotic consequences for one of the human activities in **Table 1.2**.
4. The land that your school is built on was probably once a field or farmland. Identify some biotic or abiotic consequences that might have resulted from building your school.

ACTIVITY LINK
Activity 1.15, on page 60

Table 1.2 Examples of Human Activities and Their Consequences for Ecosystems

Human Activity	Possible Abiotic and Biotic Consequences
Construction of roads and buildings	• Surface soil is removed, killing soil organisms and plants that were rooted in the soil. • The shape or slope of the land is changed, resulting in different patterns for drainage of rainwater. • Farmland that is taken over to build roads and buildings can no longer be used to grow crops and livestock.
Dam-building	• The courses of rivers and streams are changed so that water will flow to the specific place chosen for the dam. • Land is flooded to create lakes in places where none existed before. • Huge numbers of living things are killed. • Huge numbers of living things are displaced and must find new places to live. (This includes humans, too.)
Manufacturing and consumption of goods	• Soil and plant life are removed to make space to build factories and landfill sites for the solid wastes that the factories produce. • Factories consume energy to make products. • Production process creates wastes that can enter and pollute air, water, and soil. • Stores that sell goods consume energy to operate. • The packaging, transportation, and consumption of goods generate wastes that must be disposed of. • Disposal and recycling of wastes consumes energy.

Introduced species can affect the health of ecosystems.

In 1889, the only place you would find European starlings was in Europe. Then someone released 100 of these foreign birds in New York. Now there are more than 200 million of them in North America. Their dramatic success has had disastrous consequences, though. Large populations of starlings destroy grain and fruit crops. They also out-compete native birds for nesting sites. Because the native birds can't breed, their numbers decrease. Refer to **Figure 1.11**.

European starlings are an introduced species in North America. An **introduced species** is a kind of plant, animal, or other organism that lives in a place where it is not found naturally. The species has been introduced there, either deliberately or accidentally, by human activities.

Introduced species often thrive in their new ecosystems, because there are so few limiting factors to keep their populations from growing too large. So they often survive and reproduce better than native species do. As a result, native species cannot get the resources they need, and their numbers tend to decrease. Not all introduced species are harmful, though. Apples, corn, and many other food crops grown in Ontario are introduced species. Limiting factors such as consumers and disease keep populations of these species in balance.

introduced species: any species that has been introduced into and lives in an ecosystem where it is not found naturally

▲ **Figure 1.11** Starlings tend to congregate in very large groups, causing deafening noise and leaving behind great accumulations of dung.

Introduced Species Can Affect Species Diversity

Diversity refers to the "diverse-ness" or "different-ness" of things. So **species diversity** is the number and variety of different species of living things in an area. Species diversity in an ecosystem tends to decrease when an introduced species becomes well-established. For example, purple loosestrife lives in balance with other plants in its European ecosystems. But in Canada, this introduced species is deadly to other kinds of plants. Purple loosestrife, shown in **Figure 1.12**, quickly takes over a wetland ecosystem. It soaks up much of the water of the ecosystem and easily out-competes native plants. Loss of the native plants reduces food and nesting sites for waterfowl, and the ecosystem soon becomes choked off to other wildlife. As time passes, a healthy multi-species ecosystem changes to one that consists almost entirely of purple loosestrife!

species diversity: the number and variety of different species of living things in an area

▲ **Figure 1.12** Purple loosestrife is called the beautiful killer. This introduced species has invaded many aquatic ecosystems across Canada, turning them into terrestrial ecosystems with very little species diversity.

LEARNING CHECK

1. Use pictures or words to explain the following terms: introduced species, species diversity.
2. Introduced species such as apples and corn have less impact on ecosystems than European starlings. Explain why.
3. Use a graphic organizer to summarize the biotic and abiotic parts of an ecosystem that are affected by purple loosestrife.

Activity 1.13

Inquiry Focus

ONTARIO'S MOST WANTED—NOT!

1. These are some of Ontario's most destructive introduced species. Choose one, and use Internet and library resources to find out how and why this species is a threat to species diversity. Analyze the information you gather for reliability and bias. (Turn to Science Skills Toolkit 8: How to Do a Research-Based Project, for help with assessing information for reliability and bias.)
2. Create your own "Wanted" poster to communicate your findings.

WANTED

Asian Long-Horned Beetle: Small but dangerous insect wanted on numerous counts of forest destruction.

Eurasian Watermilfoil: Aquatic plant wreaking havoc on Ontario lakes. A real slippery character.

Sea Lamprey: A parasitic fish wanted for sucking the guts out of native fish species. Well-known to authorities.

Zebra Mussel: A striped suit seems just right for this critter, wanted for vandalism in the Great Lakes area.

Pollutants from human activities can travel within and between ecosystems.

watershed: any area of land (either natural or human-made or both) that drains into a body of water

You live in a watershed. So do all the other living things in and around your community. A **watershed** is any area of land that drains into a body of water. The area of land can be a natural area, or it can be a city, or both. The body of water that the land area drains into could be a pond, a river, a lake, or the ocean. **Figure 1.13** shows a watershed.

All watersheds include land and water. In other words, all watersheds connect terrestrial ecosystems with aquatic ecosystems. This connection means that what we do on land or in water can affect the land or water around us. For instance, if you pour old juice down the sink or flush a toilet, the wastes go somewhere. Where might that be?

- If you live on a farm or in a rural area, the wastes flow into a septic tank and then into a septic field on your property. There, they seep into the ground. Some of those wastes are taken up by the roots of plants. Some are consumed by soil organisms. And some seep deeper into water flowing under the ground, where they are carried to other bodies of water.
- If you live in a city or in a small urban area, the wastes flow into a system of pipes. These pipes lead to larger pipe systems that direct the wastes either to waste-water treatment plants or directly into bodies of water.

▼ **Figure 1.13** You share your watershed with thousands, perhaps millions, of other living things. So everything you do affects each of those living things.

Watersheds provide:
- water people use for drinking, cooking, and washing
- water used by other living things for drinking, washing, and cooling off
- places to live
- irrigation for farmland
- water for use by industries for cooling and cleaning equipment
- recreation areas for swimming, boating, snowmobiling, relaxing
- beauty

In both cases, the wastes from your watershed feed sooner or later into a body of water of some kind. In weeks, months, or years, trace amounts of some of those wastes will appear elsewhere. This "elsewhere" might be a place nearby or a place farther away. It even might be a place in another province or another country.

Go to **scienceontario** to find out more

LEARNING CHECK

1. What is a watershed?
2. How do watersheds connect terrestrial and aquatic ecosystems?
3. Why do we need to be careful about the things we wash down the drain and flush down the toilet?

Literacy Focus

Activity 1.14

A WATERSHED MIND MAP

Use the diagram below to make a mind map with "Watershed" as your starting term. In your mind map, include all the ways that the watershed is being used. Use a different-coloured pen to include other ways that watersheds are used but which are not shown in the diagram

Making a DIFFERENCE

As captain of her high school's 2007 Envirothon team, Dayna Corelli helped develop an award-winning proposal that was reviewed by the city of Sudbury. Dayna's team was concerned that water was wasted during lawn-watering and street-cleaning. They were also worried that the city's sewer pipe system could not handle heavy rainfall. This could cause storm sewers to overflow and dump raw sewage into the Great Lakes. Since reviewing the proposal, and recommendations from other groups, Sudbury has restricted lawn-watering, made street-cleaning more efficient, and improved the sewer system.

Dayna also has been a member of her school's Environmental Club and coordinated its recycling program. She campaigns to promote recycling, energy efficiency, water conservation, and anti-idling. "I am committed to continuing to make my community more environmentally friendly," she says.

What changes could your municipality make to improve water quality and energy conservation?

Rebekah Parker has motivated other people to help the environment. In 2007, she created the "Living Out Loud" conference for students in the Waterloo Region. More than 60 students took part to discuss environmental and social issues. Rebekah also started an Eco-Carnival in her community. She encouraged other teens to run environmental activities for elementary students at the carnival. While in high school, Rebekah also led the school's Roots and Shoots Club and organized her community's Car Free Festival.

Rebekah believes young people should pick one issue they are passionate about and get involved. "Don't be afraid to get your hands dirty. I think students can be more inspired to work toward saving a local conservation area if they actually have the chance to go and spend a day there."

How could you motivate people in your community to help the environment?

Investigation 1C

Human Activity in a Local Ecosystem

In what ways are human activities changing ecosystems in your community?

Skill Check

✓ Initiating and Planning
✓ Performing and Recording
✓ Analyzing and Interpreting
✓ Communicating

What To Do

1. With your class, gather stories from newspapers, TV, radio, and the Internet about activities that affect ecosystems in or near your community. Here are some examples of issues that might affect local ecosystems.
   - Diverting water for a construction project lowers the water level. This reduces the number of places where water birds nest and fish reproduce.
   - Adding fertilizers to a local field changes the make-up of the soil. This affects the kinds of plants that can be grown in the field.
   - The shoreline of a lake is being developed for a new recreation area. This can affect the numbers and kinds of organisms that can live by the lakeshore.
   - Chemical pesticides are used to kill a harmful introduced species. The chemicals may leach into rivers and lakes.
   - The growing human population is filling a landfill site with wastes faster than originally planned. The site must be expanded, or another solution must be developed.

2. Work in a group. Choose several ecosystems to study. Mark these ecosystems on a map of your community.

3. Your teacher will assign one of these ecosystems to your group.

4. Write a hypothesis to explain how a human activity that affects this ecosystem could threaten its health.

5. With your group, create a list of sources of information to help you learn about the activity and predict possible consequences for the ecosystem. Analyze all information for reliability and bias before using it. Create a "References Cited" page to record your sources of information.

6. Create a presentation to communicate your ideas and your findings.

What Did You Find Out?

1. What did you learn about the ways that human activities are changing ecosystems in your community?

TOPIC 1.5 HOW DO HUMAN ACTIVITIES AFFECT ECOSYSTEMS? • MHR 59

Activity 1.15

Research Focus

I REMEMBER WHEN...

The ecosystems in and around your community have experienced a great deal of change over the last few decades. Many of these changes are due to human activities. In this activity, you will interview seniors or Elders who have lived in your community for a long time to learn what changes they have observed.

What You Need

writing materials

audio visual equipment, such as a video camera, tape recorder, or camera (optional)

What To Do

1. Your teacher will make arrangements for you to interview a local senior or Elder, either at your school or in the community.

2. Prepare a list of questions you would like to ask for this interview. Here are a few examples of questions you could ask.
   - How long have you lived in this area?
   - What was it like when you were a child (or when you first came here)?
   - How has the city (or town, or village, or area) changed since then?
   - Are there more or fewer animals around than there were in the past?
   - What kinds of work and other activities did people do in the past? How did their work and other activities affect the land and living things in the area?
   - Have you noticed changes in local ponds, rivers, or lakes? What has caused them?

3. Show your list of questions to your teacher for approval before you begin your interview. Your teacher will discuss guidelines for the interview process with you before you begin.

4. Take notes during the interview. You could also use audio visual equipment, such as a video camera, tape recorder, or camera, to capture the interview.

5. Use the material you gathered in your interview to prepare a newspaper article, a blog, or a film documentary of your research findings.

What Did You Find Out?

1. Did local ecosystems change as much as you expected over the years? Explain.

2. What part of your research surprised you the most? Why was this the case?

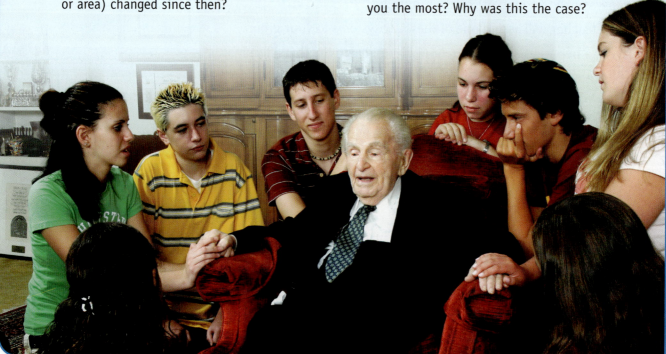

Topic 1.5 Review

Key Concepts Summary

- We cannot always accurately predict the consequences of our actions.
- Introduced species can affect the health of ecosystems.
- Pollutants from human activities can travel within and between ecosystems.

Review the Key Concepts

1. **K/U** Answer the question that is the title of this topic. Copy and complete the graphic organizer below in your notebook. Fill in four examples from the topic using key terms as well as your own words.

   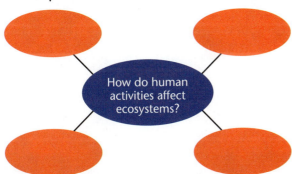

2. **K/U** Refer to **Table 1.2**. What biotic and abiotic parts of ecosystems can be affected by human activities?

3. **A** Make a flowchart to show the effects that a plastic bottle of water might have on the biotic and abiotic parts of ecosystems. In your answer, consider the manufacturing process as well as the disposal of the plastic water bottle.

4. **K/U** Explain how an introduced species can affect species diversity.

5. **A** a) Use a cause-and-effect map to show why people should not release exotic pets, such as snakes, hedgehogs, or tarantulas, into the wild if they do not want to keep them any longer.

   b) Describe at least one responsible action that people could take to provide exotic pets they no longer want with the care and support they need.

6. **C** Make a labelled drawing to show an example of how watersheds connect terrestrial and aquatic ecosystems.

7. **T/I** Scientists have confirmed that the blood of certain shark populations contains common prescription drugs. Using your knowledge of how pollutants travel within and between ecosystems, make a hypothesis that could account for this phenomenon.

8. **T/I** People sometimes build dams to divert water from rivers, streams, and lakes. The photo below shows one of the reasons that people divert water. Water diversion results in lower water levels and affects waterfowl nesting areas and fish reproduction.

   a) Identify at least two other reasons that people would divert water.

   b) Research the effects of water diversion on waterfowl and fish in rivers and streams.

   c) Summarize your research in a PMI chart or another graphic organizer of your choice.

The dams located at Niagara Falls divert water to generate electricity at the hydroelectric plant located there.

Topic 1.6 — How can our actions promote sustainable ecosystems?

Key Concepts

- We must understand and commit to sustainability.
- We must understand the link between biodiversity and sustainability.
- Our actions can maintain or rebuild sustainable ecosystems.
- You can choose actions that benefit ecosystems now and for the future.

Key Skills

Inquiry
Literacy
Research

Key Terms

sustainability
biodiversity
equilibrium

It began with a single city–Sydney, Australia–in 2007. As 2 million people and 2000 businesses shut down their lights for one hour, they sent a message into the world, to all who might hear: "We are concerned about climate change, and we hope our simple act inspires you to reduce activities that release excess carbon and other substances that contribute to it."

The world heard. As part of the second Earth Hour, Sydney was joined by people from 35 countries and more than 400 cities around the globe. Those people included students from 925 Ontario schools and their families. Across the province, demand for electrical energy dropped 5.2 percent during Earth Hour. At the University of British Columbia, demand dropped a full 100 percent—for the whole day!

The following year, 2009, was even more successful. People from 88 countries and more than 4000 cities took part. The province of Ontario, minus Toronto, saw an overall decrease of 6 percent in electrical energy used. Toronto, meanwhile, decreased its usage by 15.1 percent.

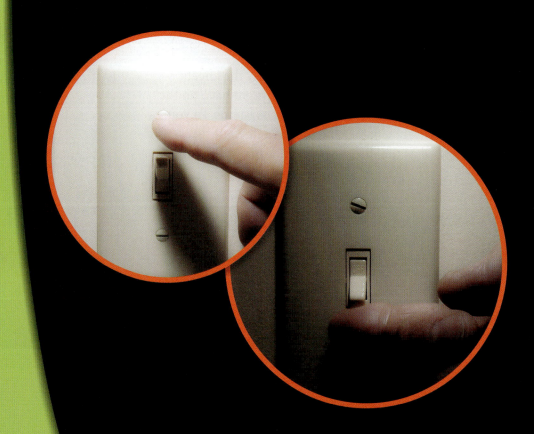

Starting Point Activity

Almost half of Ontario's energy drop came from Toronto and the surrounding region. The results are shown in the graph below.

1. Many cities and towns kept their own Earth Hour data. Find out the data for the community where you live, plus three other communities in other parts of the province.
2. Earth Hour 2008 asked people to turn off only their lights for one hour. What other things around the home and in the community could be turned off as well?
3. Some people have said that the Earth Hour event has little or no effect on dropping energy use worldwide.
   a) How could the event have a greater effect?
   b) Even if the event has little or no effect on energy use, what other benefits might it have?

Earth Hour Results for Select Cities in Ontario in 2008

City	Percent drop	Megawatts saved	Number of homes the saved energy would power
Milton	15%	2.85	2 580
Newmarket	14%	12	12 000
Aurora	10.2%	6.6	6 600
Halton Hills	9.4%	4.8	4 800
Toronto	8.7%	262	262 000
Oakville	8.2%	14	14 000
Clarington	7.7%	2	2 000
Uxbridge	7.6%	0.3	300
Whitby	7.5%	7.6	7 600
Ajax Pickering	6.7%	10.7	10 700
Burlington	6.6%	11.5	11 500
Markham	6%	22.2	22 200
Port Perry	5.4%	0.3	300
Oshawa	4.8%	6.9	6 900
Mississauga	3.6%	29	29 000
Vaughan	3.6%	13.7	13 700
Richmond Hill	3.6%	14.5	14 500
Brampton	3.2%	13.3	13 300
Brock	3.1%	0.2	200

We must understand and commit to sustainability.

sustainability: maintaining an ecosystem so that present populations can get the resources they need without risking the ability of future generations to get the resources that they will need

The word "sustainability" can mean different things, depending on who uses it and in what situation. In terms of what you have been learning in this unit, one way to think about the word is this: **Sustainability** is maintaining the abiotic and biotic parts of an ecosystem so that present populations can get the resources they need without risking the ability of future generations to get the resources that they will need.

Sustainability helps to ensure that populations stay within the carrying capacity of their ecosystem. It is a way of believing, and thinking, and acting that takes into account the effects that our actions will have on the ability of future generations of all living things to live a good life.

LEARNING CHECK

1. In terms of this unit, what does sustainability mean?
2. What resources do we humans use that future generations will need as well?
3. Provide two examples of ways that we can ensure that future generations will have what they need to live a good life.
4. Provide three examples of your own actions that affect ecosystems. Classify these actions as sustainable or unsustainable.

64 MHR • UNIT 1 SUSTAINABLE ECOSYSTEMS AND HUMAN ACTIVITY

Literacy Focus

Activity 1.16

REFLECTING ON RESPONSIBILITIES

Read and listen to the words below, spoken by the wise of the Haudenosaunee, who are also known as the Iroquois.

1. As you read and listen to these words, how do you feel:
   - about yourself?
   - about others who are dear to you?
   - about your past, present, and future?
   - about your connection to our shared planet Earth?

   Share your feelings with others or record them only for yourself, as you wish.

2. Read and listen again to the last line: "We must consider the effects our actions will have on their ability to live a good life." This line helps to guide us in the way we feel, think, and act toward Earth as we live our lives. Non-Aboriginal people today describe this guidance as "sustainability." Use a word map to show how you would explain the meaning of this term.

We acknowledge one another, female and male. We give greetings and thanks that we have this opportunity to spend some time together.

We turn our minds to our ancestors and our Elders. You are the carriers of knowledge, of our history.

We acknowledge the adults among us. You represent the bridge between the past and the future.

We also acknowledge our youth and children. It is to you that we will pass on the responsibilities we now carry. Soon, you will take our place in facing the challenges of life. Soon, you will carry the burden of your people. Do not forget the ways of the past as you move toward the future. Remember that we are to walk softly on our sacred Mother, the Earth, for we walk on the faces of the unborn, those who have yet to rise and take up the challenges of existence.

We must consider the effects our actions will have on their ability to live a good life.

We must understand the link between biodiversity and sustainability.

There are at least 2 million species (kinds) of organisms on Earth. These are just the species we know about. Some scientists estimate there could be as many as 100 million more species yet to be discovered. There is truly a great diversity of species of living things on Earth.

Now think about ecosystems. An ecosystem's size and the number of populations it supports is limited only by the abiotic and biotic parts of that ecosystem. Any ecosystem may contain tens, hundreds, thousands, and more smaller ecosystems. There is truly a great diversity of ecosystems on Earth.

biodiversity: all the diversity of species that live in an ecosystem, as well as all the diversity of ecosystems within and beyond that ecosystem

There is a word that describes the great diversity of Earth's species and the great diversity of Earth's ecosystems at the same time. That word is biodiversity. **Biodiversity** is all of the diversity of species that live in an ecosystem, plus all of the diversity of ecosystems within and beyond that ecosystem. So biodiversity is all the different kinds of living things in a certain place, as well as all the different kinds of places within that place and elsewhere.

Linking Biodiversity with Sustainability

So how are biodiversity and sustainability linked? You already know the answer when you remember that everything is connected. A sustainable ecosystem must maintain a state of balance between its diverse living parts and its non-living parts. This state of balance is called **equilibrium**. An ecosystem that is in equilibrium (in balance) tends to have a high degree of biodiversity. Such an ecosystem tends to be a sustainable ecosystem.

equilibrium: a state of balance in an ecosystem

LEARNING CHECK

1. Explain how biodiversity refers to species as well as ecosystems.
2. How does the term "equilibrium" apply to ecosystems?
3. Draw a picture to represent a sustainable ecosystem. Include the ideas of species diversity, biodiversity, and equilibrium.

Inquiry Focus

Activity 1.17
LOOK FOR THE LINKS

The pictures show food webs in a forest ecosystem and an arctic ecosystem.

1. Compare the abiotic parts of these two ecosystems. For instance, how does the amount of sunlight in each ecosystem compare? What about temperature, water, and shelter?

2. Compare the biodiversity in these two ecosystems. Why would one ecosystem have a higher biodiversity than the other?

3. A rainforest has a much higher biodiversity than a desert. And yet both ecosystems can be in equilibrium. In other words, both ecosystems can be equally sustainable. What does that tell you about the link between biodiversity and sustainability?

Our actions can maintain or rebuild sustainable ecosystems.

We can harm ecosystems when we forget to think about the consequences of our activities. But we also have the power to heal ecosystems. Many communities have adopted programs to maintain ecosystems that are in danger of becoming unhealthy and to rebuild ecosystems that have already become so. The pictures on these two pages give a sample of the many ways that human activities are helping ecosystems.

A healthy wetland is a hotbed of biodiversity. Alfred Bog in southern Ontario is one such place. Thanks to the efforts of concerned citizens and government officials, more than 70 percent of this wetland is managed as a nature reserve, keeping it safe from mining and other activities that would harm it.

Elk were once native to Ontario, but by the late 1800s, these majestic animals were gone as a consequence of growing human settlements and over-hunting. Efforts to restore elk to Ontario have been in place since the mid-1990s. Four populations are now re-established in the areas of Sudbury, Bancroft/North Hastings, Lake of the Woods, and the north shore of Lake Huron.

This beetle, originally from Europe, is one of the species of insects used in the fight against purple loosestrife. Use of living things to control introduced species is called biocontrol. This can be effective in reducing the population size of an introduced species. However, it rarely can remove the invader entirely.

LEARNING CHECK

1. Describe two ways that humans are helping ecosystems.
2. Converting a strip mall to condominiums instead of building on a field is an example of smart growth. Explain why.
3. What might be some unintended consequences of using an introduced species for the purpose of biocontrol?

ACTIVITY LINK
Activity 1.18, on page 73

Go to **scienceontario** to find out more

Farmers and home-owners in rural areas, as well as in cities, often set up special boxes to provide places for birds to establish nests. People who love and respect birds and their role in ecosystems put up nest boxes to make up for the trees that have been logged to clear space or provide timber for various products.

Urban sprawl happens as cities with growing populations increase their size by spreading into natural areas and farmland. A strategy called smart growth helps by concentrating growth in the centre of a city, rather than in outlying areas. Homes and businesses intermingle, while green spaces are preserved. Smart growth also enhances public transit, which reduces traffic pollution.

INVESTIGATION LINK
Case Study Investigation, on page 76

You can choose actions that benefit ecosystems now and for the future.

Consumers have power. What choices do you make about the products you will and will not buy? What reasons lie behind, or motivate, your choices?

Volunteers inspire by their commitment and example. Where do you, or can you, volunteer your time? Who benefits from your willingness to share a part of yourself?

Change in society starts with change in individuals. Each one of us has tools and gifts that can help us bring about change. At the start of this unit, you thought about this statement:

"Instead of waiting for change that might never come, many people are choosing to become the change they are waiting for."

Citizens have responsibility. In what ways are you a citizen of your community? Your province? Your country? Your planet? What responsibilities do you have as a citizen?

Over the course of this unit, you have seen ways that humans can harm ecosystems as well as help them. Each of us holds in our thoughts, in our hands, and in our hearts the ability to create a more sustainable future. How do you use this ability now? How will you use it in the future?

Making a DIFFERENCE

In Grade 9, Yvonne Su found out that her Newmarket school was not recycling, because it lacked the resources. So she and some friends and teachers decided to tackle the problem themselves. They started a recycling and environmental club. Yvonne has been engaged in environmental activities ever since.

"As Grade 9s, my friends and I didn't know where to turn to learn more about our planet. But after speaking to some teachers, we found out that our greatest resources were right in front of us—our science classes."

The more Yvonne and her friends learned, the more they wanted to share their knowledge. They organized campaigns at their school about environmental issues. They then took their campaigns to schools across Canada.

What changes could be made at your school to help the environment?

Students at a Toronto school have adopted a local aquatic ecosystem. Over the past 10 years, students in Chaminade College's science classes and environmental club have worked on projects to restore the Black Creek. "The Black Creek was once a pristine, cold-water trout and salmon fishery, which, with some effort, can be returned to its former glory," says science teacher Tino Romano, who is moderator of the environmental club.

The school's efforts include an ongoing project to decrease erosion of the banks of the creek. Students have planted aquatic and terrestrial plants in and around the creek. They also have removed garbage from the creek and the area around it. The school also has its own fish hatchery. Each year, students raise thousands of brown trout and 100 Atlantic salmon in the hatchery and release them into the nearby Humber and Credit Rivers. Through their efforts, Chaminade students hope to re-establish spawning grounds for brown trout.

Is there a local stream that your school could "adopt" and help restore?

Research Focus

Activity 1.18

TOWN COUNCIL MEETING

When it comes to environmental decision making, not everyone has the same opinion. People come from different backgrounds and have different life experiences. As a result, people's viewpoints on the same environmental issue may differ dramatically.

What To Do

1. As a class, identify a local project that will have an effect on the environment if it is carried out. Examples include building a new golf course, introducing a sport fish that is not native to local streams, or paving over a city green space. You can also create your own scenario to use for this activity.

2. Agree on four to six citizens who will present their point of view at a town council meeting concerning this project. The citizens should all have different points of view about the project. For example, the project may be to open a mine near a fishing lake. Citizens presenting their point of view might include an unemployed miner, a local business person, the head of a company that uses minerals from the mine, a local angler, the Minister of Tourism, and a resident concerned about water quality.

3. Form groups so that each group represents one citizen. You do not have to agree with the point of view of your group's citizen, but you must present it fairly.

4. Use your library, the Internet, and other sources to research the issues concerning the project. Focus on questions that your citizen would likely want answers to. For example:
   - What economic or social benefits will the project bring that will affect me?
   - What impacts will it have on the soil, vegetation, and water that I might care about?
   - Use an organizer such as a PMI chart to organize the results of your research.
   - How might the direct and indirect effects of the project affect my future and the future of those I care about?
   - Do the impacts of this project sit well with my beliefs and values?

5. Once you have collected your information, prepare a presentation for the town council (played by your teacher) that explains your views and concerns about the project. Explain why you do or do not support the project. Use charts, tables, or graphs to display data where helpful.

6. Present your point of view to the council, allowing time for questions from the council and other citizens (your classmates).

7. Based on the information presented, try to come to a joint decision, as a town, to approve, abandon, or modify the project.

What Did You Find Out?

1. Give an example of how the knowledge you gained in this unit helped you research and present your citizen's position.

2. What other issues were raised during the presentation? Identify any that you thought were interesting or important.

3. Do you think the process you modelled is an effective way to make decisions about local projects that have an environmental impact? Explain.

Case Study Investigation

Securing a Bright Future for Songbirds

ELECTRONEWS

SATURDAY, JUNE 2, 2114

TORONTO NOVA
TORONTO, ONTARIO

Toronto's Oldest Citizen Remembers Songbird Past

Toronto resident Jared Riozk is celebrating a milestone. Born at the end of the 1990s, Riozk is celebrating his 115th birthday today. Living with his 58-year-old granddaughter, Riozk enjoys quiet times with his family and an occasional game of Masiko, which he got hooked on when it came out in 2065. Although he doesn't get out much these days, Riozk spends a lot of time recalling a greener, feather-filled Toronto. "There were a lot of songbirds when I was young," Riozk recalls. "Cardinals, chickadees, goldfinches, orioles—you name it." Riozk smiles a bittersweet smile. "People tried to preserve the areas where songbirds lived and bred, but it was too late. Urbanization pushed its way through the last of the natural ecosystems in this city, and pushed the songbirds out right along with them." Asked what he would do differently if he could go back in time, Riozk replied, "I don't know. Not think someone else would take care of it, I guess. Take action. Make a difference." Riozk stares at the empty blue horizon as he speaks. The last songbird was spotted in Toronto 22 years ago.

Cardinals are common visitors to urban Ontario green spaces.

Unless adequate conservation measures are put in place, human expansion may make birdsong a rare sound in urban areas such as Toronto.

The Science behind the Story

Every day, multiple futures lie before us. But we also have opportunities to influence these futures, to create the world we wish to see. Today, urban sprawl plays a large role in the destruction of many natural ecosystems. Native songbirds rely on the plant life in these ecosystems for food, shelter, and other resources. Without these ecosystems, the future of songbirds may be bleak in these urban areas.

Pause and Reflect

1. How can urban sprawl affect songbird populations?

How can we influence the future of songbirds?

Private citizens can make a difference when it comes to preserving songbird habitat in urban areas. In the greater Toronto area, for example, 80 percent of the natural green spaces are privately owned. This means that native backyard plantings and water sources can provide songbirds with the resources they need to thrive in this area. Not only does the number of songbirds increase, but so does their diversity. Even native plantings on balconies can help provide a stop-over sanctuary in the city that songbirds can rely on year after year.

Pause and Reflect
2. How can private citizens help preserve songbird habitat in urban areas?

What projects are already forging a brighter future?

Project CHIRP! encourages people to restore natural vegetation to private and public green spaces in southern Ontario. CHIRP stands for **C**reating **H**abitat in **R**esidential **A**reas and **P**arkland. The project hopes to inspire gardeners to plant native plants in their gardens, providing food, water, and shelter for local songbirds. It was started by Christina Sharma, an Etobicoke, Ontario, gardener who turned her love of native plants and songbirds into an environmental program that makes a difference.

Pause and Reflect
3. How does Project CHIRP! make a difference for urban songbirds?

Inquire Further
4. Find out which native plants provide food and shelter for songbirds where you live.
5. Design a garden that would increase the number of songbirds that visit your school grounds.
6. Investigate other projects that restore wildlife populations where you live.

Project CHIRP! shows people how to plant native plants in gardens such as this one. Such gardens provide food, water, shelter, and other resources for songbirds.

Investigation 1D

Skill Check

✓ Initiating and Planning

✓ Performing and Recording

✓ Analyzing and Interpreting

✓ Communicating

Investigating a Local Environmental Project

Do you have a friend who is part of a group that is working to preserve a local bog? Is your neighbour involved in a spring riverbank clean-up each year? Are you involved in a local frog count? In this investigation, you will research an environmental program or project to find out how it is linked to the sustainability of a local ecosystem.

What To Do

1. With your class, identify sources of information that could be used to discover local projects that promote the sustainability of a terrestrial or aquatic ecosystem in your area.

2. With your group, use these resources to choose a local project that interests you. Ask your teacher for permission to research this project before you continue.

3. As a group, write a list of questions that you would like to answer about your chosen project. Your goals are to learn more about the project in general and find out why it is important to the sustainability of the ecosystem it is targeting.

    Below are examples of questions that your group could ask.

    - Is there a program that is working to green the grounds of your school? What impact has this program had on the local ecosystem? What other plans could be put in place to help this program succeed?

    - Is there a project in your community to reduce the amount of pollution released into a nearby river or lake? How has this aquatic ecosystem become more sustainable as a result of the project? What other actions could be taken to improve the sustainability of the ecosystem?

    - Has the implementation of an Environmental Farm Plan (EFP) changed practices on a farm near your community? What changes have been made? How have these changes affected the sustainability of nearby terrestrial and aquatic ecosystems?

4. Use your research to design a website, poster, or brochure to inform other community members about the project.

What Did You Find Out?

1. How has the project you investigated improved the sustainability of a local ecosystem?

2. Suggest another way the project could make the ecosystem more sustainable. How would your suggestion do this?

Topic 1.6 Review

Key Concepts Summary

- We must understand and commit to sustainability.
- We must understand the link between biodiversity and sustainability.
- Our actions can maintain or rebuild sustainable ecosystems.
- You can choose actions that benefit ecosystems now and for the future.

Review the Key Concepts

1. **K/U** Answer the question that is the title of this topic. Copy and complete the graphic organizer below in your notebook. Fill in four examples from the topic using key terms as well as your own words.

2. **K/U** Revisit the word map you made in Activity 1.16 that described what sustainability meant to you. Now that you have finished the topic, add any new understandings or insights to complete your picture.

3. **T/I** Many towns and cities in the country ban the use of pesticides in parks and on home lawns. This is one example of a human activity that can work in support of, rather than against, ecosystems.
   a) Identify at least three other human activities that you think work in support of ecosystems.
   b) Conduct a risk-benefit analysis in the form of a PMI chart to assess one of these activities. Based on your analysis, would this action be something worth doing to promote sustainable ecosystems?

4. **C** Choose a different activity from the ones you identified in question 3. Create a poster or write an opinion piece such as an editorial or blog to persuade people why they should undertake that activity.

5. **K/U** What does it mean to say that an ecosystem is "in equilibrium"?

6. **A** Refer to the photo of the pond ecosystem below. Suppose that a poisonous substance leaked into the water and killed most of the fish. Explain how this change in equilibrium might affect the sustainability of the ecosystem.

   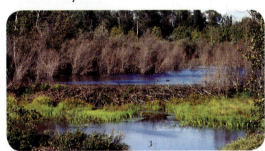

   A pond ecosystem

7. **K/U** a) What does "urban sprawl" mean?
   b) Use a cause-and-effect map to show possible effects on a city of using the strategy known as "smart growth."

8. **T/I** Use the Internet to find examples of different online campaigns that people have launched to promote sustainability. Present one example to the class and discuss whether or not you think the campaign has been effective. How might you take advantage of technology to help maintain or develop sustainable ecosystems?

SCIENCE AT WORK

CANADIANS IN SCIENCE

Lake Simcoe is one of the largest aquatic ecosystems in Ontario and is home to many different populations of organisms. It is also less than an hour away from half of Ontario's human population. Development around the lake continues to put stress on the ecosystem. The Lake Simcoe Region Conservation Authority works to restore and protect Lake Simcoe and its watershed. Chandler Eves started working for the authority as a fisheries technician six years ago. Fisheries technicians conduct field research and collect data that can be used by municipalities to make decisions about what types of development should take place in and around an ecosystem.

▲ Chandler Eves is an inventory project coordinator for the Lake Simcoe Region Conservation Authority.

What challenges do you face in your job?
The biggest challenge is unpredictability. "Humans can sometimes forget that we are not always in control. Field work is extremely unpredictable," says Chandler. Fisheries technicians have to be good at solving problems and adapting to change.

What do you find most rewarding about your job?
Chandler says the most rewarding part of his job is seeing how data he collects is used to make important decisions. For example, he might collect information about water temperatures and fish species in a section of a stream. Some species of fish can breed in warm water and some breed only in cold water. Building and development can lead to thermal pollution, making water warmer and not suitable for the breeding of some species. Information Chandler collects can be mapped, and municipal planning departments can use the maps to make decisions about where development should be allowed.

What advice do you have for students who are interested in getting into your field?
Chandler thinks students interested in working in his field should gain as much field experience as possible, as soon as possible. Find opportunities to volunteer with local conservation authorities or nature groups, he says. "If you are interested, stay interested and plan ahead to ensure you have all the course requirements to enrol in an appropriate program."

Ecology at Work

The study of ecology contributes to these careers, as well as many more!

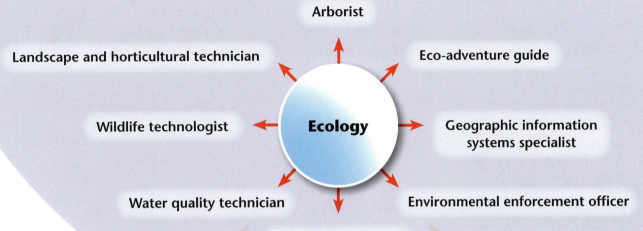

▲ Water quality technicians collect and analyze water samples and may also inspect sites for sources of contamination. They work for water treatment plants, governments, conservation authorities, and environmental consulting companies.

▲ Forestry technologists measure, survey, and map treed areas. They may be involved in reforestation, fire protection, harvesting, or tree health.

▲ Environmental enforcement officers help protect Canada's wildlife and natural resources from poachers, smugglers, and other people who break the law. They work for the federal government across the country.

Over To You

1. If you could interview Chandler Eves, what questions about his work would you ask him?
2. Why is it valuable for a fisheries technician to be a good problem solver?
3. Research a career involving ecology that interests you. If you wish, you may choose a career from the list above. What are the essential skills needed for this career?

Go to **scienceontario** to find out more

Unit 1 Summary

Topic 1.1: What are ecosystems, and why do we care about them?

Key Concepts
- Ecosystems are about connections.
- Ecosystems are made up of biotic (alive) and abiotic (not alive) parts that interact.
- Interactions between terrestrial (land) ecosystems and aquatic (water) ecosystems keep all ecosystems healthy.

Key Terms
ecology (page 11)

biotic (page 12)

abiotic (page 12)

ecosystem (page 13)

terrestrial ecosystem (page 14)

aquatic ecosystem (page 14)

Big Ideas
- Ecosystems consist of a variety of components, including, in many cases, humans.
- The sustainability of ecosystems depends on balanced interactions between their components.

Topic 1.2: How do interactions supply energy to ecosystems?

Key Concepts
- Photosynthesis stores energy, and cellular respiration releases energy.
- Producers transfer energy to consumers through food chains and food webs.
- Interactions are needed to provide a constant flow of energy for living things.

Key Terms
photosynthesis (page 20)

cellular respiration (page 20)

producer (page 22)

consumer (page 22)

food chain (page 22)

food web (page 23)

Big Ideas
- Ecosystems consist of a variety of components, including, in many cases, humans.
- The sustainability of ecosystems depends on balanced interactions between their components.

Topic 1.3: How do interactions in ecosystems cycle matter?

Key Concepts
- Abiotic and biotic interactions cycle matter in terrestrial and aquatic ecosystems.
- Photosynthesis and cellular respiration cycle carbon and oxygen in ecosystems.
- Human activities can affect ecosystems by affecting nutrient cycles.

Key Terms
decomposer (page 30)

nutrient (page 30)

nutrient cycle (page 30)

Big Ideas
- Ecosystems consist of a variety of components, including, in many cases, humans.
- The sustainability of ecosystems depends on balanced interactions between their components.
- Human activity can affect the sustainability of terrestrial and aquatic ecosystems.

Topic 1.4: What natural factors limit the growth of ecosystems?

Key Concepts
- Ecosystem growth is limited by the availability of resources.
- Abiotic and biotic factors limit populations in ecosystems.

Key Terms
population (page 42)
carrying capacity (page 42)
limiting factor (page 42)

Big Ideas
- Ecosystems consist of a variety of components, including, in many cases, humans.
- The sustainability of ecosystems depends on balanced interactions between their components.

Topic 1.5: How do human activities affect ecosystems?

Key Concepts
- We cannot always accurately predict the consequences of our actions.
- Introduced species can affect the health of ecosystems.
- Pollutants from human activities can travel within and between ecosystems.

Key Terms
introduced species (page 54)
species diversity (page 55)
watershed (page 56)

Big Ideas
- The sustainability of ecosystems depends on balanced interactions between their components.
- Human activity can affect the sustainability of terrestrial and aquatic ecosystems.

Topic 1.6: How can our actions promote sustainable ecosystems?

Key Concepts
- We must understand and commit to sustainability.
- We must understand the link between biodiversity and sustainability.
- Our actions can maintain or rebuild sustainable ecosystems.
- You can choose actions that benefit ecosystems now and in the future.

Key Terms
sustainability (page 65)
biodiversity (page 66)
equilibrium (page 66)

Big Ideas
- The sustainability of ecosystems depends on balanced interactions between their components.
- Human activity can affect the sustainability of terrestrial and aquatic ecosystems.

Unit 1 Projects

Inquiry Investigation: Investigating Compost

Nutrients are major limiting factors for growing crops. Human-made fertilizers provide nutrients, but their production and use pollutes ecosystems and the air. Composting may offer a partial solution to this problem. Through composting, food waste is changed into rich organic matter that provides nutrients for plant growth. Does compost provide enough of the nutrients plants need to thrive?

> **Inquiry Question**
> How does a plant grown with compost compare with one grown with synthetic fertilizer?

Initiate and Plan

1. Design a test to answer the Inquiry Question above. Include:
   - a hypothesis
   - a list of equipment and materials
   - a step-by-step testing method (show how you will control variables such as amount of light and water)
   - a list of dependent, independent, and controlled variables
   - safety precautions
   - a way to measure your results
   - a method for recording your results
2. Have your teacher approve your design.

Perform and Record

3. Carry out your test.
4. Summarize the growth results with a graph.

Analyze and Interpret

1. Which plant grew the most and was the healthiest during your investigation?
2. Did your results support your hypothesis? Why or why not?
3. Explain any sources of error, and list changes you would make if you were going to repeat the investigation.
4. Explain how the application of fertilizer (synthetic and compost) might affect nutrient cycles in ecosystems.

Communicate Your Findings

5. Present your results in a brief report that discusses the use of compost for growing food a) on farms and b) in home gardens.

> **Assessment Checklist**
> Review your project when you complete it. Did you...
> - ✔ generate a hypothesis? **T/I**
> - ✔ make a list of necessary equipment and materials? **T/I**
> - ✔ include a step-by-step testing method? **T/I**
> - ✔ list the dependent, independent, and controlled variables? **K/U**
> - ✔ record your results in an appropriate table and summarize them using a graph? **C**
> - ✔ present your results in a report that discusses the possible use of compost for growing food? **A**

An Issue to Analyze: Going Greener

You have learned how humans have an impact on the environment. You could probably make some changes in your lifestyle that would help protect nature.

Issue

What lifestyle changes could you make—in terms of the food you eat, how you get from place to place, or how you use heat and electricity—that would reduce your environmental impact? Are you willing to make those changes?

Initiate and Plan

1. Dr. David Suzuki has prepared a list of ten possible changes you could make, relating to food, transportation, and housing. Study the list provided by your teacher and Think/Pair/Share with a partner to choose two lifestyle changes that would reduce your environmental impact. Plan to try them for two weeks.
2. Research information that shows how your planned changes reduce your impact on the environment.
3. Set up a daily journal so you can record your experiences. Consider including the following:
   - the date
   - the change you are referring to
   - how often the change affected your activity
   - how easy or difficult it was to maintain the change
   - how you felt about making the change
4. Ask your teacher to review your ideas.

Perform and Record

5. Begin your lifestyle changes and your journal entries for two weeks.

Analyze and Interpret

1. At the end of two weeks, review your journal. Was it a struggle to make the lifestyle changes or are they becoming new habits?
2. Consider your lifestyle changes and prepare a cause-and-effect map to illustrate how your changes affect the environment.
3. Review your experience and compare it to the positive impact on the environment.

Communicate Your Findings

4. Make a decision about whether to continue the lifestyle changes and write two or three paragraphs summarizing and justifying your decision.

Assessment Checklist

Review your project when you complete it. Did you...

- [✓] select two different lifestyle changes to reduce your environmental impact? **K/U**
- [✓] use a variety of sources to research how these changes will reduce your impact on the environment? **T/I**
- [✓] keep a journal about your experience? **C**
- [✓] prepare a cause-and-effect map to evaluate how your changes affect the environment? **A**

Unit 1 Review

Connect to the Big Ideas

1. Ecosystems consist of a variety of components, including, in many cases, humans. Dr. Fritjof Capra, founding director of the Center for Ecoliteracy, wrote, "Humans are part of the web of life, not separate from it." Use words, pictures or a graphic organizer to explain this statement.

2. The sustainability of ecosystems depends on balanced interactions between their components. Choose any ecosystem and create a graphic organizer, such as a main idea web, to summarize the interactions that make the ecosystem sustainable. Be sure to include interactions between biotic parts, abiotic and biotic parts, and terrestrial and aquatic ecosystems where applicable.

3. Human activity can affect the sustainability of aquatic and terrestrial ecosystems. Imagine you are a member of a town council. A local business has requested permission to expand its buildings onto a nearby wetland. The wetland contains a large pond that is home to many species of birds, fish, and plants such as cattails. The wetland would be drained and a hotel would be constructed in the drained area. The expansion of the business would mean more jobs for the community. How would you vote on this proposal? Explain what factors you would consider when making your decision.

Knowledge and Understanding K/U

4. Create a table with three columns. In the first column, list all the key terms from this unit. In the second column, record a definition for each term, written in your own words. In the third column, sketch or draw a small picture that will help you remember the key term.

5. In a t-chart, provide five examples of terrestrial ecosystems and five examples of aquatic ecosystems. Include examples from this unit or examples of your own, and include both small and large ecosystems.

6. Create a cause-and-effect map showing how human impact on a terrestrial ecosystem can also affect an aquatic ecosystem.

7. In what ways do consumers rely on the Sun for their food?

8. Refer to the diagram below. What happens to the amount of energy available to a consumer that is farther along a food chain compared to the other consumers in the food chain?

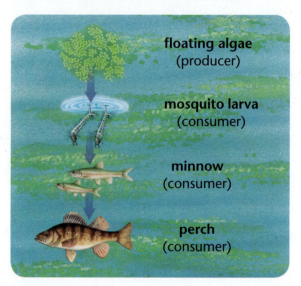

9. Summarize how photosynthesis and cellular respiration are essential for you to live.

10. Use a Venn diagram to compare these two models of energy flow in ecosystems: food chains and food webs.

11. Wolverines are strict carnivores (they don't eat plant material at all). Explain how the nutrients in a living plant might become part of the body tissue of a wolverine.

12. Some people describe cellular respiration as the reverse of photosynthesis. Explain why this is incorrect.

13. Make a drawing that shows the biotic and abiotic limiting factors in an ecosystem near your home.

14. Make a t-chart. In one column, list natural processes and human activities that add carbon dioxide to the atmosphere. In the second column, list natural processes and human activities that remove carbon dioxide from the atmosphere.

15. Explain how fertilizer run-off in a watershed could affect the biodiversity of an ecosystem.

16. Explain how the introduction of purple loosestrife to a wetland ecosystem can threaten the sustainability of that ecosystem.

Thinking and Investigation T/I

17. A pond supports a large population of minnows, which is a species of very small fish. A predatory fish population is introduced to the pond.

    a) Predict how the introduction of the predatory fish population will affect the carrying capacity of the minnow population.

    b) Do you think the introduction of the predatory fish population will affect the carrying capacity of any other populations in the pond? Justify your answer.

18. Plan (but do not actually conduct) an investigation to find answers to one of the questions below. Be sure to identify all variables, the data you will need to collect, and what method you will use to collect your data. Decide which format you will use to record your findings—for example, a chart.

    a) What are the abiotic and biotic parts of your schoolyard ecosystem?

    b) How do the abiotic and biotic parts interact with each other?

    c) What human activities have had an impact on the ecosystem?

19. Imagine you are an ecologist who has been called to investigate the reasons why the population of fish in a local stream has been declining. What tests could you do to determine if fertilizer run-off was the cause of the decline in the fish population?

20. Food and water are the limiting factors that usually have the greatest effect on population size. Hypothesize a possible reason or reasons for this relationship.

21. Use the information in the graph below to answer the following questions.

    a) Explain what happened to the moose and wolf populations between 1975 and 1980, and between 1990 and 1995.

    b) Explain why predation by wolves might be a limiting factor for moose.

    c) What other factors could influence the moose population?

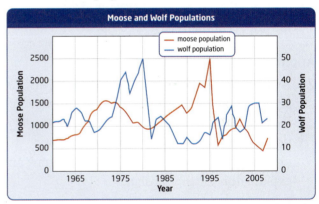

Communication C

22. If a vegetarian diet is more "environmentally friendly" than a meat-based diet, should we all be forced to become vegetarians? Explain your opinion.

23. Use a graphic organizer such as a Venn diagram to show the similarities and differences between photosynthesis and cellular respiration.

Unit 1 Review

24. Decomposers are an essential part of the cycling of matter. Imagine that all of the decomposers on Earth became extinct. Write a story or draw a comic strip that describes the effects of this extinction on producers and consumers (including humans), as well as the effect on the balance of carbon and oxygen.

25. Draw a picture or a flowchart to show how oxygen and carbon might cycle through the biotic and abiotic parts of a pond ecosystem.

26. Draw a cartoon or a storybook for a class of Grade 3 students that shows the impact of one human activity on an ecosystem and that concludes with one action everyone can take to lessen that impact. Be creative.

27. The unit opener featured song lyrics by John Mayer. Many singers and other artists use their work to convey messages about societal and environmental issues. Write your own song, rap, or poem that communicates your ideas about the impact humans have on ecosystems.

Application A

28. A gardener left grass clippings on her lawn after mowing it. She discovered that the lawn looked healthier than it did when she raked up and removed the clippings. Using your knowledge of nutrient cycles, write or draw an explanation for the gardener.

29. Explain how driving a car in Ontario could affect ecosystems elsewhere in Canada.

30. Answer the following questions about your own community.
    a) Is there new construction in or around your community? If so, how do you think this construction might affect an ecosystem near the area?
    b) Choose a species that lives in your community. List the limiting factors that might regulate the population of this species.

31. In an effort to divert waste and reduce fertilizer use, many municipalities in Ontario have adopted composting initiatives. However, these initiatives cannot always extend to schools, businesses, or apartment buildings, which all generate large amounts of wastes that can be composted.
    a) Create a graphic organizer that summarizes the pros and cons of a municipal composting initiative.
    b) Write an email or a letter to your local politician with your assessment of a municipal composting initiative and provide some suggestions about how it might be improved.

32. A ban on pesticides took effect in Ontario in April 2009. The ban prohibits the sale and use of pesticides for cosmetic use on lawns and in gardens, parks, and schoolyards.
    a) Research some effects that pesticides have on people and ecosystems.
    b) Research alternative methods for getting rid of unwanted weeds.
    c) Imagine that your neighbour was ignoring the ban and using pesticides to control dandelions on his lawn. Write a letter to your neighbourhood association or to the editor of your community newspaper explaining the impact of pesticides on people and ecosystems. Propose an alternative course of action that your neighbour could take.

33. In the unit opener, you were asked to answer how the person in the picture was acting as an agent for change in the world. Considering everything you have learned in this unit, describe a way in which you might now act, or are already acting, as an agent of change in the world.

Literacy Test Prep

Read the selection below, and answer the questions that follow it.

Populations

Ecologists define a population as a group of individuals of the same species that live together in the same place at the same time. An individual is one member of a population. If you want to identify a population, you need to know three things: the species, where the species lives, and when the species exists or existed. Populations change in size with the passage of time. The graph below shows the growth curve of a population of mosquitoes.

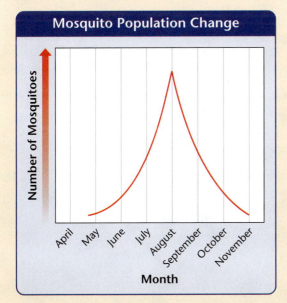

Multiple Choice

In your notebook, choose and record the best or most correct answer.

34. To identify a population, you do not need to know
    a) the species
    b) why the species exists
    c) when the species exists or existed
    d) where the species lives

35. The term that is defined in the opening sentence of the paragraph is
    a) ecologists
    b) population
    c) individuals
    d) species

36. The graph shows that the mosquito population peaks:
    a) between May and June
    b) during July
    c) during August
    d) between August and September

37. According to the graph, there is a sharp decline in the mosquito population during:
    a) spring
    b) summer
    c) fall
    d) winter

Written Answer

38. Summarize this selection. Include the main idea and two relevant points that support it.

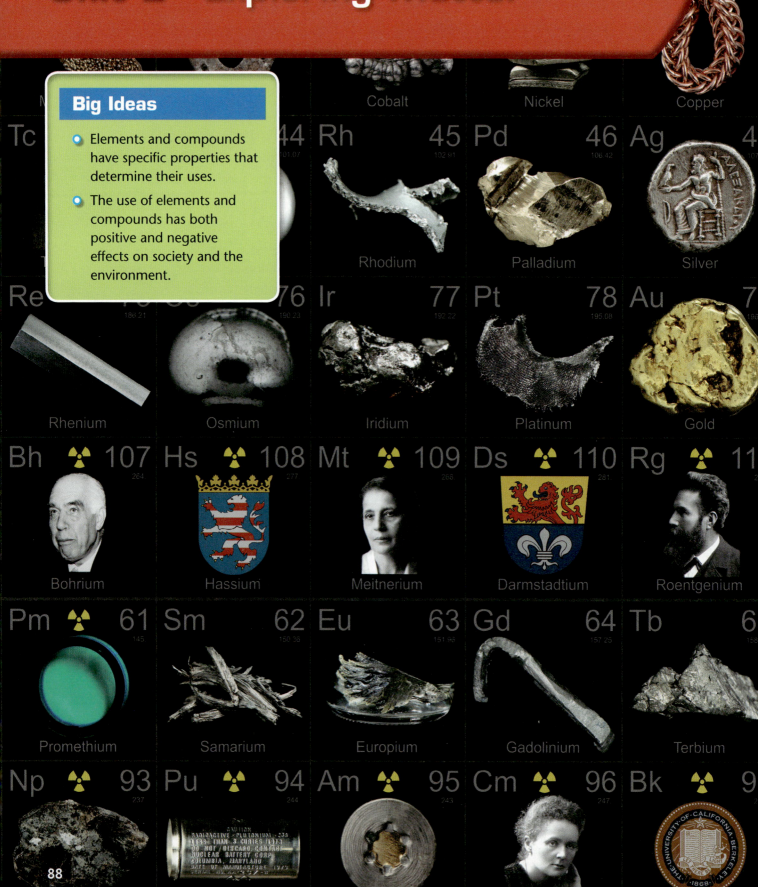

Unit 2 Exploring Matter

Big Ideas
- Elements and compounds have specific properties that determine their uses.
- The use of elements and compounds has both positive and negative effects on society and the environment.

"The Elements" by singer-songwriter Tom Lehrer

There's antimony, arsenic, aluminum, selenium
And hydrogen and oxygen and nitrogen and rhenium
And nickel, neodymium, neptunium, germanium
And iron, americium, ruthenium, uranium
Europium, zirconium, lutetium, vanadium
And lanthanum and osmium and astatine and radium
And gold and protactinium and indium and gallium
And iodine and thorium and thulium and thallium

There's yttrium, ytterbium, actinium, rubidium
And boron, gadolinium, niobium, iridium
And strontium and silicon and silver and samarium
And bismuth, bromine, lithium, beryllium, and barium

There's holmium and helium and hafnium and erbium
And phosphorus and francium and fluorine and terbium
And manganese and mercury, molybdenum, magnesium
Dysprosium and scandium and cerium and cesium
And lead, praseodymium, and platinum, plutonium
Palladium, promethium, potassium, polonium
And tantalum, technetium, titanium, tellurium
And cadmium and calcium and chromium and curium

There's sulfur, californium, and fermium, berkelium
And also mendelevium, einsteinium, nobelium
And argon, krypton, neon, radon, xenon, zinc, and rhodium
And chlorine, carbon, cobalt, copper, tungsten, tin, and sodium

These are the only ones of which
The news has come to Ha'vard
And there may be many others
But they haven't been discavard

Elements are the building blocks of matter. You and the rest of society depend on elements for your life and your way of life.

How could using elements have positive and negative effects on society and the environment?

Unit 2 At a Glance

In this unit you will learn that elements and compounds have specific properties that determine their uses. You will also learn that the use of these elements and compounds has both positive and negative effects on society and the environment.

Think about answers to each question as you work through the topic.

Topic 2.1: In what ways do chemicals affect your life?

Key Concepts
- Everything—including you and everything around you—is made up of chemicals.
- Substances have characteristics that make them useful, hazardous, or both.
- Handling chemicals and lab equipment safely and responsibly is a part of your life at school.

Topic 2.6: What are some characteristics and consequences of chemical reactions?

Key Concepts
- Compounds and elements are changed during chemical reactions.
- The properties of substances that make them useful can also make them dangerous.
- There are less-harmful alternatives to many products we use and depend on.

Exploring Matter

Topic 2.5: In what ways do scientists communicate about elements and compounds?

Key Concepts
- Chemical symbols represent elements.
- Chemical formulas are used to represent the types and numbers of atoms in compounds.

Topic 2.2: How do we use properties to help us describe matter?

Key Concepts
- Physical properties describe how matter looks and feels.
- Chemical properties describe how substances can change when they interact with other substances.

Topic 2.3: What are pure substances and how are they classified?

Key Concepts
- Pure substances are elements and compounds.
- Elements include metals and non-metals.

Topic 2.4: How are properties of atoms used to organize elements into the periodic table?

Key Concepts
- Elements are made up of atoms, which are made up of subatomic particles.
- Elements are arranged in the periodic table according to their atomic structure and properties.
- Elements in the same family (group) share similar physical and chemical properties.

Looking Ahead to the Unit 2 Project

At the end of this unit, you will do a scientific investigation about the effects of a common chemical reaction: the rusting of metals. Read pages 156–157. With tips from your teacher, start your project planning folder now.

Get Ready for Unit 2

Concept Check

1. In two minutes, jot down in your notebook all the words you can think of that describe matter. Share your list with a partner and exchange words that you did not have on your individual lists.

2. Examine the beach scene below, and write one example of each of the following in your notebook:
   a) matter in its solid state
   b) matter in its liquid state
   c) matter in its gas state
   d) evaporation

3. Use the words in the box below to answer each question.

   > rocky road ice cream   fruit punch
   > helium

   a) A pure substance is made up of only one kind of particle. Which item in the box is a pure substance?
   b) A mechanical mixture is made up of two kinds of particles that can be seen as separate. Which item in the box is an example of a mechanical mixture?
   c) A solution is made up of two or more kinds of particles but appears to be made up of only one kind of particle. Which item in the box is an example of a solution?

4. Copy the table below into your notebook. Identify each property as physical or chemical. Find examples of matter in the picture of the beach scene below that have these properties (you can use the same example more than once).

Examples of Physical and Chemical Properties

Clue	Physical or Chemical Property?	Substance
a) Smells sweet		
b) Does not react with water		
c) Feels rough		
d) Is red		

5. Read each statement below and determine whether it describes the particles that make up air or the particles that make up water. Write your answers in your notebook.
   a) particles are close together
   b) particles are spread far apart
   c) particles move freely about one another

Inquiry Check

Solubility is a measure of the ability of a substance to dissolve in another substance. An investigation was conducted to demonstrate the effect of temperature on the solubility of salt and sugar. The results are shown in this graph.

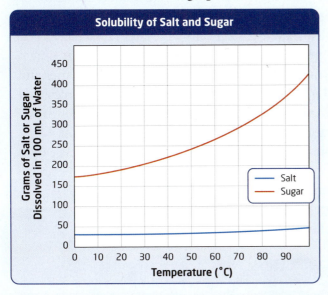

6. **Analyze** How many grams of (a) sugar and (b) salt will dissolve in water at 50°C?

7. **Interpret** Compare the solubility of salt and sugar at 80°C and identify which substance is more soluble than the other.

Numeracy and Literacy Check

8. Mass is the amount of matter in something. Mass is measured in units such as milligrams (mg), grams (g), kilograms (kg), or tonnes (t). Identify which unit would be most appropriate to measure the mass of each of these objects:

   a) an aspirin   c) a dog
   b) an ice cream truck   d) a hamburger

9. Volume is the amount of space taken up by something. Volume is measured in units such as millilitres (mL), litres (L), or kilolitres (kL). Identify which unit would be most appropriate to measure the volume of each of these objects:

   a) the amount of ice cream in an ice cream cone
   b) the amount of air in a balloon
   c) the amount of water in a swimming pool

10. **Calculate** You can use the mass and volume of a substance to determine an important physical property called density. Density is the mass of an object that occupies a certain volume. You can calculate density by dividing the mass of an object by its volume:

    $$\text{density} = \frac{\text{mass}}{\text{volume}}$$

    If a sample of ice cream has a mass of 55 g and a volume of 50 mL, what is the density of the ice cream?

11. **Discuss** Draw a three-panel comic strip showing how a forensic scientist might apply knowledge of density to identify a mysterious fragment of an object found at a crime scene.

Topic 2.1

In what ways do chemicals affect your life?

Key Concepts

- Everything—including you and everything around you—is made up of chemicals.
- Substances have characteristics that make them useful, hazardous, or both.
- Handling chemicals and lab equipment safely and responsibly is a part of your life at school.

Key Skills

Inquiry
Literacy
Research

Key Terms

matter

Each day, you use substances that have been invented to make your life easier and more enjoyable. For instance, at one time, pop, juice, and other mass-market drinks were sold in glass bottles or steel cans. Due to expense, steel cans were later replaced with cans made from aluminum. Soon glass gave way to plastic. Next time you reach for a drink in a plastic bottle, think about the matter and energy that went into making it. For example, what resources were used? What will happen to that bottle when you finish your drink?

Actually, what happens to the bottle might surprise you. As you can see on these two pages, many different products can be made from recycled PET plastic. (PET is short for a substance called **p**ol**y**ethylene **t**erephthalate [PAW-lee-ETH-eh-leen TER-ef-THAL-ate]. It's the type of plastic that is branded with the numeral one in recycling logos for plastic.)

Starting Point Activity

1. What characteristics make glass, steel, aluminum, and plastic suitable for use as drink containers?
2. Why have plastic containers for drinks become so widely used?
3. What do you usually do with plastic drink bottles that you use (if you use them)?
4. Are you surprised by any of the items that are made using recycled soft drink bottles? Explain.
5. Ontario is one of the few provinces in the country that has curbside pickup for plastics and other recyclables. (In most other provinces, people have to take recyclable materials to depots themselves.) And yet statistics in 2004 showed that Ontario had the second-worst rate of recycling PET plastic in the country. Why do you think this is so? What could be done to change this situation?

scouring pads

paint brush

waste receptable

kayaks

veterinary cone

netting

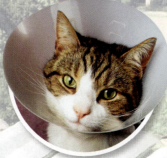

The pictures in the circles are just seven examples of products that can be made from the plastic used to make drinking bottles.

Everything—including you and everything around you—is made up of chemicals.

Literacy Focus

Activity 2.1

CHEMICAL-FREE! (OH, REALLY?)

Consider the three facts below.

- A popular coffee company says it uses a process to remove caffeine without the use of chemicals. The process involves soaking the coffee beans in water to dissolve the caffeine and wash it out of the beans.

- A company that makes environmentally respectful products makes a cleaning cloth that kills bacteria and other germs. The cloth contains tiny bits of silver. The company says, "Silver is a metal, not a chemical."

- Many gardeners proudly proclaim their lawns and gardens are chemical-free. Their results depend on methods that include the use of natural fertilizers such as manure and nutrient-rich compost.

Now work together in small groups to answer these questions. You can refer to books or other information sources if you like.

1. What do you think "chemical-free" means?
2. Is it possible for any product to be chemical-free? Explain.
3. What is a chemical?

Before you got to school today, you probably touched or used any or all of the following items: toothpaste, soap, water, food, paper, fabric, concrete, grass or snow (depending on the time of year), metal.

When you or a family member cleans your home, you probably use some or all of the following items: detergent, vinegar, window cleaner, scouring powders or liquids, furniture polish.

To maintain a car, truck, or bus in good working condition, a mechanic uses some or all of the following items: lubricating oil, engine oil, degreasing liquids, windshield-washer fluid, brake fluid, transmission fluid.

◀ **Figure 2.1** Elements are a certain kind of chemical that you will learn about starting in Topic 2.3. This pie graph shows the most abundant elements that make up the human body. The size of the pie wedges represents the proportion of the body that is made up of that element or elements.

Matter: The "Stuff" of the Universe

People commonly refer to many items in their daily life as "chemicals." But what is a chemical? You might think that it means something very specific to a scientist. After all, scientists are very careful about the ways that they define and use terms and ideas. However, the word "chemical" does not have a specific scientific meaning. That's because everything in the world that isn't energy is a chemical or contains chemicals. For example, do you think of yourself as being chemical-free? Think again. **Figure 2.1** shows that you are made up mostly of four chemicals of a certain type, with smaller amounts of many, many others.

When people use the word "chemical," they are really talking about matter. Anything—any *thing*—that has mass and that has volume (takes up space) is **matter**.

matter: anything that has mass and volume

LEARNING CHECK

1. Define the term "matter."
2. Use the pie graph in **Figure 2.1** to help you determine which element is most abundant in your body.
3. Refer to the photo of the plastic drink bottle on page 94. Check at home for products packaged in plastic that have the recycling code "1." What products did you find?
4. What kinds of problems in communication and understanding can result when people use the word "chemical" when they are talking about issues involving health and the environment? Describe two examples.

Substances have characteristics that make them useful, hazardous, or both.

Table 2.1 outlines the characteristics of two substances (certain kinds of matter) that are found in many homes. Notice that a substance can be both useful and hazardous at the same time.

Table 2.1 Useful and Hazardous Characteristics of Two Common Substances in the Home

Substance in the Home	Useful Characteristics	Hazardous Characteristics
ammonia (an ingredient in some cleaning products)	kills bacteria and other germs	• can burn skin and other body tissues • poisonous—can cause dangerous irritation if inhaled • releases poisonous gas if mixed with certain other substances such as chlorine
methane (a fuel—natural gas—that is used for heating, cooking, and transportation)	burns cleanly and efficiently in the presence of plentiful oxygen	• explosive • fumes can cause suffocation

Research Focus

Activity 2.2
CONSIDERING PROS AND CONS

What To Do

1. In pairs, choose one of the chemical substances from the list below. Do research to complete one row of the table. Compile the data as a class.

   gold   propane   muriatic acid   acetone
   iron   caustic soda   (hydrochloric acid)   lead
   mercury   (sodium hydroxide)   sulfuric acid   methanol

Substance	Common (or main) Uses	Useful Characteristics	Hazardous Characteristics	Special Storage and/or Disposal Requirements

98 MHR • UNIT 2 EXPLORING MATTER

Plastics: Not All Are Alike

Many stores and municipalities have banned or phased out the use of plastic grocery bags. But believe it or not, plastic grocery bags were introduced in the 1970s to provide a solution to the problems associated with paper bags. Producing a plastic bag takes less energy than producing a paper bag and does not use trees, a scarce resource. Recycling plastic takes less energy than recycling paper, and plastic bags can be reused for trash, as a lunch bag, or to pick up after your pet. More durable and lightweight than paper, plastic bags provided consumers with a solution to the problems that were identified at the time.

Since then, many hazards have been associated with the use of plastic bags, mostly because they are so durable that they won't decompose over time. The features that made plastic bags so attractive are now the ones causing most of the problems. But not all plastics are alike. For instance, when hospital workers change the sheets on a patient's bed, or collect the laundry after a surgery, they expose themselves to blood, urine, and other body fluids that put them at risk for infection. By collecting the laundry in polyvinyl alcohol (PVA) plastic bags (refer to **Figure 2.2**), the risk of infection is lowered because workers only handle the dirty laundry once.

Huh? How does the laundry get out of the bag? It doesn't have to! PVA bags dissolve in hot water, so hospital workers can load full bags of dirty laundry into the washing machine and turn it on. The bags dissolve in the water, leaving only the laundry behind.

These bags were first developed for use in hospitals, but also can be used to clean up after pets and to hold bait for fishing. Scientists hope that they may be able to further modify this plastic to be used in other applications. However, are there any risks to the environment from using dissolvable plastic? There is not yet enough data to answer this question. It must be answered in future so that people can make informed decisions about whether or not to use products made with dissolvable plastic.

▲ **Figure 2.2** These dissolvable plastic bags hold contaminated laundry.

LEARNING CHECK

1. Refer to **Table 2.1**. Describe one useful characteristic and one hazardous characteristic of ammonia.
2. Explain why hospital workers can load PVA bags of laundry right into a washing machine without having to empty the laundry out of the bag first.
3. Describe how you think a PVA plastic bag could be useful in your life.

Go to **scienceontario** to find out more

Handling chemicals and lab equipment safely and responsibly is a part of your life at school.

Making sure that you know how to handle chemicals safely in the laboratory is an essential part of your exploration of matter. In previous science studies, you have learned and practised safe techniques and procedures for handling chemicals and equipment. Use **Figure 2.3** to help refresh your memory on some of the safety icons and WHMIS symbols that you are likely to see in this unit. The "Safety in the Science Classroom" section on page xv near the start of this book has a more complete list of safety icons and WHMIS symbols. Also, as part of the WHMIS system, there are material safety data sheets (MSDS) that are available for each chemical that you will handle in the lab.

Figure 2.3 The safety icons (in red and white) and WHMIS symbols (in black and white) communicate important information about chemicals, equipment, and procedures in an experiment.

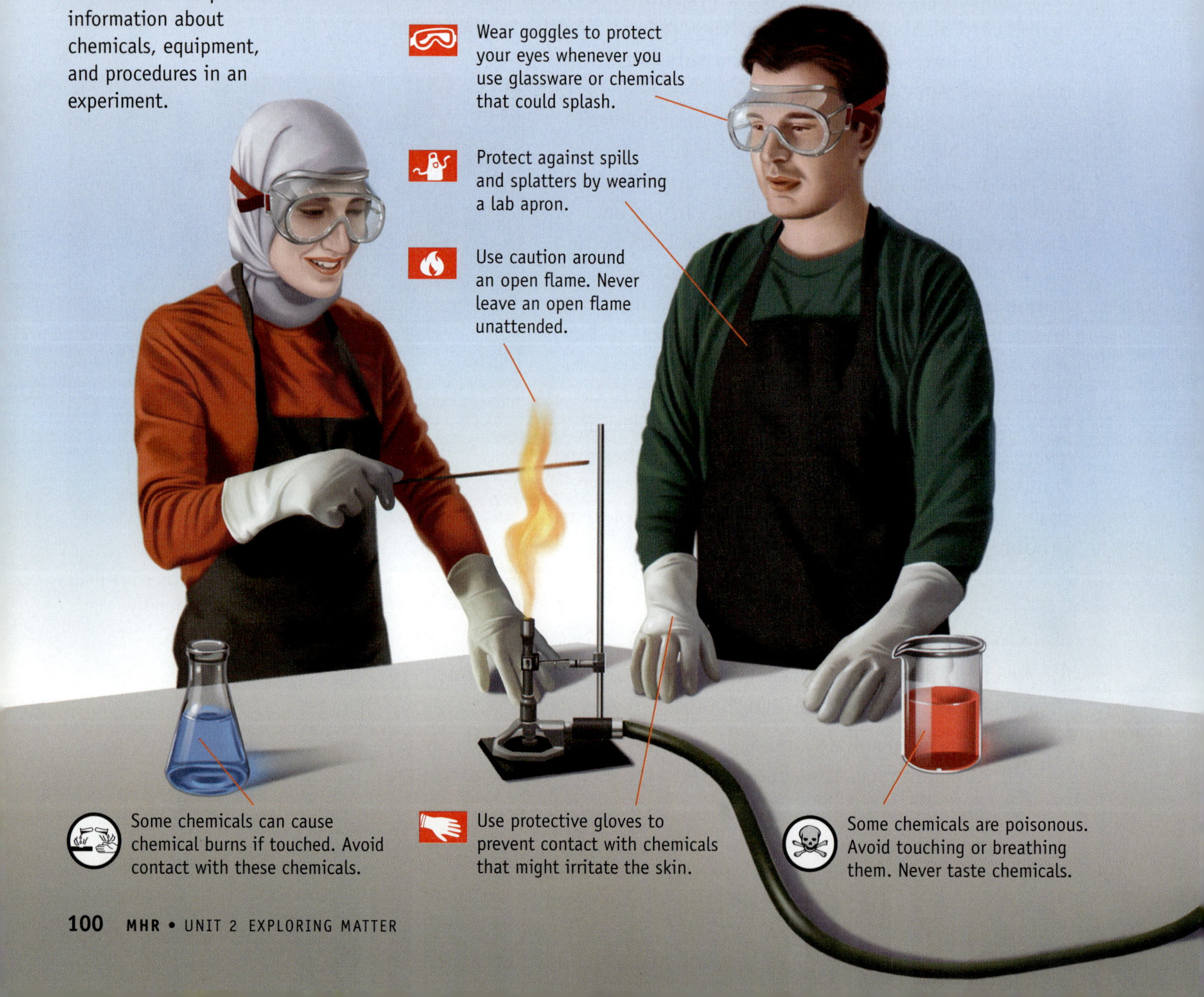

Wear goggles to protect your eyes whenever you use glassware or chemicals that could splash.

Protect against spills and splatters by wearing a lab apron.

Use caution around an open flame. Never leave an open flame unattended.

Some chemicals can cause chemical burns if touched. Avoid contact with these chemicals.

Use protective gloves to prevent contact with chemicals that might irritate the skin.

Some chemicals are poisonous. Avoid touching or breathing them. Never taste chemicals.

100 MHR • UNIT 2 EXPLORING MATTER

Activity 2.3

SAFETY FIRST

When performing an experiment, you must be able to recognize the safety icons and symbols that are used and know the precautions you need to take. Can you easily recognize all the potential hazards associated with the instructions below?

What To Do

1. Read over the list of safety icons and the list of WHMIS symbols in the "Safety in the Science Classroom" section on page xv.
2. The instructions in the column of text to the right of this column of text describe eight different lab procedures. As you read the instructions, draw the symbols that apply to each instruction. You should use every icon and symbol at least once in this activity. (An instruction might need more than one symbol.)

Instructions

A. Make sure that your lab station is clear and dry. Then plug in the electric hot plate and turn it on.

B. Do not add water to the sugar before heating the sugar.

C. Light the Bunsen burner. Then heat the test tube gently by holding it above the flame.

D. Let the steel pin cool for 10 min. When the pin has cooled, put it into the container as the teacher has shown.

E. Heat the test tube with a Bunsen burner gently at first, and then more strongly. Do not breathe the irritating ammonia gas that forms.

F. Using a medicine dropper, add the acid, one drop at a time, to the base. Be careful that you do not spill either the acid or the base.

G. Add two drops of the solution. The solution can be absorbed into the skin, so be careful that you do not get any on you.

LEARNING CHECK

1. What information is communicated by the safety icons and WHMIS symbols in this textbook?
2. Which safety icons could be used for almost any laboratory experiment. Explain your answer.
3. Write a "Caution!" statement for each of the instructions in Activity 2.3 to draw attention to one safety hazard. Here is an example for Instruction C: "Caution! Wear safety goggles to protect your eyes."
4. Make a sketch of your science lab or classroom to show the location of the emergency exits, eyewash stations, fire extinguishers, and any other emergency equipment.
5. List any safety rules that your teacher has given you that apply specifically to your classroom.

Strange Tales of Science

MINDING
SCIENTIFIC INQUIRY

He helped invent the method of scientific inquiry that scientists use today. And in his spare time he created a new type of mathematics. But after he died, someone took his brain! BWAAHHAHAHAHAHAH!

Um...Okay, so they didn't take his brain, but they did take his skull, and much of the rest of his remains. Sixteen years after the celebrated French Thinker, René Descartes, was buried in Denmark, far from his place of birth, someone dug up his coffin and made off to France with his head and bones.

So... What do you think?

1. Did someone really dig up Descartes' body and take his head and bones?
2. Find out two things that Descartes did to help invent science inquiry.
3. Stating a hypothesis is one of the skills of scientific inquiry. Which of these skills have you used in your science course this year?

Topic 2.1 Review

Key Concepts Summary

- Everything—including you and everything around you—is made up of chemicals.
- Substances have characteristics that make them useful, hazardous, or both.
- Handling chemicals and lab equipment safely and responsibly is a part of your life at school.

Review the Key Concepts

1. **K/U** Answer the question that is the title of this topic. Copy and complete the graphic organizer below in your notebook. Fill in four examples from the topic using key terms as well as your own words.

   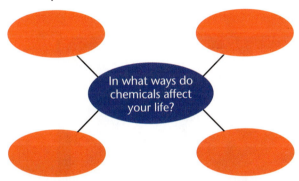

2. **T/I** The table below lists the ten most common elements in Earth's crust. They are listed in alphabetical order. Use the data in the table to make a pie graph. (The numbers have been rounded off and will add up to 100.)

Element	Approximate Percentage in Earth's Crust
aluminum	8
calcium	4
iron	5
magnesium	2
oxygen	47
potassium	3
silicon	28
sodium	3

3. **K/U** In your own words, define "matter."

4. **A** Recycling code "1" represents PET plastic. Recycling code "4" represents LDPE plastic—the type of plastic used to make plastic shopping bags. (LDPE stands for low density polyethylene.) Search the Internet to find out what products are made from LDPE.

   a) List four products that are made from LDPE.

   b) List at least two products that are made from recycled LDPE.

   c) Describe how your life would be different if you eliminated all products made with LDPE from your life.

   d) Describe how your life would be different if you eliminated all products from with plastic from your life.

5. **C** Explain using words or a picture how a common product such as toothpaste, deodorant, shampoo, dish soap, perfume, soap, hair mousse, or hair gel could affect your life, the environment, or both.

6. **C** Use a concept map to explain to a Grade 3 student what the word "chemical" means.

7. **A** Some cities in the province of Ontario have banned the sale of bottled water in certain city-run facilities such as arenas and community centres. List the pros and cons of such bans, and then write a brief paragraph expressing your opinion of banning bottled water.

Topic 2.2 How do we use properties to help us describe matter?

Key Concepts

- Physical properties describe how matter looks and feels.
- Chemical properties describe how substances can change when they interact with other substances.

Key Skills

Inquiry

Key Terms

physical properties
conductivity
density
lustre
solubility
texture
chemical properties
combustibility
decomposition
precipitate

Think of your favourite food. How would you describe it to someone who has never seen or tasted it before? Do you tell them how it smells? Do you talk about whether it's spicy, salty, sweet, or savory? Do you mention where people can get it and how much it costs?

You probably do some or all of these things. Maybe you refer to other features of your favourite food, such as its colour, its shape, or its texture (how it feels to the touch). In other words, you describe your favourite food by referring to its properties. Scientists describe matter the same way.

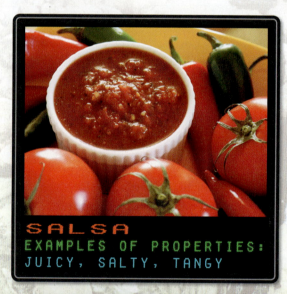

SALSA
EXAMPLES OF PROPERTIES:
JUICY, SALTY, TANGY

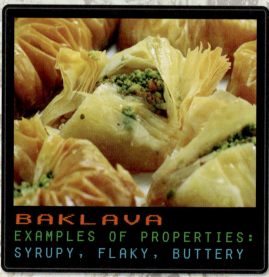

BAKLAVA
EXAMPLES OF PROPERTIES:
SYRUPY, FLAKY, BUTTERY

Starting Point Activity

1. With a partner, take turns describing three of the foods in the pictures. Come up with at least two additional properties for each of the foods.
2. As a class, list the properties used to describe the foods.
3. Use property words to describe your favourite food to your partner without using its name. Have your partner guess what it is, based on the properties you use.

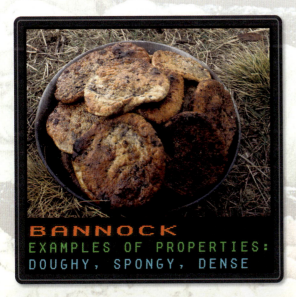

BANNOCK
EXAMPLES OF PROPERTIES:
DOUGHY, SPONGY, DENSE

CUMIN
EXAMPLES OF PROPERTIES:
FRAGRANT, MUSKY, GOLDEN

GREEN APPLE
EXAMPLES OF PROPERTIES:
TART, CRISP

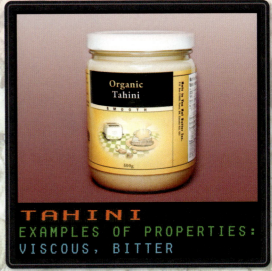

TAHINI
EXAMPLES OF PROPERTIES:
VISCOUS, BITTER

Physical properties describe how matter looks and feels.

physical property: any feature of matter that can be observed or measured without changing the type of matter it is

All matter can be described by its **physical properties**. Physical properties are features of matter that can be observed or measured without changing the type of matter that something is. **Table 2.2** lists some physical properties that are often used to describe matter.

Table 2.2 Examples of Physical Properties Used To Describe Matter

Physical Property	What is it?	Examples
	Conductivity describes how well a substance lets heat or electrical current move through it. Metals tend to be good conductors, and non-metals tend to be poor conductors.	• Copper is used to make electrical wires, because it is a good conductor of electrical current. • One reason glass is good to make windows is that it does not conduct heat very well.
	Density describes how compact a substance is, and is calculated by dividing its mass by its volume.	• Ice (solid water) floats on liquid water, because ice is less dense than liquid water. • Iron sinks in liquid water because iron is more dense than liquid water.
	Lustre describes how well the surface of a substance reflects light.	• Many people are attracted to lustrous metals such as silver, gold, and chrome because they are shiny.
	Solubility describes how much of a substance dissolves in another substance.	• Salt crystals dissolve in water to form the mixture salt water.
	Texture describes how the surface of a substance feels (its roughness, softness, or smoothness).	• Window glass has a smooth texture. • Brick has a rough texture.

conductivity: describes how well a substance lets heat or electrical current move through it

density: describes how compact a substance is, and is calculated by dividing mass by volume

lustre: describes how well the surface of a substance reflects light

solubility: describes how much of a substance dissolves in another substance.

texture: describes how the surface of a substance feels (its roughness, softness, or smoothness)

LEARNING CHECK

1. Explain what a physical property is, and give two examples.
2. Describe two physical properties of a) a pencil, b) a piece of paper, c) apple juice, and d) air.
3. Explain how the following people might depend on physical properties: a) a chef, b) a painter, and c) a carpenter.

INVESTIGATION LINK
Investigation 2A, on page 110

Inquiry Focus

Activity 2.4
LINKING PHYSICAL PROPERTIES OF OBJECTS WITH THEIR USES

In this activity, you will plan and conduct an experiment to identify the physical properties of common substances.

Safety

Before you begin work on your experiment, review pages xv and 100.

What You Need

samples of objects from your teacher (examples could include salt, copper wire, cork, aluminum foil, paper cup, Styrofoam cup, plastic paper clip, metal paper clip)

beaker

water

stirring rod

conductivity apparatus

What To Do

1. Work together in small groups.
2. Design a procedure to identify the physical properties of the substances provided by your teacher. Make sure you have your teacher approve your procedure before you begin your experiment.
3. Be sure to consider safety precautions and proper clean-up and disposal in your procedure.

   Hints:
   - Use **Table 2.2** to help you plan.
   - Some physical properties, such as density and solubility, are best measured as comparisons. For instance, ice is less dense than water. Sugar is more soluble than salt.
4. Design a table to record your observations.
5. Follow your procedure and record your results.
6. Clean up and put away all the equipment. Wash your hands.

What Did You Find Out?

1. Describe each of the substances you investigated using its physical properties.
2. Explain how the physical properties you have described make each object or substance suitable for its use.

Inquire Further

What other physical properties do you know or can you remember learning about before? (Lots of examples are possible.) Make your own version of **Table 2.2** that includes other physical properties of matter.

Chemical properties describe how substances can change when they interact with other substances.

chemical properties: describe how substances can change to produce new substances with new properties when they interact with other substances

Chemical properties describe how substances can change to produce new substances with new properties when they react with other substances. You can describe substances using their physical properties just by looking at them, but chemical properties can only be observed when substances interact.

The degree to which a substance can change is its reactivity. Reactivity can be useful in some cases and not in others. For instance, we depend on reactivity for the explosiveness of dynamite. On the other hand, we often store foods in glass containers because glass is not reactive at all. Table 2.3 lists some examples of chemical properties.

Table 2.3 Examples of Chemical Properties Used To Describe Matter

Chemical Property	What is it?	Examples
	Combustibility describes the ability of a substance to catch fire and burn in air.	• We burn wood and other fuels because of their combustibility.
	Reactivity with oxygen describes the change that can occur when a substance is exposed to oxygen.	• The flesh of some kinds of fruit turns brown when it is exposed to the oxygen in air.
	Reactivity with acids describes the change that can occur when a substance is exposed to acids.	• Some substances such as baking soda produce a gas when mixed with acids such as vinegar.
	Reactivity with other substances describes the change that can occur when one substance reacts with other substances.	• When some substances, are mixed together, they form a solid, called a **precipitate**, which is a new substance.
	Decomposition describes the change that can occur when a substance such as water is broken down into the parts that make it up.	• Chemical decomposition often happens when a substance interacts with energy such as electrical current or heat.

combustibility: describes the ability of a substance to catch fire and burn in air

precipitate: an insoluble solid substance that can form when certain dissolved substances are mixed together

decomposition: a kind of reactivity that can break down a substance into its parts

Activity 2.5

IDENTIFYING CHEMICAL AND PHYSICAL PROPERTIES OF SUBSTANCES

In this activity, you will observe the chemical, as well as the physical, properties of some substances. Keep in mind that physical properties can be observed *before* there is any contact with other substances, while chemical properties can only be observed when substances react with each other.

Safety

What You Need

2 small spoons

10 mL graduated cylinder

1 resealable plastic bag

Substances A, B, and C

What To Do

1. Copy the table below in your notebook.

Data Table for Recording Physical and Chemical Properties

	Physical Properties	Chemical Properties
Substance A		
Substance B		
Substance C		
Mixture of substances (after)		✕

2. Describe the physical properties of substances A, B, and C. Record your observations in the table.

3. Mix one spoonful of substance A, one spoonful of substance B, and 10 mL of substance C into a resealable bag. Quickly seal it up.

4. Mix the substances in the bag by tilting it back and forth a few times. Record the chemical properties of the substances when they are mixed.

5. Record the physical properties of the substances after mixing.

6. When you are done, follow your teacher's instructions for cleaning up. Wash your hands.

What Did You Find Out?

1. Compile your list as a class, and add any new observations that you did not record.

2. How did the physical properties of substances A, B and C before mixing compare with the physical properties of the mixture of them?

3. Why does the table have a separate row for the physical properties of each of the substances but only one place to record the chemical properties of the substances?

4. How are chemical properties different from physical properties?

LEARNING CHECK

1. Explain what a chemical property is, and give two examples.

2. Identify the types of properties in this sentence, and give reasons: "Beeswax is soft and burns with a bright flame."

3. Use **Table 2.3** to determine which chemical property best describes propane, the gas used in a barbecue tank. Why is this property both useful and hazardous for people using propane?

INVESTIGATION LINK

Investigation 2A, on page 110

Investigation 2A

Skill Check

✓ Initiating and Planning

✓ Performing and Recording

✓ Analyzing and interpreting

✓ Communicating

Safety

- Do not taste or eat anything in the classroom.
- Clean up any spills immediately.

What You Need

- samples that will include table sugar, baking soda, salt, vinegar, aluminum foil, and others provided by your teacher
- test tubes and test tube rack
- water
- stirring rod
- conductivity apparatus
- other equipment as needed

Physical and Chemical Properties of Substances in the Home

How do the physical and chemical properties of substances in the home differ? What do they have in common?

What To Do

1. Your teacher will give your group six substances to investigate. Plan a procedure to identify four physical properties and two chemical properties of each substance. Choose properties from Table 2.2 and 2.3.

2. Make a table like the one below to record the results of your tests. Include a title for your table. Be sure to include enough rows for all the samples you will be testing.

Sample	Physical Properties	Chemical Properties
table sugar		
salt		

3. Have your teacher review your procedure, observations table, and list of equipment. Do not start your tests without your teacher's approval.

4. Carry out your procedure. Make sure that you do your tests exactly as you planned them. Don't add any steps without your teacher's approval.

5. Record your observations in your table. They may include descriptions (such as colour), yes/no answers (does it react with acid?) or ratings (such as the hardness of a substance compared to another substance).

6. Share your group's results with your teacher, who will record them in a class chart. Add any information that you have not already recorded in your own table.

What Did You Find Out?

1. Analyze your observations to determine if there are certain properties that seem to distinguish one substance from the others. For some substances, it may be one particular property. For other substances, it may be a combination of properties.

2. Evaluate your tests to determine whether they were useful for telling the substances apart, based on their properties. What improvements could you make?

Topic 2.2 Review

Key Concept Summary
- Physical properties describe how matter looks and feels.
- Chemical properties describe how substances can change when they interact with other substances.

Review the Key Concepts

1. **K/U** Answer the question that is the title of this topic. Copy and complete the graphic organizer below in your notebook. Fill in four examples from the topic using key terms as well as your own words.

   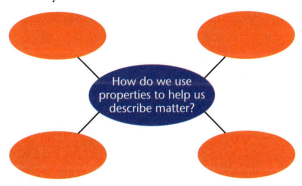

2. **T/I** Look at the photo of paper clips. Describe at least three physical properties of paper clips that make them useful.

   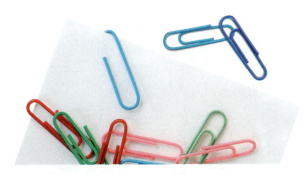

3. **K/U** Use words or a graphic organizer such as a Venn diagram to explain the difference between a physical property and a chemical property.

4. **T/I** Your teacher places a small sample of a white, crystalline substance in a beaker of water and it dissolves and disappears. Next, your teacher slowly and carefully heats a small sample of the substance and it begins to melt—you see a liquid forming. Your teacher leaves the heat on for long enough that you begin to detect a sweet odour. Then the white substance catches fire and burns.

   a) Create a table or a concept map to organize the physical properties and the chemical properties of this white substance.

   b) Describe the safety steps that your teacher would have taken, including the types of safety clothing used, to conduct all the tests safely.

   c) In this case, your teacher knows what the white crystalline substance is. (It's table sugar.) However, imagine that your teacher didn't know the identity of the substance. Explain why the tests carried out could have been dangerous.

5. **K/U** You place a stone and a piece of wood into a large tub of water. The stone sinks and the wood floats. What physical property of these two substances determines if each substance sinks or floats? Write a sentence that compares this property for the stone, water, and wood. For help, refer to **Table 2.2**.

6. **A** Homes in Ontario have insulation in the walls to prevent the loss of heat to the outside during cold winter months. Which physical property determines which substances can be used to make insulation?

Topic 2.3

What are pure substances and how are they classified?

Key Concepts

- Pure substances are elements and compounds.
- Elements include metals and non-metals.

Key Skills

Research

Key Terms

pure substance
element
compound
metals
non-metal

What do the two photos have in common? Photo A shows two people fishing with a dip net. You have used a similar tool if you have ever gone fishing or scooped a pet fish out of an aquarium. A dip net is a traditional Aborginal tool designed to separate big fish from smaller fish or water.

The device in Photo B is a piece of equipment that is often used in crime labs. The person uses the device to analyze and identify the types of matter in or on a sample from a crime scene.

The devices in Photos A and B use properties to separate mixtures of matter into their parts. The dip net uses the properties of size and state—in this case, whether something is a solid or a liquid. The crime-lab device uses different properties. But not all types of matter can be separated into parts easily, and some types of matter can't be separated at all. Why might that be?

A

Starting Point Activity

When you read the word "pure" on a product label, what do you think it means? Does "pure" mean the same thing to you as it does to someone else who looks at the same product label? Does it mean the same thing to a scientist? Classify some common materials as either mixtures or pure substances.

What To Do

1. Work in groups. Decide together how to define the words "mixture" and "pure substance." Record your definitions.

2. Make a table of observations that starts like this.

Product	Mixture or pure substance?	Reasons for your choice

3. Brainstorm a list of common products that you might find at home in the kitchen or bathroom or both.

4. Choose 10 of your brainstormed products to classify as a mixture or a pure substance. Fill out your table for each of these. Use the definitions that you developed in step 1 to help you decide how to classify each product.

What Did You Find Out?

1. Compare your products and classifications with other groups. How many mixtures were there? How many pure substances were there?

2. Which products were easy to classify? Which ones were hard? Explain.

3. At any time, did you modify your definitions or think about modifying them? Explain.

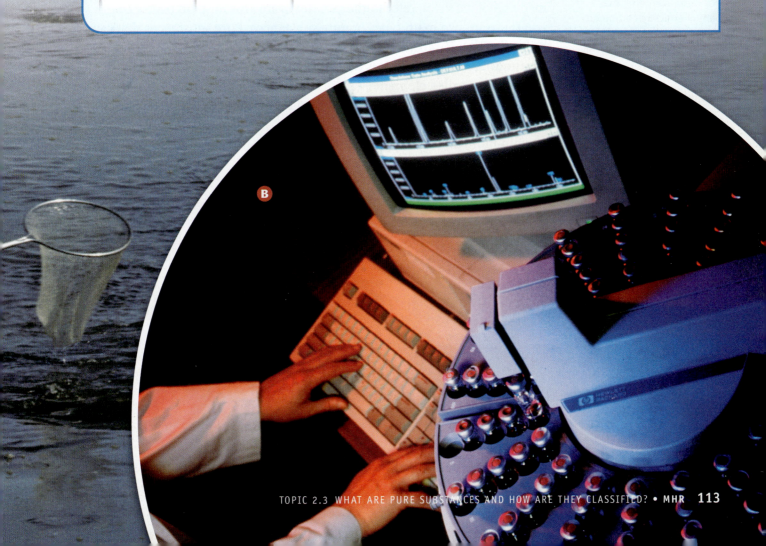

Pure substances are elements and compounds.

Matter is usually classified by its physical and chemical properties. When scientists first worked on classifying matter, they grouped matter based on how they could break it apart. Matter that could be separated into parts using differences in physical properties was classified as a *mixture*. Mixtures are made up of two or more types of matter, and these types can be physically separated.

Matter that is not a mixture is a pure substance. A **pure substance** is made up of only one type of matter, so it cannot be separated into parts physically. Table sugar (sucrose) is a pure substance. So is oxygen. Other examples of pure substances include pure water, copper, and aluminum.

Each part of a pure substance has the same properties, because each pure substance is made up of its own type of particle. The kind of particle that makes up one pure substance is different from the kinds of particles that make up other pure substances. Use **Figure 2.4** to help you understand what this means. (If you need help to recall the concept that all matter is made up of particles, refer to the Get Ready for this unit, on page 92.)

There are two main types of pure substances. Pure substances that can be broken down into smaller parts using chemical reactions are called **compounds**. Pure substances that can't be broken down using chemical reactions are called **elements**.

Why can compounds be broken down into smaller parts and elements can't? Compounds can be broken down into smaller parts, because they contain more than one type of element. For instance, look at **Figure 2.5** on the next page.

pure substance: matter that contains only one type of particle

compound: a pure substance made up of two or more elements that are chemically combined; can be broken down into elements again by chemical means

element: a pure substance made up of one type of particle that cannot be broken down into simpler parts by chemical means

▶ Figure 2.4 Copper, pure (distilled) water, and aluminum are examples of pure substances. In (A), notice that any part of copper is made up only of copper particles. See how this same idea applies to water and aluminum in (B) and (C).

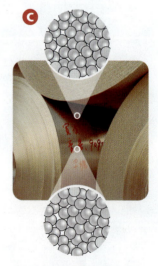

copper particles　　　distilled water particles　　　aluminum particles

In **Figure 2.5**, you can see a piece of metal, a sealed glass flask, and a white powdery solid. If you look closely at the flask, you can see a pale yellow colour inside. That is a gas. The gas is the element chlorine. The silvery-grey solid is a metal. The metal is the element sodium. These two elements, chlorine and sodium, can combine together to form the compound table salt. Energy can be used to un-combine the compound into the two individual elements again. The elements, however, cannot be un-combined further with a chemical reaction.

Chlorine combines with other elements to make other compounds, too. For example, chlorine combines with calcium to form the compound calcium chloride, which is a type of salt that is used on roads in winter. Other common compounds include table sugar and carbon dioxide.

sodium metal   chlorine gas   table salt

▲ **Figure 2.5** The compound table salt is made up of the elements chlorine and sodium, which have combined in a chemical reaction. The scientific term for table salt is sodium chloride.

LEARNING CHECK

1. Explain what a pure substance is. List two examples.
2. Create a flowchart that organizes the following terms: mixture, pure substance, matter, compound, and element. Start your flowchart with matter.
3. Use a double bubble graphic organizer to show similarities and differences between an element and a compound.
4. Look carefully at **Figure 2.4**. Notice that particles from two different parts of each pure substance are shown. Use that information to help you explain why a sample of a pure substance such as copper from Ontario has the same properties as a sample of copper from Peru.

Elements include metals and non-metals.

Research

Activity 2.6

CLASSIFY ELEMENTS

Just as pure substances can be classified as compounds or elements, elements can also be further classified. In this activity, you will investigate the properties of elements, and then classify them based on their properties.

hydrogen	fluorine	argon
helium	neon	potassium
lithium	sodium	calcium
beryllium	magnesium	iron
krypton	aluminum	gold
carbon	mercury	copper
nitrogen	sulfur	nickel
oxygen	chlorine	uranium

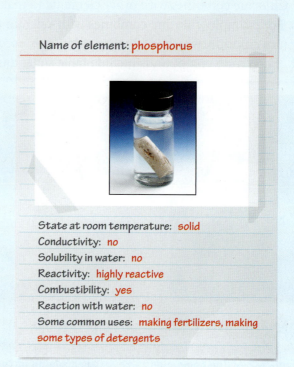

Name of element: **phosphorus**

State at room temperature: **solid**
Conductivity: **no**
Solubility in water: **no**
Reactivity: **highly reactive**
Combustibility: **yes**
Reaction with water: **no**
Some common uses: **making fertilizers, making some types of detergents**

What To Do

1. Choose one of the elements from the list so that each person has a different element. Your teacher will add other elements as needed. Do research to find key properties and uses for this element. Record the information on an index card such as the sample shown on this page. The sample shows the information you are expected to find.

2. As a class, use the information to classify the elements into groups based on their properties.

What Did You Find Out?

1. What properties did you find the most useful for classifying the elements? Why?

Distinguishing Metals and Non-metals

One way that elements can be classified is to group them into two categories: metals and non-metals. **Metals** are elements that share physical properties such as these. (Refer also to Table 2.2.)

- lustre
- malleability (ability to be bent or hammered without breaking)
- ductility (ability to be stretched into a wire without snapping)
- good conductors

Non-metals are grouped together because they do not share the properties of metals. Table 2.4 compares the properties that help in classifying metals and non-metals. Figure 2.6 shows some examples of metals and non-metals.

metals: elements that are commonly solid at room temperature, shiny, malleable, ductile, and good conductors

non-metals: elements that can be solid, liquid, or gas at room temperature, dull, brittle, not ductile, and poor conductors

Table 2.4 **Properties That Help To Distinguish Metals from Non-metals**

Substance	State at room temperature	Lustre	Conductivity	Malleability	Ductility
Metals	solid (except mercury, which is liquid)	shiny (lustrous)	good conductors	malleable	ductile
Non-metals	solid, liquid, or gas	dull (not lustrous)	poor conductors	not malleable (brittle)	not ductile

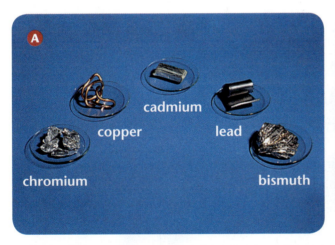 

▲ Figure 2.6 Samples of metals (A) and non-metals (B)

LEARNING CHECK

1. Refer to Figure 2.6. What properties do the metals have in common, and which are different? What properties do the non-metals have in common, and which are different?
2. Make a t-chart to summarize the properties of metals and non-metals.

INVESTIGATION LINK
Investigation 2B, on page 118

Investigation 2B

Skill Check

✓ Initiating and Planning

✓ Performing and Recording

✓ Analyzing and Interpreting

✓ Communicating

What You Need

- labelled samples of metals and non-metals
- other equipment and materials as required

Comparing the Physical Properties of Metals with Non-metals

In this investigation, you will plan and perform an experiment to compare and contrast physical properties of metals with non-metals.

What To Do

1. Work together in small groups.
2. Design a procedure to determine the physical properties of the metal and non-metal samples provided by your teacher. Make sure you have your teacher approve your procedure before you begin your experiment. Be sure to consider safety precautions and proper clean-up and disposal in your procedure.

   Consider the following points as you plan your procedure.
   - Refer to **Table 2.2** on page 106 and **Table 2.3** on page 108 to help you as you plan.
   - Your procedure should provide you with at least five physical properties for each of the samples.
   - Make a list of the equipment and materials you will need as you plan.
3. Create a table to record your observations.
4. Follow your procedure and record your results.
5. Clean up and put away all the equipment. Wash your hands.

What Did You Find Out?

1. Copy and complete the table below. Leave enough room for all of the properties that you tested.

   Observation Table for Investigation 2B: Summary of Physical Properties

	Metals	Non-metals
Properties shared by all samples		
Properties that are not shared by all samples		

2. What do all metals have in common? All non-metals?
3. Which properties (if any) can't be used to distinguish between a metal and a non-metal?

Topic 2.3 Review

Key Concept Summary
- Pure substances are elements and compounds.
- Elements include metals and non-metals.

Review the Key Concepts

1. **K/U** Answer the question that is the title of this topic. Copy and complete the graphic organizer below in your notebook. Fill in four examples from the topic using key terms as well as your own words.

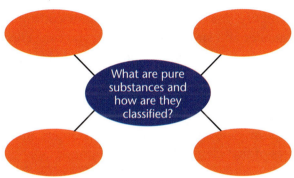

2. **T/I** Draw a labelled diagram that clearly shows the difference between the pure substance iron and a mixture of iron and sand in terms of the particles that make them up. For help, refer to **Figure 2.4**.

3. **K/U** What properties of the element aluminum make it desirable for use in constructing aircraft? What properties make it undesirable for this purpose?

4. **K/U** The flowchart below is part of a classification scheme for matter that focusses on the key concepts that you explored in this topic. Identify the category of matter that is represented by each of the letters.

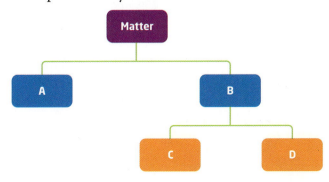

5. **T/I** Think about the substances listed in the box below and classify each one as either an element or a compound. Explain how you made your choices.

calcium carbonate	chlorine
copper	neon
table salt	hydrogen peroxide

6. **K/U** List the physical properties that most metals have in common. Think of an example of a metal and describe how its properties relate to its use.

7. **K/U** List the physical properties that most non-metals have in common. Think of an example of a non-metal and describe how its properties relate to its use.

8. **A** What property does steel (a metal) have that makes it a good choice for making car doors?

9. **A** Steel is a metal that is made from carbon, iron, and small amounts of other elements. When solid steel is heated strongly enough for it to change to a liquid, the elements of which it is made can be separated from one another easily. Based on this information, is steel a pure substance or a mixture? Explain.

10. **K/U** An element that is a non-metal can be a solid, a liquid, or a gas at room temperature. How, then, can an element that is a non-metal be distinguished from an element that is a non-metal?

Topic 2.4

How are properties of atoms used to organize elements into the periodic table?

Key Concepts

- Elements are made up of atoms, which are made up of subatomic particles.
- Elements are arranged in the periodic table according to their atomic structure and properties.
- Elements in the same family (group) share similar physical and chemical properties.

Key Skills

Inquiry

Key Terms

atom
proton
neutron
electron
nucleus
atomic number
periodic table
period
family

Elements are the building blocks of which all matter on Earth is made. But what makes up the carbon in the coal that is burned as fuel, the oxygen in the air you breathe, or the silver and gold used to make fine works of art such as those shown here?

Imagine any element—gold, for example. Imagine cutting a piece of gold into smaller and smaller pieces until you can't chop it up any further. What you are left with is a bit of gold that is so small that the most powerful microscopes in the world are needed to magnify it enough to be able to view it. Nevertheless, this extremely tiny bit of gold still has all the properties of a piece of gold the size of the Bill Reid sculpture. The smallest unit of any element that still has all the properties of that element is called an atom.

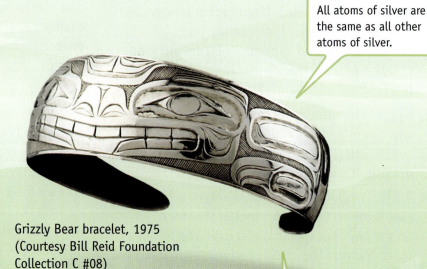

Grizzly Bear bracelet, 1975
(Courtesy Bill Reid Foundation Collection C #08)

All atoms of silver are the same as all other atoms of silver.

Atoms of gold are different from atoms of silver. As well, atoms of gold and of silver are different from atoms of all other elements.

Starting Point Activity

Use the information in the "call-out" labels around the photos to answer these questions.

1. How do atoms of one element compare to atoms of another element?
2. Imagine that you are looking at Bill Reid's gold sculpture. How do gold atoms around Raven's mouth compare to gold atoms deep in the centre of the sculpture?
3. Imagine that you are looking at Bill Reid's gold sculpture and this 3600 year old gold mask thought to show the face of the ancient king of Mycenae, Agamemnon. How do atoms of gold around Raven's mouth compare to atoms of gold around Agamemnon's mouth?
4. Describe an atom using what you have learned.

▲ The gold mask of Agamemnon.

Famed Haida artist, Bill Reid (1920–1998), created this work of art from gold. It is called *The Raven and the First Men* and is only 7 cm by 6.9 cm by 5 cm. (Courtesy Bill Reid Foundation Collection C #25)

All atoms of gold are the same as all other atoms of gold.

Any atom is a million times smaller in diameter than the thinnest human hair.

Even the thinnest piece of gold or silver that you can imagine is about 200 000 atoms thick.

Elements are made up of atoms, which are made up of subatomic particles.

atom: the smallest unit of an element that displays the properties of that element

proton: a positively charged particle that is part of the atomic nucleus

neutron: an uncharged particle that is part of the atomic nucleus

electron: a negatively charged particle that surrounds the nucleus

nucleus: the positively charged centre of an atom

An **atom** is the smallest unit of an element that has the properties of that element. **Figure 2.7** and **Table 2.5** show that atoms are made up of even smaller subatomic particles: **protons**, **neutrons**, and **electrons**. The organization of subatomic particles in an atom is called the *atomic structure*.

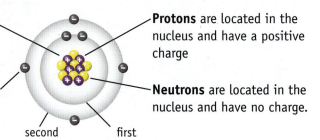

The **nucleus** is at the centre of an atom and holds its protons and neutrons.

Electrons surround the nucleus in one or more energy levels and have a negative charge.

Protons are located in the nucleus and have a positive charge

Neutrons are located in the nucleus and have no charge.

second energy level

first energy level

▲ **Figure 2.7** This model shows the subatomic particles of an atom. (The atom in this case is a carbon atom.)

Table 2.5 Subatomic Particles of Atoms

Name	Electrical Charge	Symbol	Location in an Atom	Relative Mass
proton	+	p^+	nucleus	about 1
electron	−	e^-	region around the nucleus	about $\frac{1}{2000}$
neutron	0	n^0	nucleus	about 1

The following statements are true for any atom.
- The number of protons is equal to the number of electrons.
- The number of protons that an atom of any element has is called the **atomic number**. For example, all hydrogen atoms have one proton. All oxygen atoms have 8 protons. The atomic number distinguishes the atoms of one element from the atoms of other elements. See **Figure 2.8**.
- The regions around an atom's nucleus are called *energy levels*. Atoms of each element have a different number of electrons in their energy levels. The number of energy levels ranges from one to seven. Two electrons can fit in the space of the first energy level. Eight electrons can fit in the second energy level. More electrons can fit into higher energy levels.

atomic number: the number of protons in the nucleus of an atom

Go to **scienceontario** to find out more

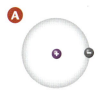 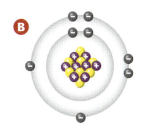

◀ **Figure 2.8** The atomic number of hydrogen (A) is 1. The atomic number of oxygen (B) is 8. Count the protons in each nucleus to confirm this.

LEARNING CHECK

1. What does an atomic number represent?
2. Use a double bubble organizer to compare and contrast the atoms shown in **Figure 2.8**.
3. Use **Figure 2.7** as a guide to draw a beryllium atom.

Inquiry Focus

Activity 2.7
BUILDING ATOMS

In this activity, you will use common materials to make models that represent your understanding of what atoms look like.

What You Need
variety of materials to represent protons, neutrons, and electrons

What To Do

1. Choose one of the elements from this list. Use the information on page 122 to build a model for an atom of your element.

Element	Atomic Number	Subatomic Particles	Element	Atomic Number	Subatomic Particles
Hydrogen	1	1 p^+, 1 e^-, 0 n^0	Sodium	11	11 p^+, 11 e^-, 12 n^0
Helium	2	2 p^+, 2 e^-, 2 n^0	Magnesium	12	12 p^+, 12 e^-, 12 n^0
Lithium	3	3 p^+, 3 e^-, 4 n^0	Aluminum	13	13 p^+, 13 e^-, 14 n^0
Beryllium	4	4 p^+, 4 e^-, 5 n^0	Silicon	14	14 p^+, 14 e^-, 14 n^0
Boron	5	5 p^+, 5 e^-, 6 n^0	Phosphorus	15	15 p^+, 15 e^-, 16 n^0
Carbon	6	6 p^+, 6 e^-, 6 n^0	Sulfur	16	16 p^+, 16 e^-, 16 n^0
Nitrogen	7	7 p^+, 7 e^-, 7 n^0	Chlorine	17	17 p^+, 17 e^-, 18 n^0
Oxygen	8	8 p^+, 8 e^-, 8 n^0	Argon	18	18 p^+, 18 e^-, 22 n^0
Fluorine	9	9 p^+, 9 e^-, 10 n^0	Potassium	19	19 p^+, 19 e^-, 20 n^0
Neon	10	10 p^+, 10 e^-, 10 n^0	Calcium	20	20 p^+, 20 e^-, 20 n^0

Elements are arranged in the periodic table according to their atomic structure and properties.

	Group 1	Group 2	Group 3	Group 4	Group 5	Group 6	Group 7	Group 8	Group 9	Group 10	Group 11	Group 12	Group 13	Group 14	Group 15	Group 16	Group 17	Group 18
Period 1	H																	He
Period 2	Li	Be											B	C	N	O	F	Ne
Period 3	Na	Mg											Al	Si	P	S	Cl	Ar
Period 4	K	Ca	Sc	Ti	V	Cr	Mn	Fe	Co	Ni	Cu	Zn	Ga	Ge	As	Se	Br	Kr
Period 5	Rb	Sr	Y	Zr	Nb	Mo	Tc	Ru	Rh	Pd	Ag	Cd	In	Sn	Sb	Te	I	Xe
Period 6	Cs	Ba	La	Hf	Ta	W	Re	Os	Ir	Pt	Au	Hg	Tl	Pb	Bi	Po	At	Rn

▲ **Figure 2.9** A simplified version of the periodic table of the elements. Short forms of the names of the elements are used. (You will learn more about the short forms in Topic 2.5.)

Figure 2.9 shows a simplified version of the **periodic table**. The place in the periodic table where each element appears depends on its atomic number and its properties.

Periods (rows on the periodic table) represent the number of energy levels that contain electrons. For instance, the elements in Period 1 have electrons in only the first energy level.

Families (columns or groups on the periodic table) represent the number of electrons in the outermost energy level. Many of the properties of elements are determined by these electrons. As you can see in **Figure 2.10**, atoms in the same family have the same number of electrons in their outer energy level. As a result, elements in the same family tend to have similar properties and react in a similar way.

Figure 2.11 on pages 126 and 127 shows a full, pictorial version of the periodic table. You can see a text version of it on page 434.

▶ **Figure 2.10** Several patterns are visible in the number and arrangements of the electrons of atoms of the first 18 elements.

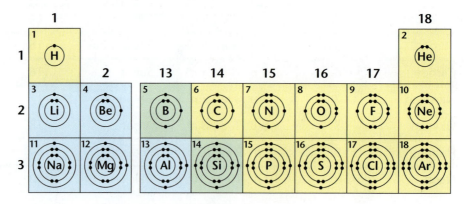

124 MHR • UNIT 2 EXPLORING MATTER

Elements in the same family (group) share similar physical and chemical properties.

When the elements are organized in order of their atomic number and atomic structure, a variety of patterns emerge. The periodic table is developed so that each family (group) of elements has similar chemical and physical properties. This occurs because the atomic structure of an element determines its chemical and physical properties.

Another pattern that emerges when elements are sorted by atomic number involves metals and non-metals. As you can see in **Figure 2.9**, all metals (shown in blue) are on the left side of the periodic table. The most reactive metals are all in group 1. The reactivity of the elements in group 1 increases as you go down the column. So the most reactive metal in this group is at the bottom. It is cesium (Cs).

The non-metals (shown in yellow) are on the right side of the periodic table. The most reactive non-metals are all in group 17. The reactivity of the elements in group 17 increases as you go up the column. So the most reactive non-metal is at the top. It is fluorine (F).

The eight elements in green in **Figure 2.9** are metalloids. They share some properties of metals and some properties of non-metals.

periodic table: a chart in which elements are listed horizontally in order of their atomic number and in which elements with similar properties are arranged vertically

period: a horizontal row of elements in the periodic table

family: a vertical column of elements in the periodic table

Activity 2.8

Inquiry Focus

PATTERNS IN THE PERIODIC TABLE

Refer to **Figure 2.9** and **2.10**, and the index cards from Activity 2.6.

1. Look closely at the properties of the elements that are near each other vertically (up and down) and horizontally (beside each other). Describe any patterns that you see in the properties.

2. Explain how the atomic structure of an element is linked to its position in the periodic table and to its properties.

LEARNING CHECK

1. What determines an element's chemical and physical properties?
2. In the periodic table, what pattern emerges when elements are grouped into families?
3. Refer to **Figure 2.10**. What is the most reactive metal in group 2? What is the most reactive non-metal in group 16?

Periodic Table

▲ **Figure 2.11** This pictorial version of the full periodic table includes photos of common uses of elements, as well as faces of important people who either discovered elements or added to our understanding of them.

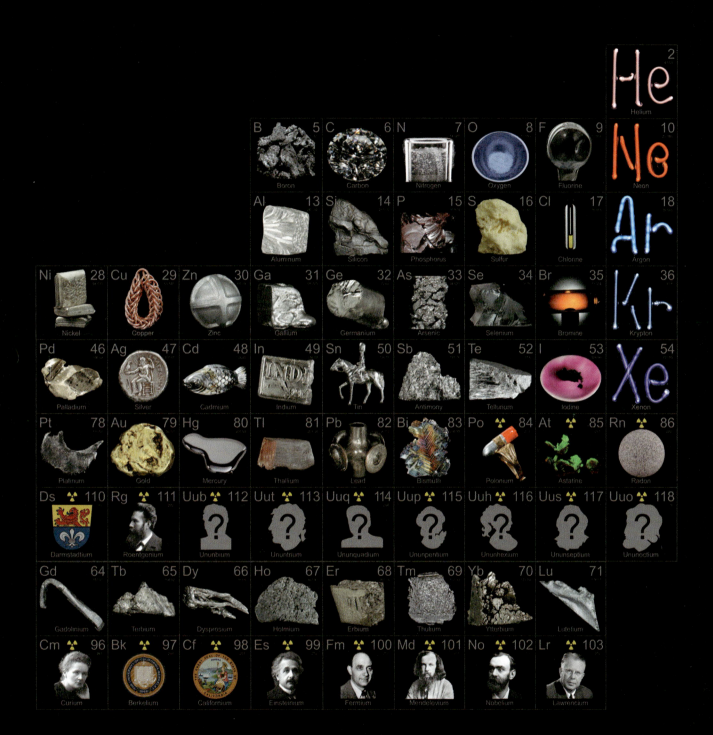

TOPIC 2.4 HOW ARE PROPERTIES OF ATOMS USED TO ORGANIZE ELEMENTS INTO THE PERIODIC TABLE?

Activity 2.9
BUILD A PERIODIC TABLE

[Periodic table template with Period 1–4 rows and Groups 1, 2, 13–18 columns, empty cells to be filled in.]

In this activity, you will use the atomic models you built in Activity 2.7 to make your own periodic table based on the organization of the protons, neutrons, and electrons in the atom. Recall that the organization of the protons, neutrons, and electrons in an atom is called the atomic structure.

What To Do

1. Work in pairs or small groups. Use your atomic models, the template above, and the information that you have learned about the periodic table to place your elements in the correct location based on the atomic structure of their atoms.

2. When you have placed all of your elements, check your periodic table with your teacher. Reorganize your elements if necessary.

What Did You Find Out?

1. What do the rows of the periodic table mean?
2. What do the columns of the periodic table mean?

Inquire Further

1. The periodic table of the elements is a human invention. In the 1800s, over a period of several decades, scientists from different countries tried to sort the elements that were known at the time. The scientist whose sorting method worked the best was a Russian chemist named Dmitri Mendeleev. He recorded properties of each element on cards. The picture shows the kind of information that Mendeleev wrote on his property cards.

    a) Find out how chemists of the time reacted to Mendeleev's ideas.

    b) Do you think that scientists in today's world still sometimes experience prejudice from their colleagues? Give reasons to justify your opinion.

Si – Silicon
Atomic Mass   28.1
Density   2.3 g/cm³
Colour   Dark Grey
M.P.   1410°C
B.P.   3265°C

Topic 2.4 Review

Key Concept Summary

- Elements are made up of atoms, which are made up of subatomic particles.
- Elements are arranged in the periodic table according to their atomic structure and properties.
- Elements in the same family (group) share similar physical and chemical properties.

Review the Key Concepts

1. **K/U** Answer the question that is the title of this topic. Copy and complete the graphic organizer below in your notebook. Fill in four examples from the topic using key terms as well as your own words.

   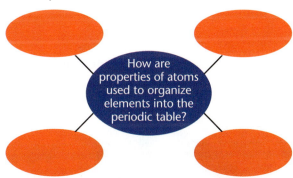

2. **K/U** Use a concept map to summarize the three parts of an atom. Include each part's charge, where each part is found, and the mass of each part compared to the other parts (relative mass).

3. **T/I** Compare an atom of sulfur with an atom of magnesium in terms of the number of protons, electrons, and neutrons. Use the table in Activity 2.7 on page 123 to help you.

4. **K/U** Why is the periodic table one of the most useful tools used by chemists and science students?

5. **K/U** List three elements found in period 2 of the periodic table.

6. **K/U** List two elements found in group 17 of the periodic table.

7. **C** The atomic number for nitrogen is 7.
   a) How many protons and electrons does an atom of nitrogen have?
   b) Nitrogen has 7 neutrons. Use the information from a) above and this new information to draw a diagram showing the structure of a nitrogen atom. (Hint: 2 electrons occupy the first energy level and then up to 8 electrons occupy the next energy level.)

8. **A** Sodium is a metal element found in group 1 of the periodic table. It is a soft metal that can be cut with a knife. Sodium reacts violently with water, as you can see in the photo below. Potassium is another element found in group 1 of the periodic table. Based on the information given for sodium and your knowledge of properties of elements in the same group of the periodic table, predict the properties of potassium.

Topic 2.5

In what ways do scientists communicate about elements and compounds?

Key Concepts

- Chemical symbols are used to represent elements.
- Chemical formulas are used to represent the types and numbers of atoms in compounds.

Key Skills

Inquiry
Research

Key Terms

chemical symbol
molecule
chemical formula

All of the compounds that exist on Earth are built from the elements on the periodic table. How many is "all"? Scientists estimate that there may be as many as 10^{200} different compounds. That number is a 1 with 200 zeros after it! The periodic table lists just under 120 elements, and only 80 of these commonly form compounds. Think of how many different structures you would be able to build if you had 80 different kinds of building blocks. The compounds that are made up of just carbon and hydrogen atoms number in the millions!

A canister containing hydrogen

A sample of carbon

Starting Point Activity

The photos in the circles show products that are entirely carbon, entirely hydrogen, or compounds that result when carbon and hydrogen combine chemically. Use the photos and your imagination to discuss possible answers to these questions.

1. How do the physical properties of hydrogen and carbon alone compare with the physical properties of the substances that are made from them?
2. List three chemical properties that many of the carbon and hydrogen compounds appear to have in common.

pencil "lead" (pure carbon)

diamond (pure carbon)

hydrogen-powered car (pure hydrogen)

gasoline (a compound of carbon and hydrogen)

natural gas (a compound of carbon and hydrogen)

candle wax (a compound of carbon and hydrogen)

plastic (a compound of carbon and hydrogen)

acetylene (a compound of carbon and hydrogen)

Chemical symbols are used to represent elements.

chemical symbol: letters used to represent the names of elements

Scientists use shorthand when they talk about elements, much the same way that you use shorthand if you send a text message. **Chemical symbols** are short forms used to represent the names of elements in compounds. You have already seen them used on the periodic table.

Chemical symbols are either one or two letters. Often the first letter of the element name is used—for instance, C for carbon and N for nitrogen. When scientists ran out of letters, they began using the first letter of the element name and a second letter from the name—for instance, Ca for calcium and Zn for zinc.

Some of the elements were discovered in ancient times and have Latin names. (For a long time, Latin—the language of ancient Rome—was the language used by most scientists to communicate their ideas.) As a result, some elements have chemical symbols that are based on their Latin name. For example, the symbol for the element gold is Au. What's the link? Au is the short form of the word *aurum*, which is the Latin word for gold. **Table 2.6** shows some other examples of chemical symbols that come from the Latin names for elements.

Table 2.6 Some Chemical Symbols That Come from the Latin Names for Elements

Element Name	Chemical Symbol	Latin Name	Meaning of Name
silver	Ag	argentum	Latin for "silver"
mercury	Hg	hydrargyrum	Latin for "liquid silver"
tin	Sn	stannum	Latin for "tin"
potassium	K	kalium	Latin for an Arabic word, *al-qalyah*, meaning "plant ashes"
iron	Fe	ferrum	Latin for "grey"
lead	Pb	plumbum	Latin for "lead"
sodium	Na	natrium	Latin for "sodium"
copper	Cu	cuprum	Latin for "Cyprian" (metal from the island, Cyprus)

Go to **scienceontario** to find out more

Elements: Atoms and Molecules

Almost all elements exist as atoms. A few elements, however, exist as molecules. **Table 2.7** lists these. A **molecule** is a type of particle that is made up of two or more atoms that are joined together by what's called a chemical bond. Chemists use a chemical formula as shorthand to describe a molecule. A **chemical formula** uses both chemical symbols and numbers. The chemical symbol is written first, and the number of atoms is shown with a subscript. A subscript is written smaller and slightly below the rest of the text. **Figure 2.12** summarizes the parts of a chemical formula.

molecule: a type of particle made up of two or more atoms bonded together

chemical formula: a short form for writing the name of a compound using chemical symbols and numbers

Table 2.7 Elements That Exist As Molecules

Element	Chemical Symbol	Number of Atoms in the Molecule	Chemical Formula
hydrogen	H	2	H_2
nitrogen	N	2	N_2
oxygen	O	2	O_2
fluorine	F	2	F_2
chlorine	Cl	2	Cl_2
bromine	Br	2	Br_2
iodine	I	2	I_2
phosphorus	P	4	P_4
sulfur	S	8	S_8

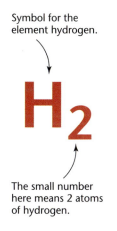

▲ **Figure 2.12** This shows the chemical formula for hydrogen.

LEARNING CHECK

1. Refer to **Figure 2.11** on pages 126 and 127. What are the chemical symbols for the following elements: helium, aluminum, tungsten, cadmium, krypton, francium, iodine, cobalt, and barium?
2. Why does silver have the chemical symbol Ag and not a symbol like Si or Sl?
3. Draw and label a diagram to explain what the symbol P_4 means. Use **Figure 2.12** as a guide.
4. Refer to **Table 2.6**. Explain how the chemical symbol for each of the elements listed is related to the Latin name for each element.
5. What chemical symbols would you use if the following new (and imaginary) elements were discovered: Asherium, Phoenixium, Wymium, Searlium, Edwardium, Weberum, Canadium, and Ontarium?

ACTIVITY LINK
Activity 2.11, on page 138

Chemical formulas are used to represent the types and numbers of atoms in compounds.

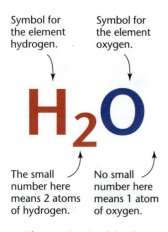

▲ Figure 2.13 This shows the chemical formula for the compound water, H_2O.

▶ Figure 2.14 These three compounds are commonly found in many homes.

When two or more atoms are joined together chemically, they form a molecule. If the atoms joined together are of the same type, then the substance is an element. If the atoms joined together are of different types, then the substance is a compound. Just as we can use chemical symbols and chemical formulas to represent elements, we also can use them to represent compounds. Refer to Figure 2.13.

While many of the substances that we use are complex mixtures, some are simple chemical compounds that are made up of just a few different elements. Figure 2.14 shows a few examples.

A Hydrogen peroxide is a compound made up of the elements of hydrogen (H) and oxygen (O). Its chemical formula is H_2O_2. This means that there are 2 atoms of hydrogen and 2 atoms of oxygen in every molecule of hydrogen peroxide. It is often used as a disinfectant for wounds.

B Lye is the common name for a compound called sodium hydroxide. Sodium hydroxide is made up of 1 atom of sodium (Na), 1 atom of oxygen (O), and 1 atom of hydrogen (H). The chemical formula is NaOH. It is the main ingredient in many kinds of oven cleaner.

C Table salt is a compound that is made up of atoms of sodium (Na) and chlorine (Cl). The chemical formula for table salt is NaCl. Sodium chloride is the type of salt compound that you put on your food.

LEARNING CHECK

1. Two different atoms are joined together to make a molecule. Is this an element or a compound? Explain your answer.
2. Explain which elements and how many atoms of each element are present in hydrogen peroxide, H_2O_2.
3. Baking soda is a compound with 1 atom of sodium, 1 atom of hydrogen, 1 atom of carbon, and 3 atoms of oxygen. What is the chemical formula for baking soda?

Activity 2.10
BUILDING MOLECULES

To understand how atoms combine to form compounds, we can build models. In this activity, you will draw and build some common molecules.

What To Do

1. Copy the table below. One entry in the second column has been filled in for you.

Chemical Formula	Number and Type of Atom Present	Drawing of Molecule
H_2O (this is water)	2 atoms hydrogen (H) 1 atom oxygen (O)	
H_2 (this is hydrogen)		
NH_3 (this is ammonia)		
CO_2 (this is carbon dioxide)		
CH_4 (this is methane)		

2. Complete the second column of the table. Use the information in this topic as a guide.

3. Using the materials provided by your teacher, build a model for each of the molecules in the table using the information from the first column.

   Hints:
   - Start with the atom written first in the chemical formula.
   - The element you start with is the centre atom. All other atoms attach to this atom.

4. Sketch your models in the third column of the table.

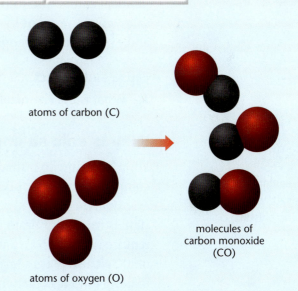

▲ An atom of the element carbon (C) combines with an atom of the element oxygen (O) to form a molecule of carbon monoxide (CO). The chemical symbols C and O are used to represent the two atoms. The chemical formula, CO, represents the molecule. Spheres of different colours can be used to model atoms and molecules.

Case Study Investigation: Salt of the Earth

The Bombay Chronicle

WEATHER 30°C

CRICKET NEWS INSIDE

THURSDAY, MARCH 13, 1930 - BOMBAY, INDIA

Gandhi Goes on Salt March to Dandi

Mohandas Karamchand Gandhi began a non-violent march to Dandi, Gujarat yesterday. The protester is perhaps better known by his other name, Mahatma Gandhi, or great soul. The march, a protest of the tax the British Empire has placed on salt in India, will be 400 km long. Gandhi also plans to break the British tax law at the end of the march in another show of protest.

In doing so, he hopes to inspire other Indians to join his ongoing peaceful opposition to British rule in India. However, Gandhi's simple act may do much more than that. News coverage of the march is poised to draw the world's gaze to the plight of the Indian independence movement. Can the steps of one man shake the British throne's hold on our nation? Only time will tell.

Large numbers of Indians joined Mohandas Gandhi to protest the British-imposed salt tax.

The Science behind the Story

We toss it on French fries, cucumber slices, and popcorn without a second thought. But common table salt, sodium chloride (NaCl), may be the most political compound found on our planet. Produced by the evaporation of salt-rich sea water or by mining it from the ground, salt was once so rare and expensive that people gave their very lives to gain control over it.

Pause and Reflect

1. What is the chemical name and formula for the salt we put on our food?

Why is salt so important?

For a long time in human history, salt was one of the most expensive compounds around. Like water, salt is essential for life. For example, the chemical reactions in your body that send signals from your brain and spine to other parts of your body require sodium (Na). Salt (NaCl) is one of the main sources of it. Salt also plays a large role in preserving food and making food taste better. For this reason, the demand for salt has always been great. Until the invention of modern mining methods, the world supply of salt was very limited, keeping prices high. As a result, those who had control of salt also controlled a great deal of power.

NaCl, commonly known as table salt, occurs naturally as cubic crystals.

Pause and Reflect

2. Why was salt so expensive in the past?
3. Describe two important uses for salt.

Inquire Further

4. The Sifto Salt Mine in Goderich, Ontario is one of the largest salt mines in the world. The mine was discovered in 1866 under Lake Huron. Create a historical graphic novel explaining how the mine was discovered and how its discovery caused the town to be struck by salt fever!
5. Find out more about Gandhi's salt march. Did Gandhi get arrested for his actions? What was the end result of his protest?
6. Salt played a major role in wars in the past. The Arabic word for peace actually means "to negotiate for salt." Find out what this word is.
7. Canadians get far too much salt in their diet.
    a) Find out what happens to your body when you eat too much salt.
    b) Think of three ways you could cut down on your salt consumption.

Because salt is affordable today, we often take it for granted.

What food do you put salt on?

Research Focus

Activity 2.11
LEARNING MORE ABOUT THE ELEMENTS AND THEIR COMPOUNDS

The chemical and physical properties of elements can change significantly when they are found in compounds. How are some of the most common elements used in the world around you? What are some of the social and environmental consequences that are linked with using them?

What To Do

1. In pairs, choose one of the following elements:

   titanium   sulfur
   iron   magnesium
   copper   calcium
   carbon   nitrogen
   gold   phosphorus
   oxygen   chlorine
   nickel   cadmium
   hydrogen

2. Use the Internet or other sources of information to find out the following for your element:
   - chemical symbol
   - sources of the element (where and how is it found?)
   - common uses of the element (both as an element and as part of a compound)
   - some social and environmental consequences of its use

3. Compile your information in a format to share with the class. Your teacher will provide you with template choices that you can select from.

What Did You Find Out?

1. Do you think the benefits of using your element outweigh the risks? Give reasons to justify your answers.

lithium metal

Onboard the space shuttle

prescription lithium compound medication

▲ The element lithium has an atomic number of 3 and the chemical symbol Li. It is very reactive, so it is not found as an element in nature. Instead, it is commonly found as part of compounds. One of these compounds is lithium hydroxide. It is used to remove carbon dioxide from the air in the space shuttle. Another lithium compound is lithium carbonate. It is used as a drug in the treatment of certain mood disorders.

Topic 2.5 Review

Key Concept Summary
- Chemical symbols are used to represent elements.
- Chemical formulas are used to represent the types and numbers of atoms in compounds.

Review the Key Concepts

1. Answer the question that is the title of this topic. Copy and complete the graphic organizer below in your notebook. Fill in four examples from the topic using key terms as well as your own words.

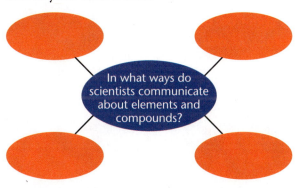

2. **T/I** You have come into contact with the elements carbon, oxygen, nitrogen, and silicon already today. Find the symbols for these elements in the periodic table in **Figure 2.9** and write the symbols in your notebook.

3. **K/U** Why is the chemical symbol Fe used for iron?

4. **K/U** Use words and a picture to show the particle that is produced when two atoms of the same type are combined and when two atoms of different types combine.

5. **K/U** List the names and formulas of the nine elements that exist as molecules.

6. **A** Explain what every letter and number represents in the chemical formula for glucose, $C_6H_{12}O_6$.

7. **C** Draw a labelled diagram to model the structure of a) hydrogen sulfide, H_2S; b) silicon hydride, SiH_4 c) chlorine, Cl_2.

8. **A** Propane is a molecule that contains 3 atoms of carbon and 8 atoms of hydrogen. Write the chemical formula for propane.

9. **A** The photograph shows equipment that passes an electrical current through the compound water. This causes a chemical reaction that decomposes water into the elements that make it up, oxygen and hydrogen. Write the chemical formulas for all the substances involved in this chemical reaction.

Topic 2.6

What are some characteristics and consequences of chemical reactions?

Key Concepts

- Compounds and elements are changed during chemical reactions.
- The properties of substances that make them useful can also make them dangerous.
- There are less-harmful alternatives to many products we use and depend on.

Key Skills

Inquiry
Literacy

Key Terms

chemical reaction

When you observe that a chemical change has occurred, you know that elements or compounds (or both) have reacted. Some substances are produced during the reaction, and some substances are consumed during the reaction. For example, think about the hot, glowing coals in the main photo. Before the reaction that started the coals glowing, there was carbon (C) in the coals and oxygen (O_2) in the air. When energy was added to start a fire, the carbon and oxygen reacted together to produce carbon dioxide gas (CO_2). Now look at all the bubbling shown in the smaller photo. Before the reaction, there was acetic acid, which you know better as vinegar (CH_3COOH), and there was sodium bicarbonate, which you know better as baking soda ($NaHCO_3$). After the two substances react, there is sodium acetate ($NaC_2H_3O_2$), water (H_2O), and carbon dioxide (CO_2).

Starting Point Activity

You can learn a lot about the interaction of substances from one of the most common substances in the home: vinegar. Most of these mini-activities take a few days for observation. Your teacher will tell you how to proceed with them. (You can also do some or all of them easily at home.)

1. For each of these mini-activities, what evidence of a new substance being produced can you observe?
    - Seal up a jar containing a piece of steel wool covered with vinegar.
    - Place a stainless steel spoon in a jar of vinegar.
    - Mix about 5 mL of baking soda with about 10 mL of vinegar in a small jar.
    - Place a clean egg in a jar of vinegar so it is fully covered, and seal the jar. (Always wash your hands after handling a raw egg.)

2. What will you do with the substances of the mini-activities when you are done with them? What harm, if any, might there be to the environment? How would your answer change if you were using powerful metal-eating car-battery acid?

Compounds and elements are changed during chemical reactions.

chemical reaction: any change that occurs when substances interact to produce new substances with new properties

A **chemical reaction** occurs when pure substances interact in a way that causes them to change into other pure substances. During chemical reactions, the chemical and physical properties of the pure substances change. That's how you know a chemical reaction has occurred.

We depend on some chemical reactions to produce certain types of products that are desirable to us. We depend on other chemical reactions for the energy they release while substances interact. However, chemical reactions also often produce undesirable products. Read about this in the examples below and at the top of the next page.

A Desirable Product of a Chemical Reaction

One important chemical reaction that affects your life combines nitrogen (N_2) with hydrogen (H_2) to make ammonia (NH_3). Ammonia is used to make fertilizer. As shown in **Figure 2.15**, farmers around the world use fertilizer to grow the food you eat.

▲ **Figure 2.15** Naturally occurring fertilizing compounds in soil are not nearly enough to support the world's population with food. Commercial production of fertilizer helps to grow enough food to support one-third of Earth's population.

▲ **Figure 2.16** Chemical reactions between the compounds in fireworks and the air produce the sound and light we associate with a fireworks display.

A Desirable Release of Energy from a Chemical Reaction

Some chemical reactions are useful not because of the substances they produce, but because of the heat or light that is released during the chemical reaction. For example, we use the energy produced during combustion reactions to keep us warm, cook our food, and power our cars and trucks. As **Figure 2.16** shows, we also use combustion reactions for entertainment.

An Undesirable Product of a Chemical Reaction

Most chemical reactions produce substances that we do not want, in addition to those that we do. For example, the combustion reactions that drive our cars produce gases that contain compounds responsible for acid rain. The acid in acid rain causes changes to the acidity of water in lakes and ponds, which can kill aquatic organisms. As well, the acid in acid rain reacts with the compounds in certain kinds of rock and metal. This reaction breaks down elements or compounds in the rock and metal. As a result, surfaces of bridges, buildings, and sculptures erode much more quickly than they would otherwise. **Figure 2.17** shows an example of damage caused by acid rain.

▲ **Figure 2.17** Acid rain has wide-reaching negative effects on the environment and on human-made structures. The acid in the rain reacts with the building materials, causing them to crumble away with the passage of time.

LEARNING CHECK

1. How do you know when a chemical reaction has occurred?
2. List two desirable products of a chemical reaction.
3. Which undesirable products are formed by the combustion reactions used to heat many homes and power cars?

Activity 2.12

ANALYZE SOME CHEMICAL REACTIONS

The photos here involve chemical reactions. Answer these questions about the photos.

1. Describe the chemical reaction taking place in each photo.
2. Do you think the reactions are useful? Explain your answer.
3. Discuss the social effects, the environmental effects, or both, of the chemical reactions shown.
4. Describe two more chemical reactions that could be analyzed using the first three questions. Then answer them.

Literacy Focus

Combustion

Rusting

Cellular respiration

Food spoilage

Photosynthesis

The properties of substances that make them useful can also make them dangerous.

Many of the substances that we use at home and at work have properties that make them both beneficial and dangerous. For instance, chlorine compounds are included in many cleaners because of their bleaching ability and strong disinfectant properties. The term disinfectant really means "poisonous." Chlorine kills bacteria and moulds because one of its useful properties is that it is poisonous.

Chlorine compounds are also used on a large scale to purify drinking and swimming pool water. Here, again, the term purify refers to the poisonous property of chlorine. We use chlorine to kill bacteria and other microbes in drinking and pool water.

So although chlorine compounds are widely used, they must be used with great care because of the hazardous nature of their properties. That's why many products sold for use in the home have hazard symbols and safety warnings on their packaging. **Figure 2.18** shows the meaning of the symbols that are commonly used on products for the home. This information comes from Health Canada to inform all Canadians about the safe way to use chemical products.

Figure 2.18 Many home products that are potentially dangerous have safety warnings on their labels. Hazardous Household Product Symbols (HHPS) are designed to be easy to understand. The shapes that outline the pictures are designed to look like road signs.

LEARNING CHECK

1. Use a t-chart to summarize the properties of bleach that make it both useful and harmful.
2. Sketch the HHPS symbol for (a) a flammable product and (b) an explosive container.
3. What HHPS symbol would you see on a container of bleach?

Literacy Focus

Activity 2.13

WHAT'S ON A LABEL?

Work in small groups. Your teacher will provide your group with a common product used in the home or the label from a common product. Examine the labelling on this product. Answer the following questions.

1. What is the intended use of this product?
2. What ingredients (substances) are found in the product?
3. What, if any, hazard-related information is included on the product labelling?
4. What, if any, safety-related instructions are provided?
5. If your product does not have a HHPS on it, design a suitable warning or safety symbol that could be used for it.
6. How does the warning and safety information for home products compare with WHMIS symbols? (Turn to page xv to review the WHMIS symbols.)

You will be asked to examine and analyze a label from products such as these in this activity.

There are less-harmful alternatives to many products we use and depend on.

If you gathered up all of the cleaning products in your home, you would find that all or most of them have some kind of warning or first aid information on them. We continue to use them because the benefits of their properties seem to outweigh the risks, or because we are unaware of the risks. However, there are alternatives with similar properties that do a similar job with fewer risks to health and the environment.

As people become more aware of the hazards of using some chemical substances, they are less willing to use more-hazardous cleaning products. Now, there are many commercially available cleaning products made with substances that are often less harmful to people and to the environment. Another, less expensive alternative is to use readily available substances that many people already have at home in the kitchen and bathroom cupboards. These are safer-to-use chemical substances that you probably already have in your kitchen cupboards at home. For most of the alternative cleaning products listed in **Table 2.8**, lemon juice can be added to improve the odour.

Note: In most cases, alternative cleaning substances are safer than traditional ones. But "safer" does not mean the same thing as "safe." Any substance—even water—can be dangerous under the right (or, rather, wrong) circumstances. There is no such thing as a chemical product that is 100% safe. All substances should be treated with care, thought, and respect.

Table 2.8 Some Alternatives to Traditional Cleaning Products

Traditional Cleaning Product	Safer-to-use Alternative
window cleaner	• a mixture of vinegar and water
furniture polish	• a mixture of white vinegar and vegetable oil
stain remover	• baking soda and water paste • hydrogen peroxide (3%) for some kinds of stains
oven cleaner	• borax and vinegar (and lots of vigorous scrubbing) • baking soda (and lots of vigorous scrubbing)
dishwasher detergent	• a mixture of baking soda and borax
fabric softener	• vinegar
toothpaste	• baking soda

Activity 2.14
WHICH WOULD YOU CHOOSE?

Inquiry Focus

In this activity, you will use a number of criteria to assess which type of cleaning product is the best choice for you.

What You Need
- commercially available cleaners
- lemons
- vinegar
- baking soda
- vegetable oil
- fabric
- "dirt"

What To Do
1. Choose one pair of household products: one commercially available and one alternative.
2. Make a table in your notebook similar to this one. Write the name of the commercial product and alternative product in the two right columns.

Test	Commercial Product	Alternative Product
effectiveness		
price		

3. Test each of your two products for their functionality. Test them as you would use them at home. For example, if they are glass cleaners, use them to clean glass. Assign each of the products a value from 5 (great) to 1 (awful) on their ability to do their job.
4. Ask your teacher for the price of each of the products. Assign the products a value from 5 (inexpensive) to 1 (very expensive) based on their cost.
5. With your partner, determine two other criteria to use to assess the products. Assess the products on these two criteria, and record the values from 5 (great) to 1 (awful).
6. Clean up your equipment. Wash your hands.

What Did You Find Out?
1. Which of your two products do you think is the "best"? Explain your answer.
2. Why did you choose the two criteria you added to assess your products?
3. How do your results compare with the rest of the class? Are commercially available products "better" than alternative products? Explain.

LEARNING CHECK

1. Why do we use cleaning products that can be hazardous?
2. Explain why using baking soda would be a safer way to clean an oven than using a more-hazardous product, but would not be considered 100 percent safe.

Investigation 2C

Identifying an Unknown Gas

Skill Check

Initiating and Planning

✓ Performing and Recording

✓ Analyzing and Interpreting

✓ Communicating

When chemical reactions take place, a gas is often produced. During this investigation, three different gases will be collected. You will identify the gas produced based on its chemical properties.

Caution! Your teacher may perform part or all of this investigation in a different way in order to ensure the safest possible environment for your class.

What To Do

1. Work with a partner to perform each of the following tests. Record your observations as you complete each step.

2. Be sure to clean up your work station as you complete each part. Place each substance in the appropriate waste container, as directed by your teacher.

Part 1: Test for Hydrogen Gas

3. Obtain 5 mL of hydrochloric acid in a test tube, a piece of mossy zinc, and a wooden splint. Have the wooden splint nearby.

4. One partner holds the test tube at a 45° angle, using a test-tube holder, and then slides the zinc down the side of the test tube into the acid. A reaction should begin. Trap some of the gas in the tube using a rubber stopper.

5. **Test for Hydrogen:** Your teacher will show you how to light the splint. The other partner brings the flaming splint close to the mouth of the test tube. Hydrogen gas will ignite and burn rapidly down the test tube with a "whoop" sound.

6. Extinguish the wooden splint.

Safety

- Put on safety goggles and a lab apron.
- Be cautious when testing for gases.
- Be careful when handling the burning splints.
- Make sure the splints are properly extinguished immediately after being used.

What You Need

10 mL dilute hydrochloric acid

4 test tubes

test tube rack

mossy zinc

rubber stopper

test tube holder

2 wooden splints

5 mL 3% hydrogen peroxide

yeast

marble or limestone chip

5 mL limewater

balloon

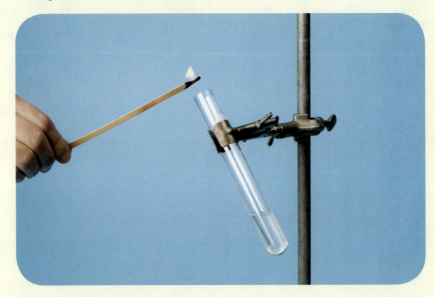

Part 2: Test for Oxygen Gas

7. Obtain 5 mL of 3% hydrogen peroxide in a test tube, some yeast, and a wooden splint.

8. One partner adds the yeast to the hydrogen peroxide. A reaction should begin. Trap some of the gas in the tube using a rubber stopper.

9. **Test for Oxygen:** Your teacher will show you how to light the splint and produce a glowing ember. The other partner brings the ember to the mouth of the test tube and inserts the glowing ember into the test tube. If oxygen is present, the glowing ember will burst into a bright flame.

10. Extinguish the wooden splint.

Part 3: Test for Carbon Dioxide Gas

11. Obtain 5 mL of hydrochloric acid in a test tube, a small piece of marble or limestone, and a second test tube containing 5 mL of limewater.

12. One partner holds the test tube with the acid at a 45° angle, and slides the piece of marble down the side of the tube into the acid. The other partner places the balloon over the top of the test tube. The balloon will inflate with any gas that is produced.

13. **Test for Carbon Dioxide:** Keep the new gas inside the balloon by twisting the balloon closed. While the balloon is still twisted closed, attach it to the mouth of the test tube containing limewater. Once attached, invert the balloon-covered test tube so the limewater will mix with the gas in the balloon. Then return the test tube to an upright position. If carbon dioxide is present, the limewater will turn white and milky.

If oxygen gas is present, a flame will form from the ember.

Slide the marble or limestone chip down the side of the test tube.

What Did You Find Out?

1. Describe the three gases by their physical properties. Could you use their physical properties to tell them apart?

2. Which chemical property was used to tell the three gases apart?

3. Which of the three gases could be used in a fire extinguisher? Explain your answer.

Making a DIFFERENCE

Adrienne Duimering was 14 when she saw statistics about fatal fires in Canada. Many of the fires were preventable, and Adrienne wanted to know what products were available to help prevent fires. She discovered that fire retardants (chemicals that fireproof flammable fabrics) can be expensive and toxic to the environment. Adrienne decided to study fire retardants for her 2007 science fair project. She wanted to find an inexpensive, environmentally friendly fire retardant. Adrienne tested the fire-retarding abilities of sodium bicarbonate (baking soda) and ammonium sulfate (a common fertilizer) on common materials. She found that both compounds were effective retardants. Her project won a silver medal at the Canada-Wide Science Fair.

Are there products in your home that could be replaced with safe, less expensive alternatives?

In 2004, when Sarah Mediouni was 12, she and her friends started a campaign to get Orangeville, Ontario to ban the use of pesticides.

During their campaign, Sarah and her friends met with their mayor and town council. They presented the officials with statistics and a 2004 report by the Ontario College of Family Physicians that linked pesticide exposure to cancer and other illnesses in children. They also presented petitions signed by 300 young people and more than 400 adults. Sarah helped deliver flyers advertising local seminars on ways to care for lawns without using pesticides. A pesticide by-law came into effect in Orangeville in 2007.

Sarah says it felt great to have an impact on the issue. "It made me realize that a small group of kids can actually have an impact on something important."

In what ways can you have a valuable impact or make an important contribution in your community?

Topic 2.6 Review

Key Concept Summary

- Compounds and elements are changed during chemical reactions.
- The properties of substances that make them useful can also make them dangerous.
- There are less-harmful alternatives to many products we use and depend on.

Review the Key Concepts

1. **K/U** Answer the question that is the title of this topic. Copy and complete the graphic organizer below in your notebook. Fill in four examples from the topic using key terms as well as your own words.

   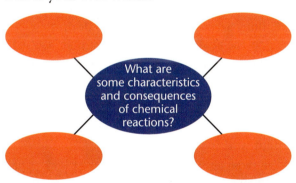

2. **A** Why are chemical reactions important to us? Answer this question by providing examples of chemical reactions that you encounter in your daily life.

3. **K/U** Are the chemical products of a reaction always something we can use? Explain why or why not.

4. **A** Commercial fertilizers are produced to help farmers try to grow enough food to feed everyone in the world. Fertilizers are also used by some homeowners in cities to keep their lawns and gardens lush and healthy. However, not all the fertilizer is used by the plants, and excess fertilizer runs off into water systems and makes its way into waterways. This run-off damages the water systems. Do you think fertilizer use should be banned in all cities? Support your opinion with appropriate evidence.

5. **T/I** A reaction that produces a gas occurred and the gas was collected. Several tests were performed. A flaming splint was brought close to the gas and nothing happened. A glowing ember did not relight when placed in the gas. When mixed with limewater, the limewater turned cloudy.

   a) In your notebook, draw a t-chart with the headings "Evidence" and "Inferences." Record each observation in the "Evidence" column and then complete the "Inferences" column.

   b) Identify the gas that was collected. Explain how you know.

   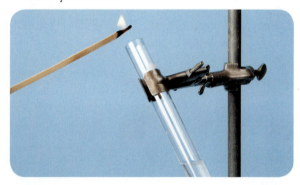

6. **A** The compound calcium carbonate (lime) is sometimes added into a lake that has been affected by acid rain. The compound reacts with acid in the lake, causing the lake water to become less acidic. Liming a lake, as this process is called, is not a permanent solution to the effects of acid rain. Explain why this is the case.

SCIENCE AT WORK

CANADIANS IN SCIENCE

Florell Essibrah

Florell Essibrah's lab is a bakery, and her chemicals include flour, egg, sugar, and butter. She mixes them together to make cake batter. Nearby are cookies decorated with the colourful icing that she has made. Her cake, cookies, and other baked goods each will taste, smell, and feel different. The taste, smell, and texture depend on the chemical and physical properties of the ingredients that go into making them. Like any chemist, a baker is keenly aware of the properties of the substances that go into making a final product.

When did you decide you wanted to pursue a career as a baker?

Florell Essibrah attended E.C. Drury Secondary School for the Deaf. While in high school, she completed a co-operative education program that involved working at a bakery, where she learned from an internationally trained pastry chef. The co-op program helped her decide to pursue a career in baking. She registered as a baker in the Ontario Youth Apprenticeship Program and will attend George Brown College to continue her education.

What do you like most about working in a bakery?

"It can be very busy, but I like to be busy. I like to have a lot of work to do," says Florell, who uses American Sign Language. "I enjoy making desserts and cookie icing the best."

What skills do bakers need?

Bakers handle large amounts of many different ingredients and use and work with large equipment, including mixers and ovens. Good organizational and planning skills, therefore, are essential. Creativity, attention to detail, and a keen eye for making precise, accurate measurements are also valuable assets.

Bakers work for food manufacturers, grocery and other food stores, hotels, catering businesses, and other specialty baking and pastry businesses.

Put Science To Work

The study of chemistry contributes to these careers, as well as many more!

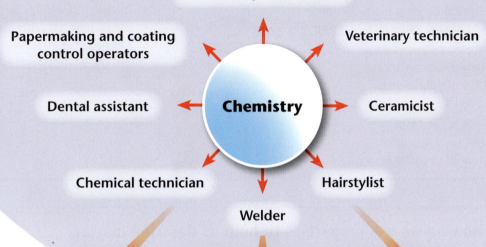

▲ Chemical technicians help prepare and conduct chemical tests. They may work in research and quality control laboratories, in the manufacturing industry, or companies that make products such as foods.

▲ Welders use special equipment to permanently join metals together to make or repair metal parts. Welders often work as mechanics in the automotive industry, as well as in aerospace and pipeline construction.

▲ Hairstylists apply chemicals, such as dyes, tints, and bleach, to treat and colour hair. They also use chemicals to straighten or curl hair.

Over To You

1. How does baking involve chemistry?
2. Why does a baker need good organizational and planning skills?
3. Research a career involving chemistry that interests you. What are the essential skills needed for this career? What would you need to do to pursue this career?

Go to **scienceontario** to find out more

Unit 2 Summary

Topic 2.1: In what ways do chemicals affect your life?

Key Concepts
- Everything—including you and everything around you—is made up of chemicals.
- Substances have characteristics that make them useful, hazardous, or both.
- Handling chemicals and lab equipment safely and responsibly is a part of your life at school.

Key Terms
matter (page 97)

Big Ideas
- The use of elements and compounds has both positive and negative effects on society and the environment.

Topic 2.2: How do we use properties to help us describe matter?

Key Concepts
- Physical properties describe how matter looks and feels.
- Chemical properties describe how substances can change when they interact with other substances.

Key Terms
physical property (page 106)
conductivity (page 106)
density (page 106)
lustre (page 106)
solubility (page 106)
texture (page 106)
chemical property (page 108)
combustibility (page 108)
precipitate (page 108)
decomposition (page 108)

Big Ideas
- Elements and compounds have specific properties that determine their uses.

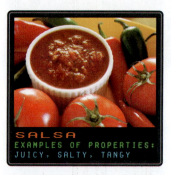
SALSA
EXAMPLES OF PROPERTIES: JUICY, SALTY, TANGY

Topic 2.3: What are pure substances and how are they classified?

Key Concepts
- Pure substances are elements and compounds.
- Elements are metals and non-metals.

Key Terms
mixture (page 114)
pure substance (page 114)
compound (page 114)
element (page 114)
metal (page 117)
non-metal (page 117)

Big Ideas
- Elements and compounds have specific properties that determine their uses.

154 MHR • UNIT 2 EXPLORING MATTER

Topic 2.4: How are properties of atoms used to organize elements into the periodic table?

Key Concepts
- Elements are made up of atoms, which are made up of subatomic particles.
- Elements are arranged in the periodic table according to their atomic structure and properties.
- Elements in the same family (group) share similar physical and chemical properties.

Key Terms
atom (page 122)
proton (page 122)
neutron (page 122)
electron (page 122)
nucleus (page 122)
atomic number (page 122)
periodic table (page 124)
period (page 124)
family (page 124)

Big Ideas
- Elements and compounds have specific properties that determine their uses.

Topic 2.5: In what ways do scientists communicate about elements and compounds?

Key Concepts
- Chemical symbols are used to represent elements.
- Chemical formulas are used to represent the types and numbers of atoms in compounds.

Key Terms
chemical symbol (page 132)
molecule (page 133)
chemical formula (page 133)

Big Ideas
- Elements and compounds have specific properties that determine their uses.

Topic 2.6: What are some characteristics and consequences of chemical reactions?

Key Concepts
- Compounds and elements are produced and consumed during chemical reactions.
- The properties of substances that make them useful can also make them dangerous.
- There are less-harmful alternatives to many products we use and depend on.

Key Terms
chemical reaction (page 142)

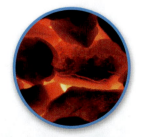

Big Ideas
- Elements and compounds have specific properties that determine their uses.
- The use of elements and compounds has both positive and negative effects on society and the environment.

Unit 2 Project

Inquiry Investigation: Rust Formation

Rust forms when a metal reacts with oxygen in the air. This chemical reaction is called corrosion. Corrosion damages metal in vehicles and supports for roads and bridges, and exposed metals in buildings.

In this project, you will investigate one factor affecting the formation of rust in iron, steel, and aluminum nuts and bolts.

> **Investigate Question**
> What factors influence the corrosion of certain metals?

Initiate and Plan

1. As a class, list the factors that influence the corrosion of metals.
2. Choose one factor (independent variable) to test its contribution to rust formation (dependent variable). Predict its effect on each of the metals.
3. Design your test. Include:
   - equipment and materials
   - step-by-step testing method
   - safety precautions
   - a procedure for recording results
   - criteria for measuring and judging results
4. Have your teacher approve your investigation.

Perform and Record

5. Conduct your investigation and record the results.

Analyze and Interpret

1. Describe the results of your investigation. Did the factor tested affect the formation of rust on each metal? Describe how.
2. Did the results match your prediction? Explain why or why not.
3. Compare your results with those of other classmates who tested the same factor.
4. What changes would you make if you were going to repeat your investigation?

Communicate your Findings

5. Present your test procedure in a written report. Present your results in a chart.
6. Explain what your results suggest about the corrosion of different metals.

Assessment Checklist
Review your project. Did you…

- ✓ Select only one factor (independent variable) for testing? **K/U**
- ✓ Select proper equipment and conduct your test safely? **T/I**
- ✓ Use the same procedure to measure the rust formation for each metal? **T/I**
- ✓ Keep all the other factors constant? **T/I**
- ✓ Explain any changes you would make if you were going to repeat the investigation? **T/I**
- ✓ Record and present your results clearly? **C**
- ✓ Explain what your results suggest about the formation of rust in different metals? **A**

An Issue to Analyze: Evaluating the Use of Road Salt

Road salt melts ice on winter roads, making driving safer. Road salt also contributes to corrosion of vehicles, roads, and bridges and contaminates land and aquatic ecosystems.

Issue

Federal and municipal authorities and the private sector release about 5 million tons of chloride salts into the Canadian environment annually. A half million tons is from road salt on Ontario roads. Should this use be regulated? Explain your reasoning.

Initiate and Plan

1. Create questions to give you enough information to evaluate this issue. Include:
   - the benefits of using salt on winter roads
   - the ways road salt gets into land and aquatic ecosystems
   - the effect of road salt on those ecosystems
   - some possible ways of reducing the impact of road salt (consider what other countries do)
2. Research and list possible sources of the information you need.
3. Choose a graphic organizer or other method to organize your findings.
4. Present your plan for your teacher's approval.

Perform and Record

5. Gather and record information to answer your questions.

Analyze and Interpret

1. Prepare a PMI chart showing positive points for use of road salt alongside negative effects of road salt.
2. Is road salt necessary to keep winter roads safe? Justify your conclusion.
3. How serious is the damage that road salt does to the environment? Give examples.
4. How could regulation help the environment? State your position on the issue and justify it.

Communicate your Findings

5. Present your findings to the class, using a poster, slide show, or a written or oral report.
6. Include a graphic organizer to show how road salt finds its way into different ecosystems.

Assessment Checklist

Review your project. Did you...

- ☑ research how road salt is used in winter to minimize driving hazards? **K/U**
- ☑ consult credible resources to find out about possible environmental damage? **T/I**
- ☑ use graphic organizers to record your research? **C**
- ☑ prepare a PMI chart showing benefits road salt alongside negative effects? **T/I**
- ☑ clearly state your position on the issue and justify your position? **C**
- ☑ note at least one technique or alternative to reduce the use of road salt? **A**

Unit 2 Review

Connect to the Big Ideas

1. Elements and compounds have specific properties that determine their uses. The wide use of silicon in computers and electronic devices is due to the unique properties of this element. Use a research process to discover what properties make silicon the element of choice for computers and electronic devices. Present your findings in an information pamphlet or a small poster.

2. The use of elements and compounds has both positive and negative effects on society and the environment. In recent years, citizens around the world have been encouraged to change the type of light bulb used in their homes. Incandescent light bulbs produce waste heat. Compact fluorescent light bulbs (CFL bulbs) are much more energy-efficient; they use less electricity and thus reduce greenhouse gas emissions. However, CFL bulbs are made with a very small amount of the element mercury, which is poisonous to the human nervous system. CFL bulbs are sealed so that the mercury stays inside the bulb unless it gets broken. Use a PMI chart to weigh the advantages and disadvantages of the use of CFL bulbs. Decide whether you think they should continue to be produced. Write a blog promoting your decision.

Knowledge and Understanding K/U

3. Identify each of the following properties as physical or chemical.
   a) butter melts in a frying pan
   b) honey is denser than water
   c) a limestone statue bubbles when an acid touches it
   d) aluminum is a good conductor of heat and electrical current
   e) snow is white
   f) many non-metals are brittle
   g) peroxide reacts with protein in hair

4. What chemical property of potassium is shown in the photograph below?

5. A substance dissolves in water. Describe one of its physical properties.

6. Solid water—ice—is less dense than liquid water. Therefore, ice floats on water. Describe how your life would be different if ice were not less dense than liquid water.

7. Draw and label pictures of the particles found in a pure substance and the particles found in a mixture. Describe how pure substances and mixtures are different.

8. Use a concept map to show the relationship between an element and a compound.

9. Classify each of the following as an element, a compound, or a mixture.
   a) gold
   b) air
   c) orange juice
   d) sugar
   e) salt

10. How do the properties of bleach make it suitable for cleaning? What are some hazards related to using bleach?

11. Use a Venn diagram to summarize the similarities and differences between metals and non-metals.

158 MHR • UNIT 2 EXPLORING MATTER

12. What can you determine from the atomic number of an atom?

13. Draw a diagram comparing the three subatomic particles found in an atom. Use a symbol and a charge to label each diagram.

14. Draw a diagram of a calcium atom. Place the subatomic particles in the correct position and energy level.

15. Describe the different patterns involving elements in the periodic table.

16. Identify the features of the periodic table below that are indicated by the letters A, B, C, and D.

17. What is the symbol or the chemical formula for each of the elements and compounds listed below?
    a) carbon
    b) fluorine
    c) carbon dioxide
    d) hydrogen peroxide
    e) potassium

18. Draw the HHPS symbol that you would see on a container of a poisonous household product.

19. What properties of copper make it useful for electrical wiring?

Thinking and Investigation T/I

20. While baking at home, you discover that the labels have fallen off the containers of flour and baking soda (sodium hydrogen carbonate). Describe a *safe* way in which you could use a difference in the properties of these two substances to tell them apart. (Do *not* taste the substances.)

21. You have found a chunk of a dull, grey substance while on a nature walk. Outline the steps you could use to determine if this chunk is a metal or a non-metal.

22. Do the following diagrams show atoms of the same element? Explain your answer.

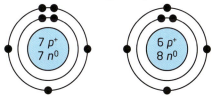

23. A company that sells bottled water calls one of its products, "Nature's Pure Source Water."

    a) Write the chemical formula for the only ingredient that you would expect to find in this product. Explain your answer.

    b) The nutrient analysis label on the product records that the product contains 25 ppp (parts per million) of magnesium, 70 ppm calcium, and 1 ppm sodium. Write the chemical symbols for these three elements.

24. Choose one of the lab tests for hydrogen, oxygen, or carbon dioxide. Use a flowchart to outline the procedure for this test.

Communication C

25. Create a collage of pictures that shows different examples of matter. Draw your own pictures or cut out visuals from newspapers or magazines.

Unit 2 Review

26. Draw models of the following compounds:
    a) water, H_2O
    b) methane, CH_4
    c) ammonia, NH_3

27. Use diagrams and words to explain the difference between an atom and a molecule to someone who is in Grade 6.

28. Batteries are used in many everyday devices. Research the chemicals present in the batteries you use in technologies such as cell phones, as well as the hazards associated with these chemicals. Devise a plan of action for the safe disposal of these batteries after they are used up. Create a school PA message, a poster, a website, or another text form to share your plan of action with others.

29. You are the manager of a cleaning service. You have recently hired some new employees, and you are training them about various health hazards they need to be aware of when doing their jobs. Design an information sheet that outlines the benefits and the hazards of using chlorine bleach.

Application A

30. A facial tissue, a sheet of paper, and a cardboard box are all made from wood fibres. Discuss how the physical properties of each material make it useful for its intended purposes. Name a chemical property that these three materials have in common.

31. In question 23, you read about a bottled water product that has the word "pure" in its name. Write a letter to the marketing manager of this company explaining how the use of the word "pure" on the label is misleading. Use your science knowledge to support your claim.

32. Identify the element at each of the following positions in the periodic table in **Figure 2.9**. Provide the full name and the symbol for each element.
    a) group 2, period 3
    b) group 14, period 2
    c) group 18, period 2

33. Silver is a metallic element with many useful properties. It is malleable and can be polished, moulded, and stretched. It is better than any other metal at conducting heat and electricity. Do some research to find out more about this precious metal and its properties. Make a t-chart listing some properties of silver and some everyday uses based on those properties.

34. A question on a quiz asks you how many energy levels in calcium contain electrons. Explain how you would use a periodic table to find the correct answer to the question.

35. What is the meaning of each of these icons?

36. The chemical symbol for the element, nitrogen, is N. However, when chemists describe the gas, nitrogen, in the atmosphere, they write N_2. Explain why they write the symbol with a subscript of 2.

37. Sodium phosphate is a compound that may be used in some cleaning products, The chemical formula for sodium phosphate is Na_3PO_4. What does the formula tell you about sodium phosphate?

Literacy Test Prep

Read the selection below and answer the questions that follow it.

Aluminum Production

Aluminum is a useful material because of its properties. Aluminum has a low density compared with other metals, so it forms very strong, lightweight alloys with other metals. It is very reactive but does not corrode. Aluminum is easy to work with, so it can be flattened and bent to the desired shape. It also is a very good conductor of heat and electricity.

Aluminum is obtained from a compound called aluminum oxide, which comes from bauxite that is mined. A large amount of electrical energy is needed to process the bauxite and aluminum oxide. Although bauxite is not mined in Canada and must be imported, our abundant hydroelectric power plants allow Canada to be one of the world's main producers of aluminum.

Each year, about 1.5 billion kg of aluminum are used to produce beverage cans in North America. This requires about 5 billion kilowatt-hours of electricity—the same amount of electricity that is used by over half a million homes each year. Recycling the aluminum in cans, however, requires only 5 percent of the energy needed to make new aluminum. Also, unlike most other metals, 100 percent of the aluminum can be recycled.

Multiple Choice

In your notebook, record the best or most correct answer.

34. Canada is a large producer of aluminum because Canada has many
 a) bauxite mines
 b) hydroelectric power stations
 c) beverage canning factories
 d) deposits of elemental aluminum

35. Aluminum is useful because of all the following properties **except**
 a) its ability to form alloys
 b) its conductivity
 c) its ability to be easily shaped
 d) its high density

36. What is the correct order in the production of aluminum beverage cans?
 a) bauxite → aluminum oxide → cans
 b) aluminum oxide → bauxite → aluminum → cans
 c) bauxite → aluminum → aluminum oxide → cans
 d) bauxite → aluminum oxide → aluminum → cans

37. The purpose of the last paragraph is to
 a) encourage the reader to recycle aluminum cans
 b) organize information about the production of aluminum
 c) inform the reader of aluminum use
 d) recommend that the reader use less energy in the home

Written Answer

38. Aluminum is used to manufacture parts for airplanes. Describe two properties of aluminum that you think are important in the production of these parts. Use specific details from the selection to support your answer.

Unit 3 Space Exploration

Big Ideas

- Celestial objects in the solar system and universe have specific properties that can be investigated and understood.
- Technologies developed for space exploration have practical applications on Earth.

"Moonshot" by Buffy Sainte-Marie

off into outer space you go my friends
we wish you bon voyage
and when you get there we will welcome you again
and still you'll wonder at it all
see all the wonders that you leave behind
the wonders humble people own
I know a boy from a tribe so primitive
he can call me up without no telephone
see all the wonders that you leave behind
enshrined in some great hourglass
the noble tongues, the noble languages
entombed in some great english class
off into outer space you go my friends
we wish you bon voyage
and when you get there we will welcome you again
and still you'll wonder at it all

Science helps us explain the universe, technology helps us explore it, and imagination links the two together.

How does the human imagination help us investigate space?

Unit 3 At a Glance

In this unit you will learn about objects in the solar system and the universe. You will find out about their specific properties and how these properties can be investigated. You will also learn about the technologies developed for space exploration, and how they are used in space and on Earth.

Think about answers to each question as you work through the topic.

Topic 3.1: What do we see when we look at the night sky?

Key Concepts
- We see stars that we organize into patterns.
- We see celestial objects of the universe.
- We see objects separated by immense distances.

Topic 3.5: How do we benefit from space exploration?

Key Concepts
- We develop technologies that shape our lives.
- We are challenged to think and act locally, globally, and universally.
- We gain a deeper appreciation for ourselves and our home planet.

Earth and Space Science: Space Exploration

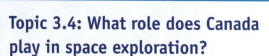

Topic 3.4: What role does Canada play in space exploration?

Key Concepts
- Canada contributes people and technology to explore space.
- Canada helps build the future of space exploration.

Topic 3.2: What are the Sun and the Moon, and how are they linked to Earth?

Key Concepts
- The Sun is our nearest star.
- Interactions of Earth and the Sun make life possible.
- The Moon is our nearest neighbour in space.
- The Sun, Moon, and Earth interact to create eclipses.

Topic 3.3: What has space exploration taught us about our solar system?

Key Concepts
- The four inner Earth-like planets are small and rocky.
- The four outer "gas giant" planets are large and ringed.
- Rocky chunks of various sizes make up the rest of the solar system.

Looking Ahead to the Unit 3 Project

At the end of this unit, you will do a project. The **Inquiry Investigation** challenges you to find a way to purify waste water for human consumption in space. The **Issue to Analyze** examines the costs and benefits of three different space technologies used to carry equipment into space. Read pages 232–233. With tips from your teacher, start your project planning folder now.

Get Ready for Unit 3

Concept Check

1. Match each term with its correct definition below.

   a) planet
   b) meteorite
   c) Moon
   d) star
   e) Earth
   f) comet

   i. Earth's natural satellite
   ii. emits light
   iii. much of its surface is covered with water
   iv. has a tail consisting of gas and dust
   v. orbits stars and reflects light
   vi. stony or metallic matter that has fallen to Earth

2. With a partner, brainstorm ways in which stars differ from planets. Organize your comparisons in a table like the one below. Alternatively, make a Venn diagram to compare and contrast stars and planets.

   Comparing Stars and Planets

Characteristics of Stars (like our Sun)	Characteristics of Planets

3. Use a word from the box below to answer each of the following questions. Write your answers in your notebook.

   a) Which term describes Earth's turning on its axis?
   b) Which term describes Earth's movement around the Sun?
   c) Which term means the same thing as the answer to b)?

revolution	rotation	orbit

4. It takes 11.86 Earth-years for Jupiter to revolve around the Sun. Each Jupiter day is 9 h, 50 min, and 30 s long.

   a) Is Jupiter's day longer or shorter than Earth's?
   b) How often would Earth have winter in 11.86 Earth-years?
   c) How often would Jupiter have winter in 11.86 Earth years?

5. Each of the following space technologies is illustrated below. Match each technology to its corresponding illustration.

   a) *Discovery* space shuttle
   b) *Atlas V* rocket
   c) the *Hubble Space Telescope*
   d) the *International Space Station*

Inquiry Check

Scientific and technological advances have enabled humans to adapt to life in space. However, astronauts face many challenges while living in space. For example, some of the needs that must be met while living on the *International Space Station* include:

   a) elimination of human waste
   b) regular exercise to maintain muscle and bone density
   c) a long-term supply of drinkable water

6. **Think Critically** Using jot notes, add three to five more challenges to the list.

7. **Analyze** How might some of these challenges be overcome? Choose one challenge from the list above and one from your own list, and discuss your recommendations in a small group.

8. **Think Critically** The average stay for an astronaut on the *International Space Station* is six months. Why do you think astronauts remain on the *International Space Station* for this period of time?

Numeracy and Literacy Check

Astronomers commonly use scientific notation to express the sizes of objects in space and distances between them. For example, the diameter of the Moon is 3475 km, or 3.475×10^3 km.

9. **Convert** The diameter of Earth is 12 756 km. Express this measurement in scientific notation.

10. **Compare** How much larger is the diameter of Earth compared to the diameter of the Moon?

11. **Writing** While risky and expensive, space exploration technology is beneficial. For example, the technology used for fuel pumps in space shuttles is also used to make better artificial hearts. Write a blog expressing your opinion on the money spent by the federal government for space exploration (see the table below). Support your viewpoint with examples.

Federal Spending in Canada in 2004

Area	Money Spent (millions of dollars)
Environmental programs	900
Defence	9 800
Health care	99 000
Space exploration	308

Topic 3.1 What do we see when we look at the night sky?

Key Concepts

- We see stars that we organize into patterns.
- We see celestial objects of the universe.
- We see objects separated by immense distances.

Key Skills

Inquiry
Numeracy

Key Terms

constellation
universe
gravitational pull
orbit
solar system
star
galaxy
astronomical unit (AU)
light-year

You probably have many ways to describe and share what you see in the sky. Imagine that you and others share your ideas about what you see with friends. Imagine that the friends pass on this information to others. Imagine that this sharing of information takes place over months, years, centuries, and longer. This process of sharing is the act of storytelling. For thousands of years, the sky has inspired us to share stories, to pass on ideas, explanations, guidance, and wisdom. What kinds of stories do we tell? As many as there are stars in the sky. We tell:

- ancient stories about the shapes, patterns, and events that our ancestors saw when they gazed at the heavens
- scientific stories told by astronomers to explain what we see when we look at and beyond Earth
- graphic stories, told with rock walls, animal hides, paper, and pixels, to help us communicate our wonder of and connection to the beauty of space
- visionary stories told by science-fiction authors to amuse us and teach us about our past, our present, and our future
- inspirational stories told by Elders, seers, and spiritual leaders—past and present—to link space, Earth, and all creation in our hearts, minds, and souls

Starting Point Activity

In the book, *The Hitchhiker's Guide to the Galaxy*, we are told: "Space is big. Really big. You just won't believe how vastly, hugely, mind-bogglingly big it is."

What do you see when you look into the mind-boggling hugeness of the sky? What objects, shapes, events, and patterns do you see or recall seeing? For instance, have you ever watched the Moon's changing face from night to night? Have you seen connect-the-dot animals and other objects in the stars? Share what you have seen, what you know, and what you would like to know about space.

We see stars that we organize into patterns.

Activity 3.1

ESTIMATE THE NUMBER OF STARS

The jar in the photo has as many beans as there are stars that are visible to the naked eye in the night sky of the northern hemisphere. How many are there? Share ideas to come up with a method to estimate the number of beans in the jar. (Hint: What if you counted only a portion of the beans?)

Numeracy Focus

The shape of the sky is like the shape of an upside-down bowl, with the rim of the bowl being the horizon. The stars at night look like dots of light painted on the inside of this bowl. This model of the sky—the upside-down bowl—is called the celestial sphere. The word "celestial" means sky.

If you were to watch the stars for a whole night, you would see them appear to move from east to west across the celestial sphere, just as the Sun does during the day. But the stars are not really moving across the celestial sphere. Earth's rotation causes the illusion of their movement.

Look at **Figure 3.1**. Notice the east-to-west movement of stars labelled on the diagram. Notice also the movement around the label indicating north. As you look north, the stars look like they are rotating around a single point in the sky. This point is a star. This star happens to be lined up with the North Pole of Earth's axis. So it seems to stay fixed in place while other stars circle around it. The name of this fixed star is Polaris, which means pole star. You might know it better by its common name: the North Star.

Go to **scienceontario** to find out more

▶ **Figure 3.1** The stars in the night sky seem to move from east to west. In the northern hemisphere, stars seem to rotate around a star that is fixed (unmoving) in the sky.

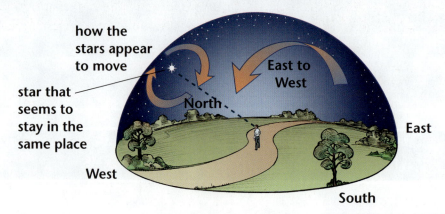

170 MHR • UNIT 3 SPACE EXPLORATION

Navigating the Night Sky with Polaris

Polaris is useful because it can help people in the northern hemisphere find direction at night. But how can you find Polaris? You can use **constellations**, which are patterns formed by other stars. For instance, find Polaris in **Figure 3.2**. Polaris is the last star in the handle of the constellation called Ursa Minor (Little Bear). Ursa Minor is also called the Little Dipper, because the seven stars that make up this constellation look like the shape of a small ladle or dipping spoon.

constellation: a shape or pattern of stars in the night sky made by imagining that stars are joined together by make-believe lines

◀ **Figure 3.2** Polaris is the last star in the handle of the Little Dipper (part of the Little Bear).

Figure 3.3 shows Ursa Minor and a few other constellations that circle around Polaris. These constellations are called circumpolar, because they travel around (circum-) the pole star, Polaris. Circumpolar constellations are always visible in Canada. They never go below the horizon.

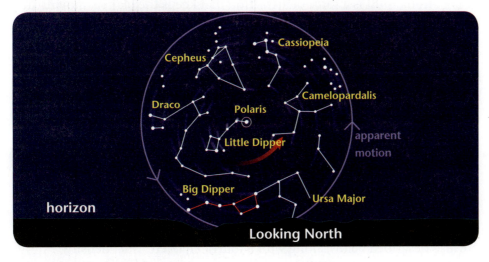

◀ **Figure 3.3** The circumpolar constellations are visible all year long in Canada. Cepheus was a mythical king. Cassiopeia was a mythical queen. Camelopardalis is the Giraffe. Ursa Major is the Great Bear. Draco is a mythical dragon.

LEARNING CHECK

1. Describe the direction the stars seem to move in the sky.
2. Explain why stars do not move the way they seem to.
3. Identify the star that stays fixed while other stars circle it.
4. Refer to **Figure 3.3**. Would you always be able to see these constellations on a clear night? Justify your answer.

INVESTIGATION LINK
Go to Investigation 3A, on page 180.

We see celestial objects of the universe.

The celestial objects that we see in the sky include our Moon, planets, stars, and collections of stars. With the help of telescope technology, we also can see other moons, other planets and planet-like objects, and ever-more distant collections of stars. All the celestial objects that we see in the sky, and all that there is to be seen, make up what we call the **universe**.

universe: everything that exists, including celestial objects such as stars and planets as well as all the matter and empty space surrounding them

> **Inquiry Focus**
>
> ### Activity 3.2
> **DRAW ORBITS**
>
> Use a mathematical compass to draw a model of a circular orbit that involves the following pairs of celestial objects. Use labels to identify the objects. Explain how you decided which object should be at the centre of the orbit.
>
> a) Earth and the Moon   b) Earth and the Sun

Our Solar System

Our Sun's gravity exerts a powerful pulling force on the planets. This **gravitational pull** is a force of attraction that keeps the planets moving in a circular pattern around it. This circular pattern is called an **orbit**, and the planets are said to *revolve* around the Sun, which means that they move in an orbit around the Sun. Most of the planets also have moons that revolve around them. The Sun, the planets, the moons, and other objects that orbit the Sun due to its gravitational pull make up the **solar system**. (You will explore the solar system in Topic 3.3.)

gravitational pull: the force of attraction that any two masses have for each other

orbit: the circular path of one object around another object

solar system: the system of planets, including Earth, moons and other objects that orbit (revolve around) the Sun

star: a massive ball of superheated gases that radiates heat and light

Stars and Their Characteristics

Our Sun is a star. A **star** is a ball-shaped mass of superheated gases that produces and gives off light, heat, and other kinds of energy. **Figure 3.4** shows that stars vary in size, colour, and temperature. These characteristics are also described below, along with a fourth characteristic: density.

- **Size:** Some stars are millions of kilometres in diameter, while others may be only 20 km across.
- **Colour:** Some stars have a reddish, orange, or yellow colour, while others are bluish, white, or bluish-white.
- **Temperature:** Reddish-coloured stars have relatively cool surface temperatures of 3000°C, while bluish-coloured stars can be as hot as 55 000°C. (Our yellowish star is about 6000°C.)
- **Density:** Some stars have such low density they would float on water. Others are so dense a gram of the star would crush the CN Tower.

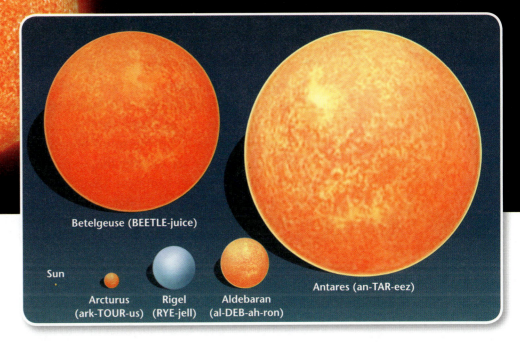

Galaxies: Collections of Stars

Stars do not exist on their own. A collection of many billions of stars held together by gravity is called a **galaxy**. There are billions and billions of galaxies in the universe. Our solar system is located in the Milky Way galaxy. It gets its name from the way it looks—like a hazy or milky path across the night sky. It appears this way because you are actually gazing into the dense centre of our galaxy from your viewpoint on Earth, near the outer edge of the Milky Way. You saw a picture of the Milky Way galaxy at the start of this topic.

In addition to stars, and any solar systems that a star might have, galaxies also contain masses of gas and dust. The gas is made up mainly of hydrogen atoms. The dust is not like the dust we have on Earth. Instead, space dust is made up of atoms and fragments of atoms. Most of a galaxy, however, has no matter of any kind. It is just empty space. (It's not called space for nothing. On second thought, maybe it *is*!)

▲ **Figure 3.4** Compared with other stars, our Sun is considered to be an average-sized star. Notice that some stars can be quite a bit larger than our Sun is.

galaxy: a collection of many billions of stars, plus gas and dust held together by gravity

LEARNING CHECK

1. Explain how gravitational pull affects planets.
2. Describe what a star is.
3. Explain how stars and galaxies are related.
4. Choose three stars in **Figure 3.4**. Design a table to compare their size and colour.

ACTIVITY LINK

Activity 3.4, on page 176

We see objects separated by immense distances.

On the celestial sphere model of the sky, stars, planets, and other objects look like they are on a flat surface close together. This is just an illusion, though. Most objects in the universe are so far apart that it is hard to imagine how far apart they are. (Remember about space: "You just won't believe how vastly, hugely, mind-bogglingly big it is.")

Distances in space are so large that well-known units such as kilometres are almost meaningless. Using kilometres to measure the distance between Earth and Jupiter would be as silly as using millimetres to measure the distance between Ottawa and Kenora. The units just aren't suitable. (By the way, there are about 2 000 000 000 mm between Ottawa and Kenora. See?)

Measuring Distances in the Solar System

Astronomers have created their own unit to measure distances between planets in the solar system. One **astronomical unit (AU)** is equal to the distance between the Sun and Earth. In familiar units, that distance is 150 000 000 (one hundred and fifty million) km. Using AUs, the Sun-Earth distance is much easier to express. It's just 1 AU, as shown in **Figure 3.5**.

Using AUs, the distance from the Sun to the last planet of the solar system (Neptune) is about 30 AU. This means that you would have to travel the distance between Earth and the Sun 30 times to reach Neptune.

Two planets, Mercury and Venus, are closer to the Sun than Earth. So their distance in AUs is expressed as a decimal fraction. The distance from the Sun to Mercury is 0.39 AU. The distance between the Sun and Venus is 0.72 AU.

astronomical unit: a measurement equal to the distance between the Sun and Earth (150 million km)

▶ Figure 3.5 Using AUs to measure distances in the solar system is much simpler than using kilometres.

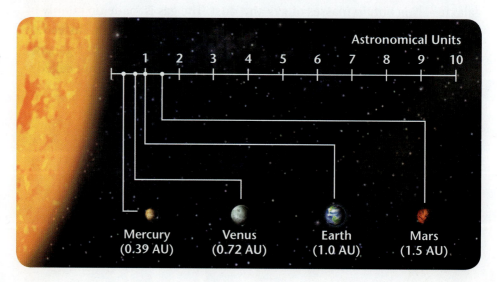

Distances involving Galaxies

Distances involving galaxies are much larger than those involving solar systems. In the movies, a spacecraft can travel across a galaxy in less time than it takes its crew to execute a daring rescue plan. Even with our more current advanced technology, however, the distances between stars and galaxies are simply too great to be covered in a human lifetime.

Because these distances are so great, even AUs are inadequate to measure them. Instead, astronomers use a unit called the light-year. A **light-year** is the distance that light travels in one year. Light travels faster than anything in the known universe—a mind-bogglingly fast speed of 300 000 km/s. In one year, light can travel about 9.5×10^{12} km (9.5 trillion or 9 500 000 000 000 km). That might sound like a lot—and it is—but it's nothing in terms of distances in the universe. Most stars and galaxies, such as the one shown in **Figure 3.6**, are hundreds, thousands, and even millions of light-years away!

light-year: a measurement equal to the distance that light travels in one year (about 9.5×10^{12} km)

◂ **Figure 3.6** The Andromeda galaxy is the nearest large galaxy to our own and the most distant object that is still visible to the unaided eye. It is 2.3 million light-years away.

LEARNING CHECK

1. Compare an astronomical unit (AU) to a light-year.
2. Explain why kilometres are inadequate for measuring most distances in the universe.
3. Refer to **Figure 3.5**. Calculate the distance from Mars to the Sun in kilometres. Show your work.

Numeracy Focus

Activity 3.3
CHOOSE YOUR UNITS

Name the most suitable units to measure each of these distances:
1. your home to your school
2. your home to the nearest large city
3. Canada to Saturn
4. the Sun to Polaris

ACTIVITY LINK
Activity 3.5, on page 177

Activity 3.4
CLASSIFY GALAXIES

Scientists have invented a system to classify the millions of galaxies into just a few types. In this activity, you will invent your own system.

What To Do

1. Examine the galaxy photos. Pay attention to the similarities and differences in the features that you see. Record these in a table with headings like this:

Galaxy Number	Description of Features

2. Use your completed table to help you make up a system to classify galaxies.

What Did You Find Out?

1. What features did you use to invent your classification system?
2. How does your system compare with others in the class? Decide which systems work best, and explain why you think so.

Galaxy M108   Galaxy M32   Galaxy M82   Galaxy M87

Large Magellanic Cloud   Galaxy M83   Galaxy M86   Galaxy M81

Activity 3.5

BUILD CONSTELLATIONS IN 3-D

The stars in constellations look like they are all the same distance away, but this is an illusion. In this activity, you will build a 3-D model to show the real positions of the stars in the Big Dipper.

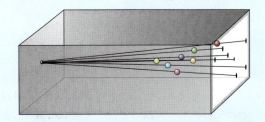

What You Need
- shoebox (or other small box)
- string, scissors, glue, and tape
- Big Dipper diagram (from your teacher)
- seven small beads with holes

What To Do
1. Glue the Big Dipper diagram to the inside of one end of the box. Poke a hole through the box at each star.
2. At the other end of the box, poke one hole through the middle. This hole will be Earth.

3. Cut seven pieces of string. The string pieces should be a few centimeters longer than the length of the box.
4. Thread each piece of string through a bead. Each bead will be a star of the Big Dipper.
5. Thread the end of one piece of string through the hole at Earth. Tape the end to the outside of the hole.
6. Make sure the bead is threaded on the string. Then thread the other end of the string through a hole on the Big Dipper diagram. Tape the end of the string to the outside of the box.
7. Look at the table. Find the distance from Earth to Star 1. Glue the bead (star) to the string at the correct distance from the "Earth" hole.
8. Repeat steps 5, 6, and 7 for each piece of string.

What Did You Find Out?
1. In the sky, the seven stars of the Big Dipper look like they are the same distance from Earth. How does this model show this is just an illusion?
2. Why do the stars of the Big Dipper look like they are the same distance from Earth? (Hint: Look back on page 170 and read about the celestial sphere.)

Inquire Further
The names of the stars of the Big Dipper are Arabic. In fact, many stars that are visible in the night sky have Arabic names.

- Find out why many stars that are visible in the night sky have Arabic names.
- One of the stars in the Big Dipper is actually two stars. What is this eighth star in the Big Dipper? How is it linked to ancient tests of eyesight?

Stars of the Big Dipper and their Distances on the 3-D Model

Star in the Big Dipper	Distance from Earth on model (in cm)
Alkaid (Star 1)	1.8
Mizar (Star 2)	2.2
Alioth (Star 3)	1.7
Megrez (Star 4)	3.3
Phecda (Star 5)	2.4
Merak (Star 6)	2.3
Dubhe (Star 7)	1.8

Case Study Investigation: A Tale of the Bear

Welcome to the Science Links Planetarium. Today, as part of our Sky Tales series, we are honoured with a traditional story about the Great Bear in the night sky, known to many as Ursa Major. Many stories have been told about Ursa Major. They have been shared by various cultures in the past and still are today. This is one of them.

Madjikiwis and Nanabush, a tale of the Anishinabeg

Madjikiwis was young, courageous and cared deeply for his family who had nurtured and taught him. As a young Anishnabe man he learned the skills of a hunter and he loved spending time in the forest where each moment offered better understanding of the four orders of all life and the way people depended on them.

His speed was great and skills excellent but in his heart he felt something was missing in his understanding which kept him from finding his own place in his world. His family was good and his needs were sufficiently met but what was the outcome of their efforts to be? Each passing of the long Winter brought another reawakening of Spring when all life once more was renewed but where could each of these steps lead?

Madjikiwis could not find the answer.

Finally, near the end of a fine Summer, he gathered his hunting equipment and left home for hill country. After traveling for days he came to the home country of the Bear and began to feel concern for how he might be received since his provisions were almost gone and he had no gifts to offer. As darkness fell and mists floated above the valleys Madjikiwis became fearful that his journey had been unwise as his preparations had been hasty and maybe a bit careless. He stopped walking and prepared a bough lean-to as the sky filled with stars.

Just then Nanabush, a spirit-guide (able to take the form of different animals) adopted the image of Bear high above Madjikiwis and spoke in a language the young man could understand. All of the usual night time sounds stopped as Nanabush explained to Madjikiwis the place of

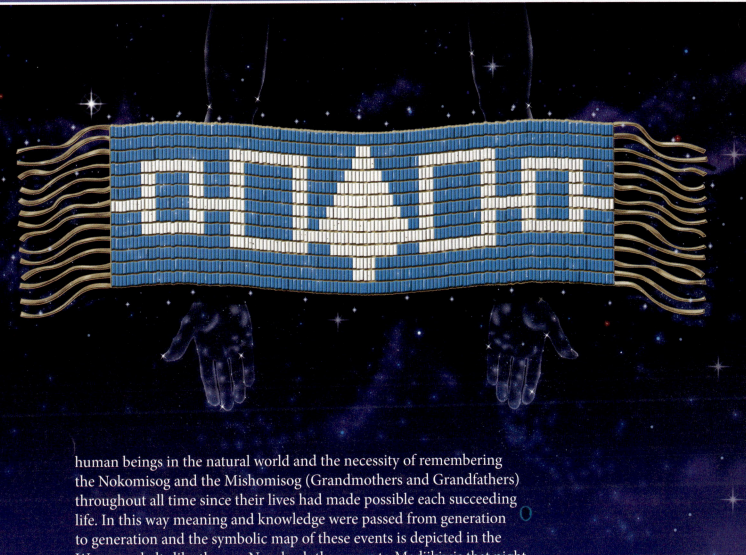

human beings in the natural world and the necessity of remembering the Nokomisog and the Mishomisog (Grandmothers and Grandfathers) throughout all time since their lives had made possible each succeeding life. In this way meaning and knowledge were passed from generation to generation and the symbolic map of these events is depicted in the Wampum belts like the one Nanabush then gave to Madjikiwis that night. Slowly the comforting sounds of the forest at night returned and the young man drifted off in a comfortable sleep.

From that day Madjikiwis gained in wisdom which he then shared with his people through care for and teachings of the Wampum belts.

High above in the heavens the constellation of the Bear symbolizes this gift of understanding and meaning which has provided understanding for thousands of years even to this day—or night.

Investigate Further

1. Create the next story for the Star Tales planetarium series. Your story can be in the form of a narrative, graphic novel, song, mural, or other creative work about a constellation of your choice.

Investigation 3A

Skill Check

Initiating and Planning

✓ Performing and Recording

✓ Analyzing and Interpreting

✓ Communicating

What You Need
- Star-Finder Wheel template (from your teacher)
- cardboard
- scissors
- tape

Make a Star-Finder Wheel

How can you use a model of the night sky to identify and find constellations on any day and at any time?

What To Do

1. Follow the directions on the Star-Finder Wheel template to make your Star-Finder Wheel.
2. Set the Wheel to today's date.
3. Use the Wheel to help you name five constellations that you would be able to see in the sky tonight at 10 P.M. (Only two of your choices can be circumpolar.)
4. Use the Wheel to determine if you would be able to see these same constellations at 12 A.M., 2 A.M., and 4 A.M.
5. Design a table to record the five constellations you named and whether you would be able to see them at 10 P.M., 12 A.M., 2 A.M., and 4 A.M.

What Did You Find Out?

1. Would you be able to see the two circumpolar constellations that you named any night of the year? Explain why or why not. Then use the Wheel to verify your answer.
2. During which seasons of the year would you be able to see the other three constellations you named at 10 P.M.?

Inquire Further

You can use the Big Dipper to tell time. Use the library or the Internet to find out how. Or ask your teacher for an activity that you can use to find out instead.

Topic 3.1 Review

Key Concepts Summary
- We see stars that we organize into patterns.
- We see celestial objects of the universe.
- We see objects separated by immense distances.

Review the Key Concepts

1. **K/U** Answer the question that is the title of this topic. Copy and complete the graphic organizer below in your notebook. Fill in four examples from the topic using key terms as well as your own words.

   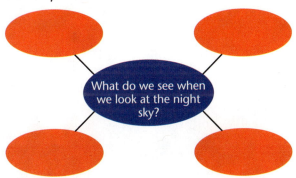

2. **C** Copy the graphic organizer below into your notebook. Complete it to show the relationship among these key terms: stars, constellations, planets, solar systems, space dust, and galaxies.

   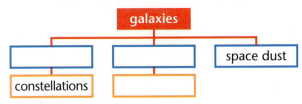

3. **A** Space dust is composed of extremely small particles. Scientists believe this dust is composed of heavy elements such as carbon, magnesium, iron, and calcium. The particles of dust pose no threat to planets such as Earth. However, they can chip away at the solar panels on spacecraft. Explain why space dust presents a serious problem for astronauts orbiting Earth on the International Space Station (ISS).

4. **C** Draw a Venn diagram to compare an astronomical unit (AU) to a light year. In your diagram, identify when astronomers would use astronomical units (AU) and when they would use light years.

5. **T/I** Look at the graph shown here and answer the questions. Identify which star is the farthest from Earth and which star is closest to Earth. Justify your answer.

   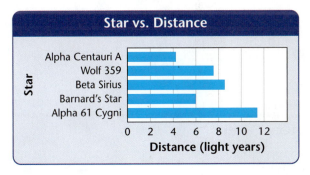

6. **K/U** Use the term "gravitational pull" in a sentence.

7. **T/I** Which star below would exert the greatest gravitational pull on its orbiting planet? Explain your reasoning.

8. **A** Imagine that you are lost in a desert at night. The desert is in the northern hemisphere. Explain how could you determine the four directions.

Topic 3.2

What are the Sun and the Moon, and how are they linked to Earth?

We know more about the Sun and the Moon than any other objects in our solar system. One reason for this is that they are the largest objects in our sky, so they are easy to observe and become familiar with. Another reason is that we have sent more satellites and probes to observe and study the Sun and the Moon than all other objects in the solar system combined.

It makes sense for us to know so much about the Sun and Moon. These objects not only inspire the stories we tell, but also affect weather systems, the ways we keep time, and the very existence of life on Earth.

Key Concepts

- The Sun is our nearest star.
- Interactions of Earth and the Sun make life possible.
- The Moon is our nearest neighbour in space.
- The Sun, Moon, and Earth interact to create eclipses.

Key Skills

Inquiry
Research

Key Terms

magnetosphere
aurora
solar eclipse
lunar eclipse

Starting Point Activity

1. What do you know about effects of the Sun on Earth? For instance, what is a link between the Sun and our seasons? What other Sun-Earth effects can you describe?

2. What do you know about effects of the Moon on Earth? For instance, what is the link between the Moon and tides? What other Moon-Earth effects can you describe?

The Sun is our nearest star.

▼ **Figure 3.7** Information about our star, the Sun, is displayed in this photo-based graphic organizer.

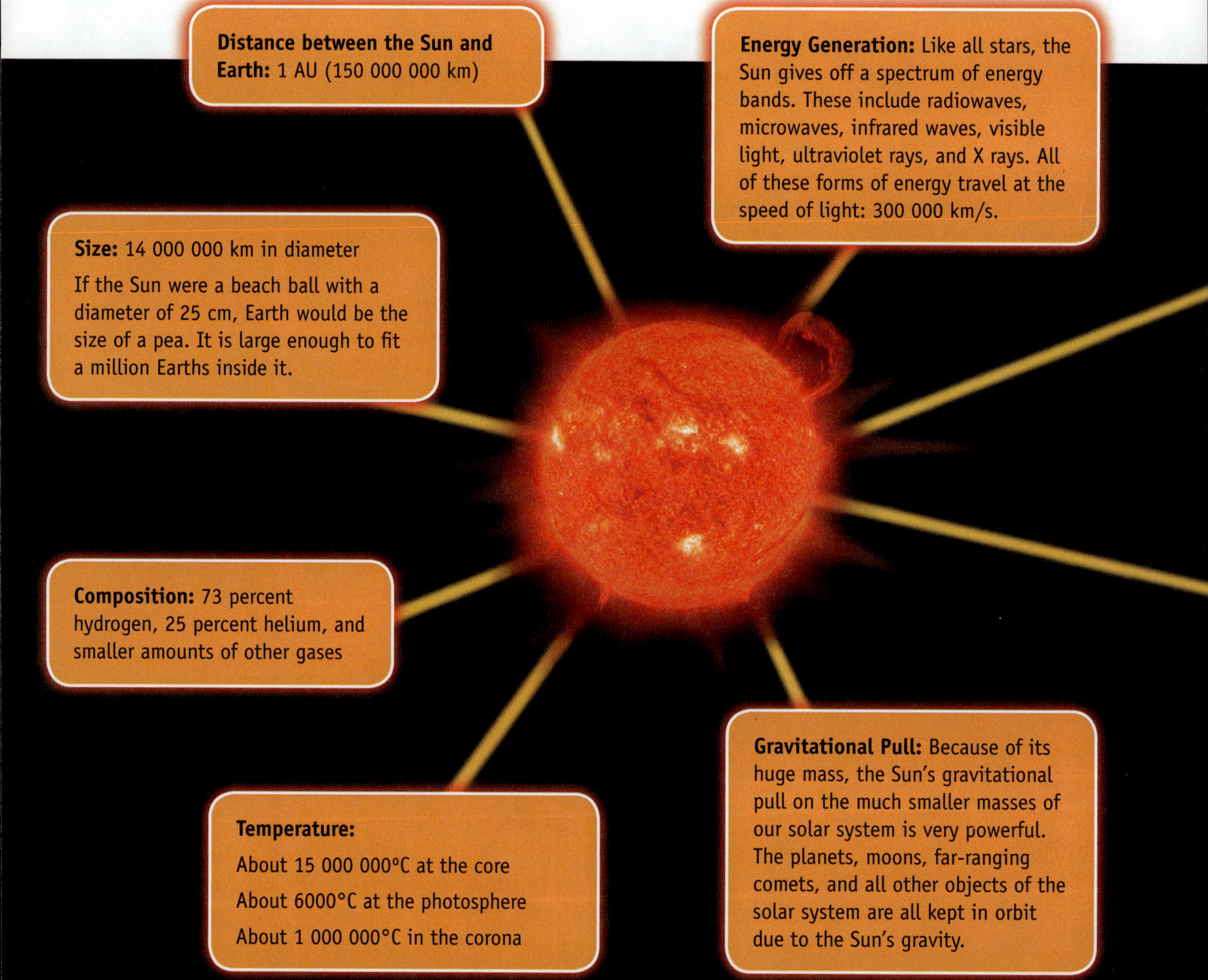

Our Sun, shown in **Figure 3.7**, was born about 5 billion years ago, and will last for another 5 billion years. It contains more mass than 300 000 Earths combined. Strong gravitational forces pull this great mass tightly together, creating great pressure and heat. Under these extreme conditions, hydrogen atoms collide violently with each other, and they combine (fuse). During this process, called nuclear fusion, two hydrogen atoms fuse together to form a helium atom, and great amounts of energy are released.

Distance between the Sun and Earth: 1 AU (150 000 000 km)

Energy Generation: Like all stars, the Sun gives off a spectrum of energy bands. These include radiowaves, microwaves, infrared waves, visible light, ultraviolet rays, and X rays. All of these forms of energy travel at the speed of light: 300 000 km/s.

Size: 14 000 000 km in diameter

If the Sun were a beach ball with a diameter of 25 cm, Earth would be the size of a pea. It is large enough to fit a million Earths inside it.

Composition: 73 percent hydrogen, 25 percent helium, and smaller amounts of other gases

Gravitational Pull: Because of its huge mass, the Sun's gravitational pull on the much smaller masses of our solar system is very powerful. The planets, moons, far-ranging comets, and all other objects of the solar system are all kept in orbit due to the Sun's gravity.

Temperature:

About 15 000 000°C at the core

About 6000°C at the photosphere

About 1 000 000°C in the corona

How much energy is released? During fusion, each gram of hydrogen releases 90 000 million kJ of energy. That's as much energy as you would get if you ate 1 trillion pizza slices. And that's only for one gram of hydrogen. The Sun has 2.0×10^{33} g (2 000 000 000 000 000 000 000 000 000 000 000 grams) of mass. No wonder the Sun has so much life left in it!

LEARNING CHECK

1. Explain how the Sun produces energy.
2. Use **Figure 3.7** to draw the Sun. Label its temperature, size, and rotation.
3. What is the total expected lifetime of the Sun?
4. Predict why the Sun is part of the sacred traditions of many cultures.

Rotation:

The Sun makes one rotation in 26 days at its equator but takes 37 days to rotate at its poles. (Yes, the Sun rotates faster at its equator than at its poles.)

This ancient symbol for the Sun is also a symbol for the element hydrogen. (See if you can figure out why.)

Research Focus

Activity 3.6

WHAT'S COOL ABOUT THE SUN?

The Sun is much more than just facts about distance, temperature, gravity, and energy. What's cool about the Sun?

Make an idea web like the one used on these two pages. Include spokes for other facts about the Sun. Six options are given below, along with some keywords and hints to help you get started. Add other spokes for other facts that you discover and want to share.

1. Other Features of the Sun: for example, sunspots, coronal mass ejections, flares, solar wind
2. Layers of the Sun (core, photosphere, corona) and their characteristics
3. Telling Time with the Sun: for example, sundials, solar calendars
4. Technology for Studying the Sun: for example, satellites such as IBEX and SOHO
5. The Sun in Song: for instance, what songs feature the Sun in their titles and lyrics?
6. The Sun in Sacred Stories: for example, what are some stories from the sacred traditions of Aboriginal peoples, other cultures, and world religions?

Interactions of Earth and the Sun make life possible.

If Earth's orbit brought us much closer to the Sun, our planet might be a lifeless, scorched desert. If Earth's orbit were much farther away, our planet might be a frigid, icy wasteland. Fortunately, our distance from the Sun is "just right" for life. In fact, astronomers call Earth's position in the solar system the "Goldilocks zone." But being "in the zone" does not guarantee safety from the Sun's energy. Fortunately, Earth's magnetosphere and atmosphere provide the protection needed for life on our planet.

The Link between the Magnetosphere and the Atmosphere

Along with energy, the Sun also sends out streams of matter in the form of charged particles that travel through the whole solar system at great speed. These streams of charged particles are called the solar wind. The solar wind bathes Earth and all other objects in the solar system.

The highly energetic particles of the solar wind are deadly to life. Fortunately, Earth is protected from their effects by a field of magnetic force that surrounds the planet. This field of magnetic force around Earth is called the **magnetosphere**. The magnetosphere deflects the solar wind and prevents much of it from entering the atmosphere.

Charged particles can, however, enter Earth's atmosphere at the poles. At times, especially powerful outbursts of solar wind particles enter at the poles and interact with atoms in the upper atmosphere to create shimmering curtains of beautiful, coloured light. **Figure 3.8** shows one such display, which is called an **aurora**. In the northern hemisphere, it is called the aurora borealis. You might have heard it called by its common name, the Northern Lights.

magnetosphere: the area of space that contains a planet's magnetic field

aurora: light shows in Earth's upper atmosphere created by solar wind

Figure 3.8 This photo shows an aurora.

The Link between the Atmosphere and Life

Green plants and other producers need light from the Sun for photosynthesis. And consumers need plants. So without the Sun, there would be no life. But too much sunlight is not a good thing, either. Sunlight includes bands of energy that are harmful to living things. These include UV (ultraviolet) rays and X rays. Fortunately, ozone and other gases in the atmosphere interact with incoming energy from the Sun to protect us. In effect, our atmosphere acts like a filter that helps to shield living things from much of the Sun's harmful effects.

Figure 3.9 shows that Earth's atmosphere also helps to trap heat from the Sun that would otherwise escape back into space. This helps to keep Earth at just the right temperature for life to thrive. Without this atmosphere, extreme fluctuations of temperature would make life as we know it impossible. Moderate temperatures also allow water to exist in all three states on Earth—gas, solid, and especially liquid. Earth is the only known planet with this unique combination, which is a key factor for maintaining life.

Some of the energy absorbed at the surface escapes into space.

Some energy is trapped by the atmosphere and redirected back to the surface.

Some energy is reflected by atmosphere and surface.

Some energy passes through atmosphere.

Some energy reaches the surface and is absorbed.

▲ **Figure 3.9** Earth's atmosphere acts like a blanket that traps heat as it escapes back into space. Instead, the heat is redirected toward the surface again. This helps to keep temperatures within a stable range for life.

LEARNING CHECK

1. Refer to **Figure 3.9**. Explain how the atmosphere protects Earth.
2. In your notebook, create a graphic organizer that explains an aurora.
3. Predict what would happen to life on Earth if the magnetosphere disappeared. Justify your answer.

ACTIVITY LINK
Activity 3.9, on page 194

The Moon is our nearest neighbour in space.

Our calendar and holidays are linked to the appearance of the Moon in the night sky. The Moon also affects the lives of animals. Some aquatic animals, for instance, mate or lay their eggs when the Moon becomes more full or less full. Even some of our language comes from words for the Moon. The word month (think of it as "moonth") is one example. The word lunar, which means moon, gives us the word lunatic.

Phases of the Moon

The Moon looks like it shines with a light of its own, but looks can be deceiving. Moonlight is really sunlight that reflects from the Moon's surface.

The reflected light we see is always from the same side of the Moon. The reason involves how the Moon rotates (spins) and how the Moon orbits. It takes the Moon 27.3 days to make one full orbit around Earth. It also takes the Moon 27.3 days to make one full rotation. The rotation rate and the orbiting rate match, so we always see the same side of the Moon from Earth. The other side of the Moon—called "the dark side of the Moon"—always faces away from us, so we never see it.

The lit-up side of the Moon is always fully lit up, but we can't always see the whole lit-up side. Instead, we see changes in the amount of lit-up surface during a month: the phases of the Moon. **Figure 3.10** shows how the Moon, Earth, and Sun are arranged to produce the phases of the Moon.

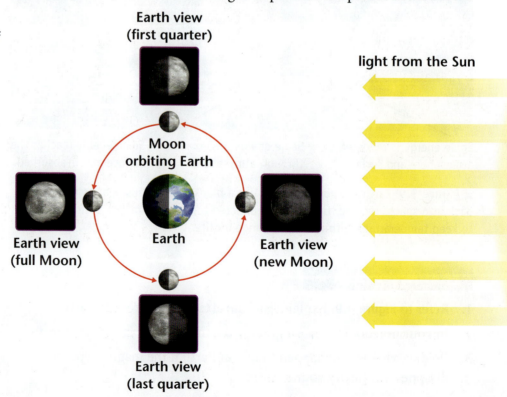

▶ **Figure 3.10** The phases of the Moon are caused by the amount of lit-up surface that we can see from Earth as the Moon orbits our planet.

Activity 3.7

MORE ABOUT THE MOON

There's plenty about the Moon to explore, even if only from Earth. Here are some things you could investigate. Present your findings in a format of your choosing, or use the template provided by your teacher.

1. What superstitions are associated with the Moon?
2. How many vehicles (rovers) have travelled on the Moon, who drove them, and what was their source of power?
3. Many surface features have names with water in them, such as Sea of Crisis, Lake of Dreams, and Swamp of Decay. But there has never been water on the Moon. What's up with that?
4. Many craters are named for people in Earth's history such as Pythagoras, Abu Arrayhan Muhammad ibn Ahmad al-Biruni, H. G. Wells, Dmitri Mendeleev, and Marie Curie. Who are they, and why are they famous or important?
5. What are some songs and poems that have been written about the Moon? What role does the Moon play in them?

MONDO MOON

Average distance from Earth: 384 400 km (0.003 AU)

Size (Diameter): 3475 km (almost one-quarter of Earth's)

Temperature Range: −170°C to 100°C

Time for One Rotation: 27.3 days

Time for One Orbit: 27.3 days

Atmosphere: none

Symbol: ☾

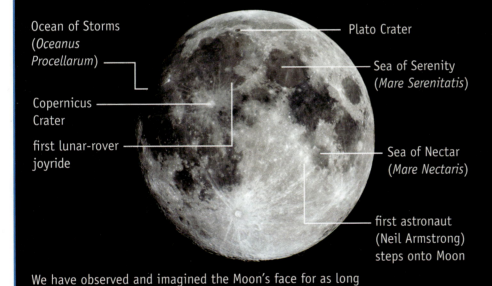

We have observed and imagined the Moon's face for as long as there have been people to do so. The dream to visit and set foot upon the Moon became a reality on July 20, 1969.

LEARNING CHECK

1. Explain why we cannot see the "dark side" of the Moon.
2. Refer to **Figure 3.10**. In your notebook, draw a labelled sketch showing the phases of the Moon.
3. What conditions make it difficult to live on the Moon?

The Sun, Moon, and Earth interact to create eclipses.

An eclipse occurs when Earth or the Moon is lined up in space so that it blocks the Sun's light for a short time. There are two kinds of eclipses: solar eclipses and lunar eclipses.

Solar Eclipses

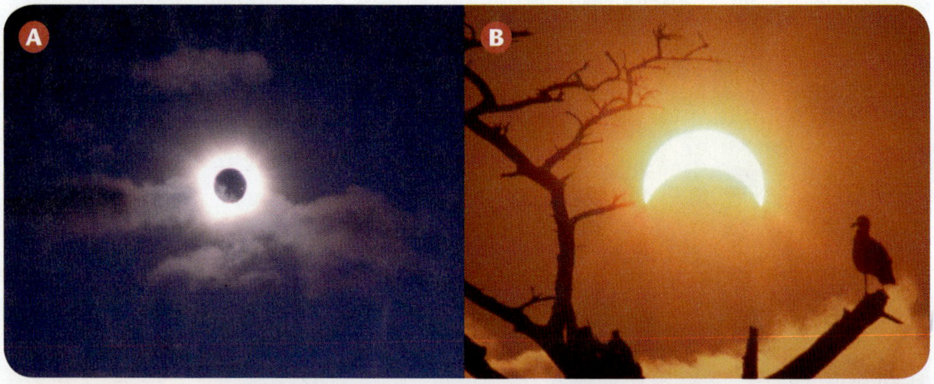

▶ **Figure 3.11** During a total solar eclipse (shown in A), the Moon covers the whole face of the Sun, leaving only a hazy, white glow of the Sun's atmosphere visible, like a halo. During a partial solar eclipse (shown in B), the Moon covers only part of the Sun's face.

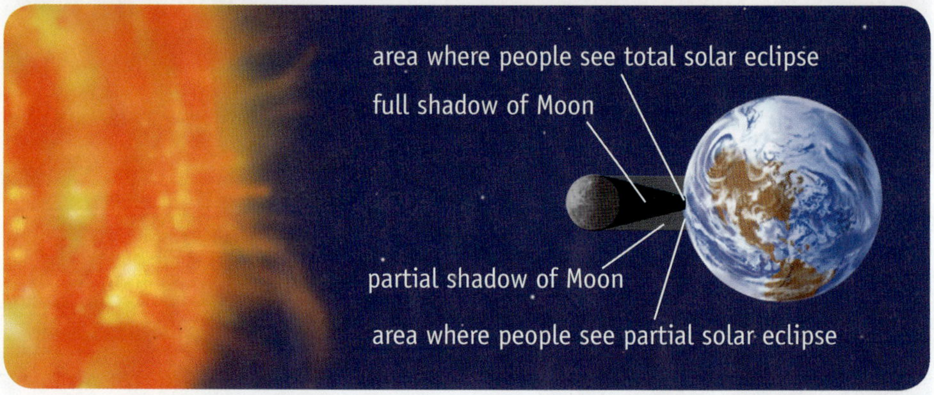

▶ **Figure 3.12** Positions of the Sun, Moon, and Earth during a solar eclipse. (Sizes and distances in the diagram are not drawn to scale.)

solar eclipse: an event where the Moon moves directly between the Sun and Earth, so that the Moon casts its shadow on part of Earth

In a **solar eclipse**, the Moon moves directly between the Sun and Earth, so that the Moon casts its shadow on part of Earth. Solar eclipses occur in the day, and the Sun's light is either totally or partially blocked. Your location determines whether you see a total or partial solar eclipse (**Figure 3.11**).

If you are standing in the area covered by the full shadow of the Moon, you see a total solar eclipse. If you are standing in the area covered by part of the shadow of the Moon, you see a partial solar eclipse. Refer to **Figure 3.12**. Many people flock to sites around the world to view solar eclipses when they occur. A solar eclipse is best viewed through a telescope or other lens equipped with a solar filter. **Danger!** *Never look directly at the Sun, either during the eclipse or any other time. The Sun's energy can damage or destroy the parts of the eye that enable you to see, causing blindness.*

Lunar Eclipses

A **lunar eclipse** occurs when Earth moves directly between the Sun and Moon, so that Earth casts its shadow on the Moon. Refer to **Figure 3.13**. During a total lunar eclipse, the Moon is covered fully by Earth's shadow. Sometimes this shadow appears red due to light bending in Earth's atmosphere. During a partial lunar eclipse, Earth's shadow only partially covers the Moon. Lunar eclipses are more common than solar eclipses. You can watch a lunar eclipse safely, because you are looking at reflected sunlight, not the direct energy of the Sun.

lunar eclipse: an event where Earth moves directly between the Sun and Moon, so that Earth casts its shadow on the Moon

◀ **Figure 3.13A** During a total lunar eclipse, the whole Moon passes through the full shadow cast by Earth. During a partial lunar eclipse, only a part of the Moon passes through the full shadow cast by Earth.

▲ **Figure 3.13B** During a total lunar eclipse, Earth's shadow totally covers the face of the Moon.

LEARNING CHECK

1. Refer to **Figures 3.12** and **3.13**. Draw the arrangement of the Sun, Earth, and Moon during a solar and a lunar eclipse.
2. Compare a total solar eclipse to a partial solar eclipse.
3. Write an announcement warning people about the danger associated with a solar eclipse.

Activity 3.8

Inquiry Focus

MODELLING ECLIPSES

Use the diagrams on these two pages to help you design models that show where the Sun, Moon, and Earth are in space during a solar eclipse and a lunar eclipse. Your model will need a light source to be the Sun, a globe for Earth, and a smaller ball for the Moon.

What Did You Find Out?

1. What is it that blocks light during a solar eclipse? During a lunar eclipse?
2. A lunar eclipse can only take place during a full moon. Use your model to explain why.
3. Use your model to explain why there isn't a lunar eclipse with every full moon.
4. Explain why more people on Earth can see a lunar eclipse than a solar eclipse.

Case Study Investigation: Solar Storms

Pittsburg Gazette

VOL. 6—NO. 22,   PITTSBURG, MONDAY, AUGUST 29, 1859

MYSTERIOUS FIRE RENDERS TELEGRAPH STATION USELESS

On the night of Sunday, August 28 fire mysteriously broke out at the Pittsburgh telegraph office, having left the entire telegraph station dead. E.W. Culgan, telegraph manager at the telegraph station in this city observed the following Sunday evening. On the night of August 28th, Culgan stated that he saw not only sparks, but also streams of fire that could not have been produced by the batteries. Much of the equipment became so hot that the hand could not be placed upon them. All lines were rendered useless. The reason for this extraordinary circumstance is not known. However, there is speculation that it may have been linked to an unusual display of coloured aurora observed in the sky that evening. This display caused dismay among many citizens.

▲ An electrical telegraph machine. Sent along telegraph lines linking different stations, telegraph messages composed of Morse code electrical signals are an indispensable form of modern, long distance communication.

The Science Behind the Story

Whatever happened on this mysterious night in August, 1859, it shut down telegraph communications across much of the world. The first clue to what was going on came from instruments recording Earth's magnetic field around the globe. The planet's magnetism had suddenly gone off the chart. The cause? A powerful storm raging on the Sun's surface. Explosions known as solar flares ejected super-hot gases from the Sun's surface. Magnetic bubbles of matter burst from the Sun's upper layer, hurling solar particles toward Earth at millions of kilometres per hour. It lasted six days and was the fiercest magnetic storm in recorded history. Classified as a solar superstorm, the magnetic storm of 1859 knocked out electrical communication around the globe. Magnetic compasses went haywire. Brilliant auroras coloured the skies from Canada to the Caribbean.

▲ Large solar flares were observed during the 1859 solar super storm.

Pause and Reflect

1. Describe how the solar superstorm of 1859 affected Earth.
2. Explain what occurs on the Sun's surface during a solar storm.

How would a solar storm affect Earth today?

A solar storm of this size has not been seen since 1859. Since then, technology has advanced in leaps and bounds. How might a solar superstorm affect modern telecommunications and global power grids? Would you believe you might not be able to send a text message, listen to the radio, or even flush your toilet? Here are just a few ways a solar superstorm could affect Earth today:

- **Communications Down:** Telecommunications could be knocked out as solar particles break down satellite solar panels and electronic equipment. In 2003, the Sun released the largest solar flare in history. If this flare had been aimed at Earth, it would have knocked out telecommunications worldwide.
- **Power Out:** Power grids across the world could be affected. In 1989, a small solar storm brought down the Hydro-Québec power grid. A solar superstorm could cause continent-wide black outs for a day or longer, affecting food storage, transportation, emergency services, and even the flow of water into your home.
- **Radio Disruption:** A solar superstorm could interfere with radio signals. This would affect aircraft communications and the Global Positioning System (GPS).

▲ Auroras are rarely seen in the southern United States, but in 1859 they were visible to people as far south as the Caribbean.

Pause and Reflect

3. Describe three ways a long-term power outage could affect your daily life.

Investigate Further

4. What are telecommunications? Choose one example of a telecommunications device and find out how it might be affected by a solar storm.
5. Learn more about the 1989 Québec power outage.
   - Why did the power grid fail?
   - For how long was the power out?
   - How was the problem fixed?
6. Find out how airline travel could be affected by a solar storm.
7. Imagine that you are a reporter for a national newspaper when a solar superstorm strikes. Write a news report covering the effects of this storm in today's world.

Activity 3.9

COLONY ON ANOTHER PLANET

One reason we explore space is to learn about other planets. If one of these planets is enough like our own, it may be a site for a future human colony. In this activity, you will explore what makes Earth so well suited for the existence of life. You will then use this knowledge to assess if humans could survive on another planet.

What To Do

1. Work with a partner to create a list of conditions that you think make Earth suitable for life. Share your list with another set of partners and add any new ideas to your list.
2. Read the description of the planet in the white box below.
3. With your partner, make a list of reasons why this planet might be able to support human life. Compare this list to your list from step 1.

Planet Loki

- The planet Loki has been discovered orbiting a nearby star, Alpha Centauri. Alpha Centauri is slightly larger than our Sun and about the same temperature.
- The distance between Loki and its star is 1.40 AU. This is almost one-and-a-half times farther than Earth is from the Sun.
- Loki is 2.3 times the size of our planet. Its mass is 2.9 times the mass of Earth, so its gravitational pull is a bit stronger.
- Loki's atmosphere is similar to Earth's, but it has about twice as much carbon dioxide.
- Water exists as solid, liquid, and gas on Loki. The planet also has a rocky crust.
- Loki's magnetic field is much weaker than Earth's.

Inquiry Focus

What Did You Find Out?

1. Use a Venn diagram or double bubble organizer to compare Earth and Loki.
2. Do you think Loki is a suitable planet to set up a human colony? Explain why or why not.
3. Do you think plants on Loki would be able to carry out photosynthesis? Why or why not?

Inquire Further

Imagine that life already exists on Loki. Construct an alien life form that might be found on Loki using art material or a computer program, or draw it on paper. Explain how it would differ from or be similar to life on Earth and why.

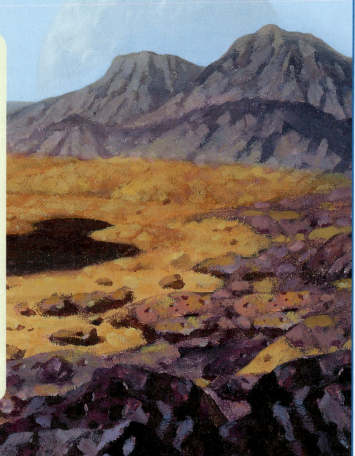

Topic 3.2 Review

Key Concepts Summary
- The Sun is our nearest star.
- Interactions of Earth and the Sun make life possible.
- The Moon is our nearest neighbour in space.
- The Sun, the Moon, and Earth interact to create eclipses.

Review the Key Concepts

1. **K/U** Answer the question that is the title of this topic. Copy and complete the graphic organizer below in your notebook. Fill in four examples from the topic using key terms as well as your own words.

   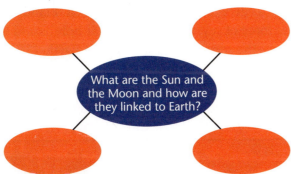

2. **C** Refer to **Figure 3.12** and **Figure 3.13A**. Use words, pictures, or a graphic organizer such as a Venn diagram to compare the similarities and differences between a total solar eclipse and a total lunar eclipse.

3. **A** The *aurora borealis,* or "northern lights," have been described in stories by many different Aboriginal peoples. Use your school's library to compare several stories about the aurora borealis told by North American Aboriginal peoples. Briefly summarize these stories in your own words.

4. **T/I** Predict why the surface of the Moon is covered with thousands of craters while there are very few craters on Earth's surface today.

5. **C** Refer to **Figure 3.10**. In your notebook, draw and label a diagram that models the phases of the moon. Write a caption for your diagram that explains why we can't always see the entire lit-up side of the Moon.

6. **T/I** Predict what might happen to Earth if the Moon broke out of its orbit.

7. **C** Use the diagram below to explain the Sun's role in sustaining life.

Topic 3.3

What has space exploration taught us about our solar system?

Key Concepts

- The four inner Earth-like planets are small and rocky.
- The four outer "gas giant" planets are large and ringed.
- Rocky chunks of various sizes make up the rest of the solar system.

Key Skills

Inquiry
Literacy
Numeracy

Key Terms

inner planets
outer planets
asteroid
meteoroid
comet

Space probes launched from Earth have visited every planet in our solar system. By venturing out into space, these robotic "eyes" have taught us much about the system of objects, our Earth included, that are linked to the star we call the Sun. Sometimes, however, we learn about our solar system when objects from *it* come to visit *us*. This is what happened, for example, 1.85 billion years ago in a part of Earth that is now called Sudbury, Ontario. A rock measuring 10 km in diameter hurtled through the atmosphere at speeds as great as 200 000 km/h. When it landed, it exploded with the force of several billion nuclear bombs and left a crater 250 km wide—the second-largest crater on Earth. (The largest is in South Africa.) Much of the wealth of material that is mined in the Sudbury area today is linked to the intense rock-changing heat from that ancient, cataclysmic explosion.

Starting Point Activity

The object that caused the Sudbury Crater was a meteorite. What do you know about meteorites? How are they related to meteors and meteoroids? Share what you know or remember about these and other solar-system words such as those below. Add any other solar-system names or words that you know or remember.

- inner planets
- outer planets
- comet
- dwarf planet
- asteroid
- asteroid belt
- moon
- solar system

110 Yards
× 11,000
= 10 km

(size of the object that struck Sudbury)

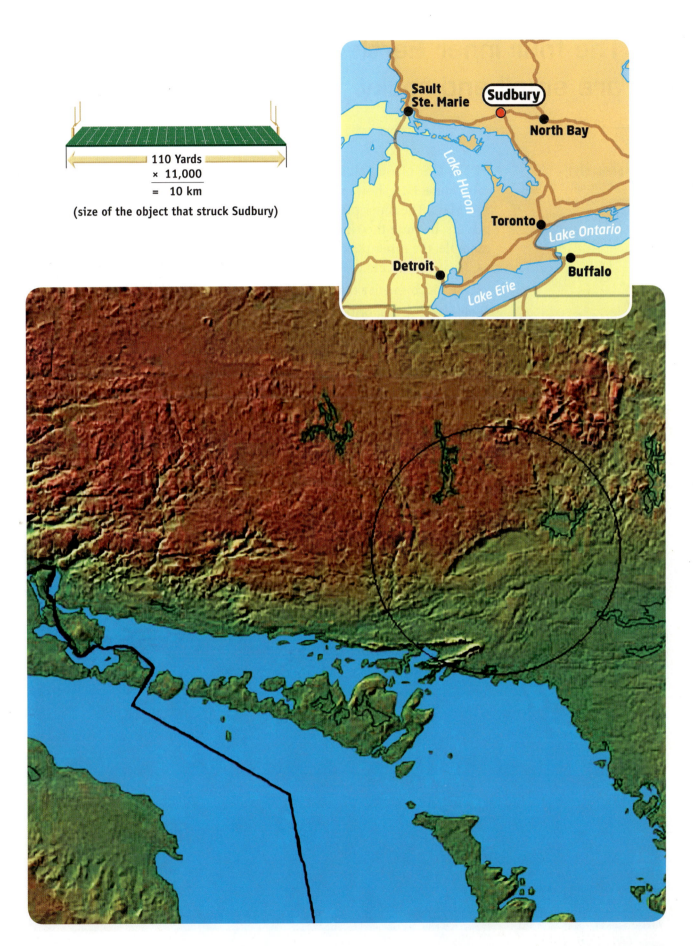

TOPIC 3.3 WHAT HAS SPACE EXPLORATION TAUGHT US ABOUT OUR SOLAR SYSTEM? • MHR

The four inner Earth-like planets are small and rocky.

inner planets: the four planets of the solar system that are closest to the Sun: Mercury, Venus, Earth, and Mars

The four planets closest to the Sun are called the **inner planets**. Figure 3.14 shows the planets in this group: Mercury, Venus, Earth, and Mars. The inner planets are also known as the terrestrial (Earth-like) planets, because they have many features in common with Earth. For example, the inner planets:

- have a rocky, cratered surface
- are much smaller than the other planets in the solar system
- have orbits that bring them much closer to the Sun than the other planets
- have no moons or very few moons compared with the other planets

LEARNING CHECK

1. Identify the inner planets in our solar system.
2. State another name for the inner planets.
3. Use a table to compare four similarities and four differences between Earth and the other inner planets.

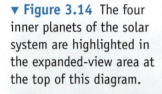

▼ **Figure 3.14** The four inner planets of the solar system are highlighted in the expanded-view area at the top of this diagram.

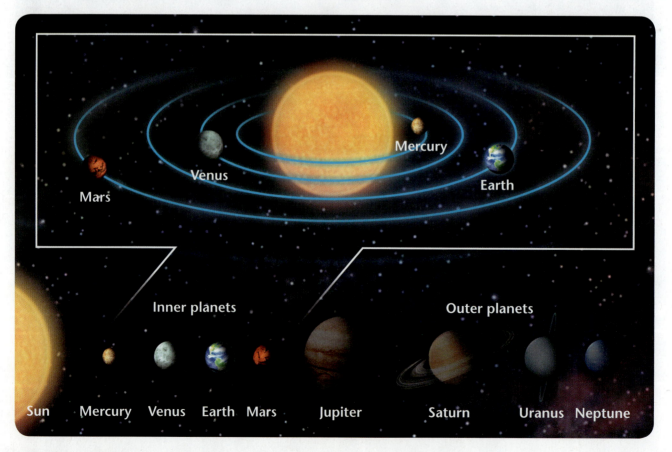

Mercury—The Smallest Planet

Distance from Sun: 0.39 AU

Size (Diameter): 4878 km (about one-third of Earth's)

Temperature Range: −184°C to 427°C

Time for One Rotation: 59 days

Time for One Orbit: 88 days (0.24 years)

Number of Moons: 0

Atmosphere: so weak and unstable as to be almost non-existent (but—as discovered in 2008 by NASA's MESSENGER probe—it includes water vapour)

Notable features: numerous craters, many named for famous Earth musicians, artists, and authors

Symbol: ☿

Name Origin: Named for the speedy Roman messenger god. (Mercury travels in its orbit faster than any other planet.)

Venus—The Hottest Planet

Distance from Sun: 0.72 AU

Size (Diameter): 12 104 km (nearly the same as Earth's, so Venus is sometimes called our sister planet)

Average Temperature: 457°C

Time for One Rotation: 243 days

Time for One Orbit: 226 days (0.62 years)

Number of Moons: 0

Atmosphere: mostly (about 97 percent) carbon dioxide, a gas that traps heat from the Sun, and that's why the temperature stays the same

Notable features: mountains, valleys, and volcanoes etched by rains of sulfuric acid

Symbol: ♀

Name Origin: Named for the Roman goddess of love and beauty. (Venus is bright, beautiful, and very noticeable in the sky at dawn and dusk.)

Earth—The Living Planet

Distance from Sun: 1.0 AU

Size (Diameter): 12 756 km

Temperature Range: −89°C to 58°C

Time for One Rotation: 23 hours, 56 minutes

Time for One Orbit: 365 days, 5 hours

Number of Moons: 1

Atmosphere: 78 percent nitrogen, 20 percent oxygen, 2 percent other gases (including carbon dioxide)

Notable features: Most of our craters are either underwater or have been eroded by weather and time. Oh, yes, and we have life!

Symbol: ♁

Name Origin: Earth is the only planet not named after a god or goddess from mythology. Earth simply refers to the material world—earth (as in soil, land, ground).

Mars—The Red Planet

Distance from Sun: 1.5 AU

Size (Diameter): 6785 km

Temperature Range: −140°C to 20°C

Time for One Rotation: 24.6 hours

Time for One Orbit: 1.88 years

Number of Moons: 2 (Phobos and Deimos, named after two children of Mars and Venus)

Atmosphere: mostly carbon dioxide (95 percent), plus other gases, including a small amount of oxygen

Notable features: rust in the soil (hence the red colour), the tallest volcano in the solar system, and methane in the atmosphere—a possible sign of life.

Symbol: ♂

Name Origin: Named for the Roman god of war and weaponry. (Mars, along with Venus and Mercury, are children of Jupiter in Roman myth.)

The four outer "gas giant" planets are large and ringed.

outer planets: the four planets of the solar system that are farthest from the Sun: Jupiter, Saturn, Uranus, and Neptune

The four planets farthest from the Sun are called the **outer planets**. Figure 3.15 shows the planets in this group: Jupiter, Saturn, Uranus, and Neptune. The outer planets are also known as the gas giants, because they have substantial gaseous atmospheres and their "surfaces" are gaseous, rather than solid. The outer planets have other features in common as well. For example, the outer planets:

- have a gassy atmosphere and no solid surface
- are very large compared with the inner planets
- have orbits that keep them far from the Sun
- have rings surrounding them
- have numerous moons

LEARNING CHECK

1. Identify the outer planets in our solar system.
2. State another name for the outer planets.
3. Use a table to compare four similarities and four differences between Earth and the outer planets.

▼ **Figure 3.15** The four outer planets of the solar system are highlighted in the expanded-view area at the top of this diagram.

200 MHR • UNIT 3 SPACE EXPLORATION

Jupiter—The Giant Planet

Distance from Sun: 5.2 AU

Size (Diameter): 142 800 km (if Earth were the size of a pea, Jupiter would be the size of an orange)

Average Temperature: −150°C

Time for One Rotation: 9.8 hours

Time for One Orbit: 12 years

Number of Moons: 63 (as of the year 2009), including the solar system's largest and the only one with a magnetosphere (Ganymede). Moons are named for mythological lovers and relatives of Jupiter.

Atmosphere: mostly hydrogen and helium

Notable features: the Great Red Spot's bone-crushing atmospheric pressure 30 000 times greater than Earth's, and a slender ring system

Symbol: ♃

Name Origin: Named for the ruler of all the Roman gods and goddesses. (Jupiter is the father of Mercury, Venus, and Mars.)

Saturn—The Ringed Planet

Distance from Sun: 9.5 AU

Size (Diameter): 120 536 km

Average Temperature: −170°C

Time for One Rotation: 10.7 hours

Time for One Orbit: 29.5 years

Number of Moons: 61 (as of the year 2009), including Titan, the only moon in the solar system with its own atmosphere. Moons are named after giants from Greek and Roman myths, as well as Inuit, French, and Scandinavian myths.

Atmosphere: mostly hydrogen and helium

Notable features: a Great White Spot (a storm that appears and disappears every 30 years), a density so small that Saturn would float on Earth's ocean, and—of course—its rings

Symbol: ♄

Name Origin: Named for the Roman god of agriculture, who was also father of Jupiter.

Uranus—The Tilted Planet

Distance from Sun: 19 AU

Size (Diameter): 51 120 km

Average Temperature: −215°C

Time for One Rotation: 12.2 hours

Time for One Orbit: 84 years

Number of Moons: 27 (as of the year 2009). Moons are named after characters from the plays of William Shakespeare and the poetry of Alexander Pope.

Atmosphere: mostly hydrogen and helium, with a small amount of methane

Notable features: a faint ring system, and the tilt of its axis is almost 90°, so it rotates "sideways"

Symbol: ⛢

Name Origin: Named for the Greek god of the sky, who was also father of Saturn. (Oh, and by the way, it's pronounced YOU-ran-us.)

Neptune—The Deep-Blue Planet

Distance from Sun: 30 AU

Size (Diameter): 49 530 km

Average Temperature: −235°C

Time for One Rotation: 16.1 hours

Time for One Orbit: 165 years

Number of Moons: 13 (as of the year 2009). Moons are named after the lesser gods and goddesses from Greek and Roman myth associated with Neptune.

Atmosphere: mostly hydrogen and helium, with a small amount of methane

Notable features: a Great Dark Spot surrounded by whitish clouds of frozen methane, and a faint ring system

Symbol: ♆

Name Origin: Named for the Roman god of the sea, who was also a son of Saturn.

Rocky chunks of various sizes make up the rest of the solar system.

The Sun contains about 99.85 percent of all the mass of the solar system. The planets and their moons contain most of the rest. What's left over, a mere 0.02 percent of the mass of the solar system, is made up of chunks of rocky material. These chunks range in size from dust-sized specks to moon-sized orbs. They include asteroids, dwarf planets, comets, and meteoroids.

Asteroids

Figure 3.16 shows that between the orbits of Mars and Jupiter lies a ring of rocky objects that orbit the Sun. Some are round, and others are irregular in shape. Some are dozens or hundreds of kilometres across, but most are much smaller. All these rocks are called **asteroids**, and their location in the solar system is known as the Asteroid Belt. Scientists know the positions of a few hundred thousand asteroids, but there are millions more.

asteroid: a rocky object, located in the region between the orbits of Mars and Jupiter, that orbits the Sun

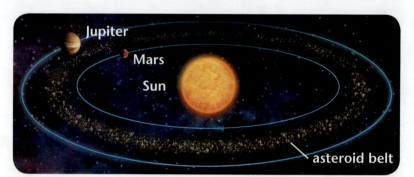

Figure 3.16 The Asteroid Belt is a ring of rocky chunks of various sizes that orbit the Sun between the orbits of Mars and Jupiter.

Dwarf Planets

The largest object in the Asteroid Belt is called Ceres. With a diameter of about 950 km, Ceres is referred to as a dwarf planet rather than an asteroid. Ceres is one of five known dwarf planets in the solar system. The others are Pluto, Eris, Haumea, and Makemake. They are located far beyond Neptune. A dwarf planet is larger than an asteroid but smaller than a planet. Unlike a planet, a dwarf planet does not have enough gravity to pull all the rocky debris around it out of the path of its orbit.

What about Comets and Meteoroids?

Comets are chunks of loosely held rock and ice that come from the outer parts of the solar system. They start their journeys toward and around the Sun when they are pulled from the outer regions of the solar system by the gravity of other objects. The orbits of comets bring them near the Sun with predictable regularity (**Figure 3.17**). For example, the orbit of Halley's Comet makes it visible to us on Earth every 76 years. Other comets have such huge orbits that they take thousands or millions of years to make one trip around the Sun.

Meteoroids are chunks of rock, metal, or both that are shed from the asteroids or comets. When a meteoroid enters Earth's atmosphere and starts to burn up, it makes a streak of light across the sky lasting a few seconds or shorter. This streak of light made by a meteoroid is called a meteor (**Figure 3.18**). Any part of the original object that survives the atmosphere and lands on Earth's surface is called a meteorite.

comet: chunks of loosely held rock and ice that are thought to come from the outer regions of the solar system

meteoroid: chunk of rock, metal, or both that is shed from an asteroid or comet

LEARNING CHECK

1. Refer to **Figures 3.17** and **3.18**. Compare a comet to a meteoroid.
2. Describe what a dwarf planet is, and state how many known dwarf planets there are.
3. In the past, many people viewed comets with suspicion and fear. Predict why people might have been fearful of comets.

Go to **scienceontario** to find out more

▲ **Figure 3.17** As a comet nears the Sun, the solar wind causes parts of the comet to vaporize and pushes them off the comet's surface. This forms what we call the tail of a comet.

▲ **Figure 3.18** The streak of light is the result of heat generated by friction between the meteoroid and Earth's atmosphere.

Activity 3.10
NEWS FROM NEOS

In the early days of our planet, nearly 4.6 billion years ago, there were many objects near Earth that were 500 km or larger in diameter. Collisions between these objects and our young planet were likely. Had there been living things on Earth in those days, a collision of this size would have wiped them out completely. Fortunately, there are no objects of this size that pose any threat to Earth now or in the future.

Near-Earth Objects (NEOs) are asteroids, comets, and meteoroids with orbits that are close to Earth's orbit. The craters that mark our planet were made by NEOs. Scientists are interested in NEOs for a few reasons. One is purely scientific. One is very practical. And one is perhaps a little alarming. Find out about any or all of the following:

- the possible link between NEOs and the origin of life on Earth
- the possible use of NEOs as a resource for raw materials
- the possible impact of a NEO with Earth, who is watching, and how it could be avoided

Record your findings in the form of a TV or radio news report.

Research Focus

A collision between a 500 km or more object and Earth likely took place more than once during Earth's early history. Such a collision would have generated enough energy to kill all life on Earth.

Activity 3.11

Numeracy Focus

BIKE ME TO THE MOON, AND BEYOND

Imagine that you can ride a bike at a constant rate of 20 km/h without ever stopping. How long would it take you to bike across the country, around the world, and to places beyond Earth? Find out in this activity.

What To Do

1. The table to the right lists the distances for several trips. For each trip, you will predict the time it would take. Then you will calculate the time. Make a table with three headings—Trip, Predicted Time, and Calculated Time—and one row for each trip.
2. Estimate and record the time you think it will take to make the trip at a speed of 20 km/h.
3. Calculate how long it would take to travel each distance. The first calculation is done as an example below.

Description of Trip	Approximate Distance
Canada, from west coast to east coast	5.20×10^3 km
Around Earth's equator	4.00×10^4 km
From Earth to the Moon	3.85×10^5 km
From Earth to Mars	5.80×10^7 km
From Earth to the Sun	1.50×10^8 km
From Earth to Jupiter	9.30×10^8 km
From Earth to Neptune	4.30×10^9 km
From Earth to the outer regions of the solar system	9.46×10^{12} km (at least)

What Did You Find Out?

1. How close were your predictions to your calculated values?
2. Which trips do you think would be practical if you could travel in a space ship at a speed of 1.0×10^5 km/h?

Description of Trip	Approximate Distance
To find the number of hours the trip will take, divide the distance by the speed. time = $\dfrac{\text{distance}}{\text{speed}}$ or $t = \dfrac{d}{s}$	$t = \dfrac{d}{s}$ $t = \dfrac{5.2 \times 10^3 \text{ km}}{20 \frac{\text{km}}{\text{h}}}$ $t = 2.6 \times 10^2$ h or 260 h
To find the time in days, divide the number of hours by the number of hours in a day. time in days = $\dfrac{\text{time in hours}}{24 \text{ hours in a day}}$	time in days = $\dfrac{\text{time in hours}}{24 \text{ hours in a day}}$ time in days = $\dfrac{260 \text{ h}}{24 \frac{\text{h}}{\text{day}}}$ time in days = 10.83 days
To find the time in years, divide the number of days by the number of days in a year. time in years = $\dfrac{\text{time in days}}{365 \text{ days in a year}}$	time in years = $\dfrac{\text{time in days}}{365 \text{ days in a year}}$ time in years = $\dfrac{10.83 \text{ days}}{365 \frac{\text{days}}{\text{year}}}$ time in years = 0.0297 years

Activity 3.12

MAP THE SOLAR SYSTEM

Inquiry Focus

You can get a better grasp of large distances in space by bringing them down to Earth. In this activity, you will use a scale to reduce the distances between planets and the Sun so they fit within the boundaries of your community.

What You Need
- map of your community
- sticky notes
- coloured pencils or markers
- ruler

What To Do

1. Copy the following table into your notebook.

Planet Distances from Sun

Planet	Distance from Sun (AU)
Mercury	
Venus	
Earth	
Mars	
Jupiter	
Saturn	
Uranus	
Neptune	

2. The planet profiles on pages 199 and 201 give the distance of each planet from the Sun. Use this data to complete the table.

3. Spread out the map of your community on a desk or tape it to the chalkboard.

4. Find the location of your school on the map. Using a sticky note and a coloured marker, label your school as the Sun.

5. Copy the following scale onto a sticky note and place it on your map: 1 AU:1 km.

6. The map of your community will have a scale on it. Find this scale and use a ruler to determine what distance represents 1 km on your map.

7. Refer to the table that you copied into your notebook. Use sticky notes and coloured markers to mark the locations of the planets on your map. Start measuring from your school.

8. Compare your map with another student and see how the locations of the different planets compare on the two maps.

What Did You Find Out?

1. What is the distance between Earth and the Sun on your map? Where is Earth located in your community?

2. Which planet is about five times farther from the Sun than Earth is? Where is this planet located in your community?

3. Did you find it more difficult to mark the inner planets or the outer planets on your map? Explain why this was the case.

4. In step 8 you compared your map with another student's map. How were your maps similar? How did they differ?

Inquire Further

Figure out how to add these other solar system objects to your map: the Asteroid Belt, the Kuiper Belt, the Oort Cloud.

Topic 3.3 Review

Key Concepts Summary
- The four inner Earth-like planets are small and rocky.
- The four outer planets are large and ringed.
- Rocky chunks of various sizes make up the rest of the solar system.

Review the Key Concepts

1. **K/U** Answer the question that is the title of this topic. Copy and complete the graphic organizer below in your notebook. Fill in four examples from the topic using key terms as well as your own words.

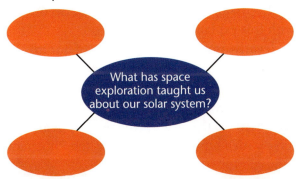

2. **K/U** Copy and complete the table below in your notebook.

Comparing Asteroids, Comets, and Meteoroids

	Asteroid	Comet	Meteoroid
Description			
Origin			

3. **C** Draw a Venn diagram to compare the similarities and differences between the inner planets and the outer planets in our solar system.

4. **C** Although Venus is sometimes referred to as Earth's sister planet, Mars has more in common with Earth than Venus does. Provide at least three facts to support this idea.

5. **T/I** In your notebook, construct a bar graph to show the data in the table below. Based on this data, list the planets in order from the smallest planet to the largest planet.

Planet	Diameter (km)
Earth	12 756
Uranus	51 120
Mercury	4878
Jupiter	142 800
Mars	6785
Neptune	49 530
Venus	12 104
Saturn	120 536

6. **A** *Voyager 2* is an unpiloted, interplanetary spacecraft designed originally to study Neptune and Uranus. To save fuel, *Voyager 2* used a planet's gravity during a flyby to slingshot it farther into space. This technique is called a "gravity assist trajectory." Predict which planet the *Voyager 2* spacecraft used as a slingshot and justify your answer.

7. **A** Refer to **Figure 3.14**. Venus is farther from the Sun than Mercury, yet it is hotter than Mercury. Explain why Venus is hotter than Mercury.

8. **K/U** Identify four features of Venus that would lead some people to refer to Venus as Earth's sister planet.

Topic 3.4

What role does Canada play in space exploration?

Key Concepts

- Canada contributes people and technology to explore space.
- Canada helps build the future of space exploration.

Key Skills

Research

For five months, come haze or shine, the Canadian-built weather station onboard the *Phoenix Mars Lander* reported the weather from its location in the north polar region of Mars. The dates were given in Earth days as well as "Sols." Each Sol was one Mars day, measured from the date of landing, May 25, 2008.

In addition to the weather station, Canada also supplied the distinctive logo for the *Phoenix* mission. Montréal's Isabelle Tremblay was the lead systems engineer for the weather station, as well as the logo's artist. Mission leader Peter H. Smith decoded the logo's symbolism with this explanation:

"Our jazzy logo cleverly unites all the elements of our mission: the planet Mars, water, and fire. The Mars image in the background shows the northern polar cap and just to the left, still inside the Arctic Circle, a droplet of water swirls out into space from our landing site. Superimposed on the water is the fiery Phoenix bird scanning the Universe with a hunter's eye. Clearly, the Phoenix is searching for something."

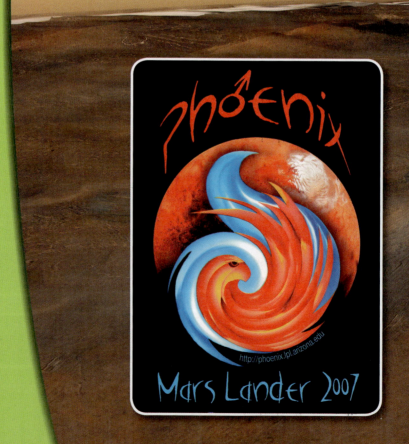

Starting Point Activity

Imagine being asked to design the logo for a lander or orbiter mission to a planet of the solar system. Think about the characteristics of the planets from Topic 3.3. Pick two or three characteristics that likely would be a focus of exploration for a mission. For instance, a mission to Venus might explore its thick carbon-dioxide atmosphere and sweltering surface temperatures. Draw the logo for your lander or orbiter. Include labels to decode the symbolism of your logo.

Canada contributes people and technology to explore space.

Did you know that Canadian scientists and engineers helped design the space capsule that carried the first American into space? Or that the oldest scientific institution in Canada was an observatory built in Toronto in 1839 to study Earth's magnetic field? Or that Canada built and launched the third satellite to orbit Earth? Canada has many notable achievements in the field of space exploration. A few of the more recent events are highlighted here.

Research Focus

Activity 3.13

WE GROW ASTRONAUTS, TOO

In early 2009, Canada had three astronauts on active duty, five who had retired, and sixteen new recruits vying for two open positions. What do you know about our first eight astronauts? Start by solving these eight puzzle-questions, and then choose one astronaut to explore in greater detail. Decide how to record your findings.

- Who knows all about Moon trees (trees grown on Earth with seeds that orbited the Moon)?
- Who has done missions in inner space (underwater) as well as in outer space?
- Who has made more trips to space than any other Canadian astronaut?
- Who was the first Canadian to walk in space?
- Who has a passion for taking photos of Earth from ground level as well as from space?
- Who was the first Canadian to operate the *Canadarm2* robotic arm?
- Who was the first Canadian to board the International Space Station?
- Who was the first Canadian trained as a mission specialist for both the space shuttle and the International Space Station?

Dr. Marc Garneau

Dr. Roberta Bondar

Dr. Steven MacLean

Col. Chris Hadfield

Dr. Robert Thirsk

Bjarni Tryggvason

Dr. Dave Williams

Capt. Julie Payette

The Maple Leaf Orbiter
CANADA'S WEEKLY SPACE JOURNAL

Seeing It and Tracking It—Even on Mars

The *Phoenix Mars Lander* includes Canadian weather technology to study Mars's polar climate. Mars may have once been a lot like Earth, with liquid water and warmer temperatures. Today, it is a world covered with dry riverbeds and endless deserts. What happened? That's one question *Phoenix* and its Canadian weather station have gone to Mars to find out. A new part of the puzzle emerged in 2008. *Phoenix* observed snow falling from clouds high above the surface. Ontario's Jim Whiteway, lead scientist for the Canadian weather station on *Phoenix,* said, "Nothing like this view has ever been seen on Mars." The *Ottawa Citizen* newspaper noted: "Trust a Canadian weather instrument to find snow. Even on Mars."

The *Phoenix* Mars Lander will help scientists better understand the similarities and differences between the two planets.

Let's Boast the MOST

Canada's first space telescope, dubbed *MOST*, was designed to study the inside of stars like our Sun. It was supposed to last one year, but more than five years later it's still going strong. Scientists from around the world book time to use *MOST*. The *MOST* team has also asked Canadian stargazers for project proposals. Recently, amateur astronomer David Garney's proposal was chosen as a *MOST* mission. Garney teaches scouts in Toronto about the night sky. His idea was to study the supergiant star, Betelgeuse. Garney says, "In a way, Canadian kids are my collaborators on this proposal, because Betelgeuse means something to them."

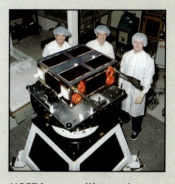

MOST is a satellite-style telescope that orbits Earth. *MOST* stands for **M**icrovariability and **O**scillations of **St**ars.

Earth, We Stand on Guard for Thee

Canadian space scientists have now entered the protection business. But instead of tracking down criminals, they are searching out asteroids that could possibly get near enough to strike Earth. Tens of thousands of asteroids have orbits that bring them near enough to Earth to warrant keeping a close eye on them. *NEOSSat* (**N**ear-**E**arth **O**bject **S**urveillance **Sat**ellite) is a small telescope-equipped satellite to monitor these asteroids. The satellite also keeps an eye on other satellites to prevent collisions.

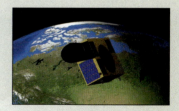

Canada's *NEOSSat*

LEARNING CHECK

1. State the purpose of Canadian technology on the *Phoenix* Mars Lander.
2. Explain why it is important for Canada to have *NEOSSat* orbiting Earth.
3. Refer to all the material on these two pages. Identify four other contributions to space made by Canadians.

Canada helps build the future of space exploration.

Space robotics play a key role in exploring planets. They also help maintain equipment and assemble spacecraft in space, as well as assist astronauts to complete many other tasks. Canadian-built and designed robotics were instrumental in building the International Space Station. They continue to be an important means of inspecting and fixing outer parts of the station. The three major components of the Mobile Service System responsible for these tasks are *Canadarm2*, *Dextre*, and the *Mobile Base System*.

Canadarm2

Canadarm2 is our main contribution to the International Space Station (ISS). The robotic arm (shown here) was installed on the ISS in 2001. The arm is larger and more flexible than its predecessor, the *Canadarm*, which retrieved and released satellites from the space shuttle. Able to reach most outer parts of the ISS, *Canadarm2* has helped build and maintain the space station. It also provides a stable platform from which astronauts perform tasks in space. The arm is a lot like a human arm, rotating at joints like those of the shoulder, elbow, and wrist. Thanks to advanced vision systems and touch sensors, *Canadarm2* can be controlled by astronauts directly from inside the station or by remote control.

Dextre

Dextre is officially known as a Special Purpose Dexterous Manipulator. It is an advanced, highly co-ordinated or "dexterous" robot that connects to *Canadarm2*. (The word "dexterous" refers to skillful use of the hands.) The robot has a pair of seven-jointed arms that sit on a set of supportive shoulders. The arms let *Dextre* perform tasks that could once only be completed by astronauts outside the ISS. These range from handling small objects to completing delicate operations that require the use of tools.

LEARNING CHECK

1. List the three Canadian-made components of the Mobile Service System.
2. Compare the design and roles of *Canadarm* to *Canadarm2*.
3. Explain how *Dextre* helps keep astronauts safe.

Activity 3.14

CANADIANS EXPLORING SPACE

Literacy Focus

Canadian organizations have contributed human and financial resources on a large scale to develop space technologies such as *Dextre* and *Canadarm2*. These investments have helped build Canada's reputation as a world leader in space exploration. Research one Canadian company, government agency, university, or college provided by your teacher. Assess its contributions to Canadian space exploration. Communicate your findings in the form of an organizer or another presentation method of your choice.

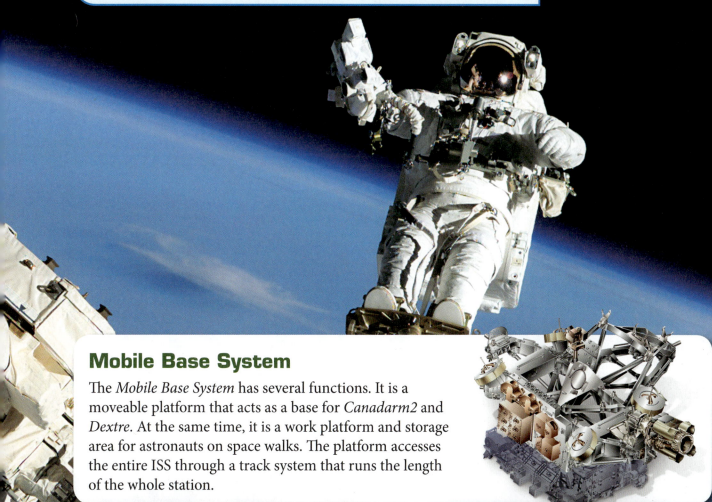

Mobile Base System

The *Mobile Base System* has several functions. It is a moveable platform that acts as a base for *Canadarm2* and *Dextre*. At the same time, it is a work platform and storage area for astronauts on space walks. The platform accesses the entire ISS through a track system that runs the length of the whole station.

Investigation 3B

You, Robot

Skill Check

Initiating and Planning

✓ Performing and Recording

✓ Analyzing and Interpreting

✓ Communicating

Because it is dangerous for astronauts to work in space, robots such as *Dextre* are designed to take over the job of performing repairs outside the space station. However, robots are not yet capable of the fine control and co-ordination that humans possess. This activity will help you get an idea of how tasks "feel" from an advanced robot's point of view.

What To Do

1. Copy the table into your notebook.

Times for Robot Simulation Trials

Shoelace test	Trial 1 time (s)	Trial 2 time (s)
Hands		
Blindfold		
Gloves		
Tongue depressors		
Pliers		

2. Work with a partner to complete this activity. Have your partner sit in a chair with shoelaces untied.

3. Time how long it takes you to tie the shoelaces.

4. Untie the shoelaces again. Time how long it takes you to tie them wearing a blindfold.

5. Remove the blindfold and repeat step 4 using heavy gloves.

6. Remove the heavy gloves and repeat step 4 with tongue depressors taped to your thumbs and forefingers.

7. Remove the tongue depressors and repeat step 4 holding a pair of pliers in each hand.

8. Switch places with your partner and repeat steps 2 to 7.

What Did You Find Out?

1. Describe how your impaired abilities affected the completion of your task.

2. Based on your results in this activity, describe your impression of how well a robot can replace a human when it comes to making repairs outside the ISS. Keep in mind that astronauts also wear gloves and spacesuits that hamper their abilities.

Safety

- Do not put the tongue depressors into your mouth.
- Be careful when handling the pliers.

What You Need

- tongue depressor
- heavy gloves
- masking tape
- two pairs of pliers
- blindfold
- shoes with laces
- stopwatch

Topic 3.4 Review

Key Concepts Summary
- Canada contributes people and technology to explore space.
- Canada helps build the future of space exploration.

Review the Key Concepts

1. **K/U** Answer the question that is the title of this topic. Copy and complete the graphic organizer below in your notebook. Fill in four examples from the topic using key terms as well as your own words.

2. **T/I** Use the Internet or print resources from your school's library to create a table like the one below to summarize the space missions undertaken by the Canadian astronauts. Give your table a suitable title.

Astronaut	Mission	Mission Date	Launch Vehicle

3. **C** Copy the following table into your notebook. Use your school's library to identify two specific examples of Canadian satellites for each category.

Observation Satellites	Communications Satellites	Exploratory Satellites

4. **C** Use a Venn diagram to show how *Canadarm2*, *Dextre*, and the *Mobile Base System* are related.

5. **A** Magnetic storms in and around Earth's atmosphere are caused by the solar wind as it strikes the magnetosphere. During magnetic storms, communications satellites can be damaged by the radiation that these storms produce. Magnetic storms can also disrupt electrical power on Earth, causing blackouts over large regions. In March, 1989, a major storm caused one third of Canada and part of New York State to lose electrical power. In 1996, Canada placed a highly specialized camera, the auroral ultra-violet imager (UVAI), aboard the Russian satellite *Interball-2*. The goal of *Interball 2* was to study the Sun's influence on magnetic phenomena around the Earth.

   a) Explain how your life could be impacted by a magnetic storm.

   b) Do you think the benefits of this Canadian-made camera outweigh the costs of developing it? Justify your answer.

6. **C** You have an opportunity to join the Canadian Space Agency as an astronaut. Write a short report describing what you would like to accomplish during your career.

7. **A** Due to concern for global environmental monitoring and protection, Earth observation is a key goal of the Canadian Space Program. RADARSAT-1 was developed as Canada's flagship satellite to pursue this goal. Predict three current environmental issues that scientists might investigate using data from satellites such as RADARSAT-1.

SCIENCE AT WORK

CANADIANS IN SCIENCE

Jason Kolodziejczak has never been to Mars, but that didn't stop him from imagining the destructive power of a Martian dust storm if such storms happened on Earth. In the world of CGI (computer-generated imagery), anything is possible. Jason uses software to create visual effects for movies and television. His Martian storm appeared in the Emmy-nominated documentary, *Planet Storm,* from The Discovery Channel. Jason also teaches visual effects at Seneca College of Applied Arts and Technology in Toronto. He studied 3-D animation at the Vancouver Film School.

▲ Jason Kolodziejczak is a digital compositor.

What challenges do digital compositors face?
Digital compositors have to work well in teams and be open to constructive criticism, says Jason. The main challenge is figuring out how to get a shot done in a limited amount of time. "Sometimes you look at a shot and think, how on Earth am I going to pull this one off!" he says.

What do you find most rewarding about your job?
Jason says he finds the combination of art and problem solving very rewarding. "Being able to see your name at the end of the movie on the big screen!" also helps, he smiles.

What advice would you give a student interested in a similar career?
Jason says students should keep in mind that math and science are part of the job. Digital compositors use complex computer programs, so strong problem-solving skills are a must. "As an artist, I don't necessarily see the underlying math of the program, but I do need to have a solid understanding of it to problem-solve."

Put Science To Work

The study of space contributes to these careers, as well as many more!

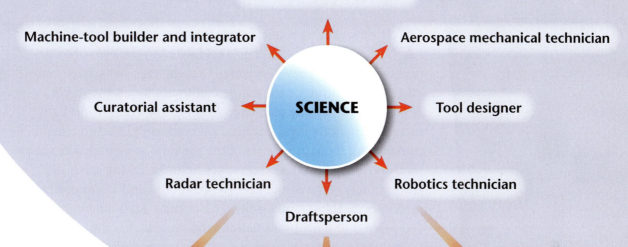

▲ Radar technicians help design, test, operate, and maintain radar satellite equipment. Technicians usually work for companies that convert radar data into pictures.

▲ A draftsperson uses computers and information from engineers to create drawings and blueprints for satellites, space probes, and other space-exploring devices.

▲ Robotics technicians build and operate automated machines. Technicians often work with engineers to build robotic parts for spacecraft and other aerospace technologies.

Over To You

1. If you could interview Jason Kolodziejczak, what questions about his work would you ask him?
2. Why would it be important for a digital compositor to have good problem-solving skills?
3. Research a career involving space science that interests you. What essential skills are needed? What else could help you find a job in this field?

Go to **scienceontario** to find out more

Topic 3.5
How do we benefit from space exploration?

Key Concepts

- We develop technologies that shape our lives.
- We are challenged to think and act locally, globally, and universally.
- We gain a deeper appreciation for ourselves and our home planet.

Key Skills

Inquiry

To protect the eyes of astronauts in space and on the Moon, NASA scientists and engineers looked to the eyes of birds of prey on Earth. They discovered that eagles, hawks, and many other raptors produce an orange-coloured fluid in their eyes. This fluid protects the eyes from harmful radiation in sunlight such as ultraviolet rays. The fluid also reduces glare and enhances contrast. All these properties of the eye fluid were highly desirable for the visor material used on space helmets. So the scientists and engineers figured out how to produce it, and it was used to manufacture them. The resulting visor material is able to block 100 percent of harmful radiation. The same technology used in space helmets is yours any time you wear sunglasses that have lenses with the brand name SunTiger® or Eagle Eyes®.

Starting Point Activity

1. In what ways do you think you benefit from space exploration? Include examples of specific products or services you use that you think came from space missions.

2. Each year, many billions of dollars throughout the world are spent on space exploration. Meanwhile, many problems here on Earth—poverty, starvation, environmental damage, to name a few—require money and expertise to help solve them. Do you think money should be spent on space exploration? Justify your opinions.

We develop technologies that shape our lives.

During a typical week, you interact with, use, and even eat dozens of things that are linked to space exploration. These products are spinoffs of space exploration. (A spinoff is a product or a technology that is originally developed for one use but is modified for other uses.) The pictures on this page barely scratch the surface of the number of space spinoffs that have helped to shape society.

1. You're off to the mall to meet some friends. You glance at your watch and hope its quartz timing crystals aren't accurate. If they are, you've just missed your bus. Unfortunately your watch is correct. After all, similar watches were used for the first human mission to the Moon.

2. Should you walk instead? Luckily you've got runners on. Maybe you'd get there faster if you knew you were supported by space exploration. Running shoes are made with shock-absorbing Moon-boot material.

3. You pause to text-message your friends that you're going to be late. Your phone is designed with miniature components created originally for the shuttle program, to save space and weight on space missions.

4. Need to supplement breakfast. You stop in at a store to buy an energy bar. The cashier scans the UPC code on the back. These codes were designed to track each of the millions of tiny pieces used to build a space capsule.

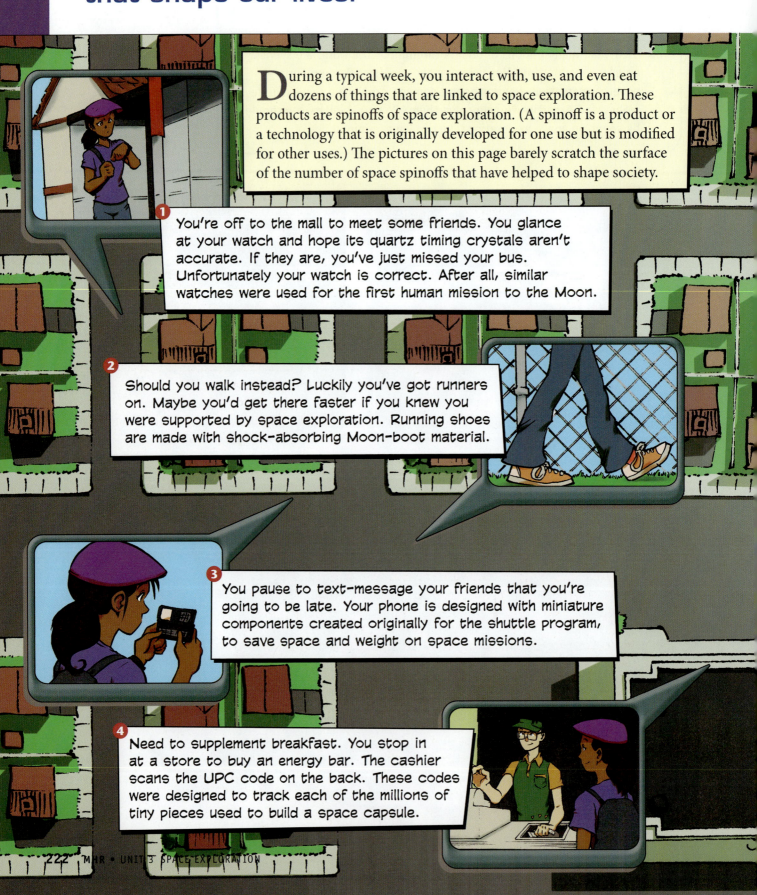

We are challenged to think and act locally, globally, and universally.

You have probably heard the statement, "Think globally, act locally." Among other things, this statement reminds us that there are consequences to the ideas we have and the actions we take in response to those ideas. Where exploring space is concerned, do we always think enough about the consequences of our ideas and actions? Do the pros of a mission always outweigh the cons? Read about the two examples here, and share what you think.

Nuclear-Powered Planetary Probes

The Cassini-Huygens (cah-SEE-NEE HIGH-genz) space probe reached Saturn and its moon system in 2004. However, its launch, shown in **Figure 3.19**, was more controversial. The probe was carrying 33 kg of radioactive plutonium as a power source. Some people asked what might happen if there were an accident, such as an explosion, during or shortly after lift-off.

NASA authorities assured everyone that the risk of radioactive material leaving the spacecraft during such an accident was very small. However, if the rocket carrying the probe had blown up in the atmosphere, over 5 billion people, and countless numbers of other organisms, would have been exposed to the radiation.

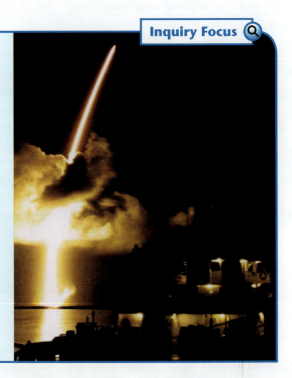

Activity 3.15

TRAVELLING BOMBS—WORTH THE RISK?

The Cassini-Huygens probe successfully reached Saturn and has greatly expanded our knowledge of the great ringed planet and several of its moons. One of those moons, Titan, is the only moon known to have an atmosphere, and contains chemicals that could support life. So understanding Titan can help us understand life back here, on Earth. Should we have allowed the probe to take off? Do the benefits outweigh the risks? What other solutions might there be?

▶ **Figure 3.19** The rocket carrying the Cassini-Huygens space probe was launched in 1997.

Terraforming—New Horizons for Humanity

Some scientists believe that it is technologically possible to transform alien, lifeless worlds or landscapes into life-sustaining ecosystems for future human colonies. Some scientists believe the technology for doing so already exists. Transforming an alien environment into one that can support Earth life is called terraforming. ("Terra" means land or earth.) At one time, terraforming was a topic only for science-fiction writers. Now, however, as the rate of our population growth increases and as our concerns for our global climate systems mount, terraforming represents one possible solution to the future of our human species. **Figure 3.20** shows what a terraformed landscape might look like.

Inquiry Focus

Activity 3.16

OFF-WORLD EARTHS—WORTH THE RISK?

Some people, perhaps many people, would say that we have not done a very good job of respecting and managing the ecosystems that sustain us and other living things on Earth. Have we learned our lessons enough to believe that we can do a better job on the Moon, on Mars, on Titan, or on other planets and moons in the universe? Do the benefits outweigh the risks? What other solutions might there be?

▼ **Figure 3.20** This picture shows an artist's vision of Mars being terraformed. Actually doing this would require, among other things, creating a water cycle, plant life to generate and sustain an atmosphere, and nutrient cycles to sustain the plants.

We gain a deeper appreciation for ourselves and our home planet.

Space exploration has influenced our sense of self—our understanding of where, when, and who we are, within and beyond the boundaries of our planet. When we look up at the night sky, we are explorers just as much as any astronaut is or has been. Travel through the words and images below and share the journey we have taken over time, from some of our earliest observations of the sky to those we are making today.

We Are Stargazers

Figure 3.21 shows a tiny piece of ivory from 32 000 years ago that records a glimpse of an ancient sky. The mammoth tusk, found in Germany, is thought to show a pattern of stars in the night sky—possibly the constellation we call Orion. It is the oldest human drawing yet known.

We Are Timekeepers

The Cro-Magnon people, ancient cousins of ours, lived in Europe over 15 000 years ago. They used their observations of the changing night sky to draw the first lunar calendar on cave walls near Lascaux, France. Refer to **Figure 3.22**. The 29 dots represent the 29 days of the Moon's phases. (It takes 29.5 days for the Moon to complete a full cycle from one new Moon to the next.)

▲ **Figure 3.21** The pose of the human figure on this ancient ivory is similar to the pose of the figure that later civilizations saw in the stars of Orion.

▼ **Figure 3.22** This painting is one of about 2000 that were made deep in caves in southern France 15 000 to 17 000 years ago.

We Are Navigators

Sailors and wanderers, both ancient and modern, rely on the predictable patterns of constellations and other celestial objects to find the way from home to there and back again.

We Are Explorers

In the mid-1900s, we decided to find out what was "out there" ourselves. In 1961, cosmonaut Yuri Gagarin became the first person in space. In 1969, American astronauts Neil Armstrong and Buzz Aldrin were the first humans to set foot on the Moon. In 2005, cosmonaut Sergei Krikalev broke the record for the longest time spent in space, spending over two years on the International Space Station. Refer to **Figure 3.23**.

◀ **Figure 3.23** Yuri Gagarin (A), Neil Armstrong (B), and Sergei Krikalev (C)—record-breakers in the history of space exploration.

We Are Voyagers

In 1977, two small probes—*Voyager 1* and *Voyager 2*—left Earth on a mission to observe and study the four outer planets before hurtling onward to the outer reaches of our solar system, and beyond. The *Voyager* probes have travelled, and continue to travel, farther from Earth than any other human-made craft in history (**Figure 3.24**). They carry with them a piece of humanity, of us—pictures, songs, and stories about the people of a small, pale-blue, life-supporting planet called Earth.

▲ **Figure 3.24** In 1990, as *Voyager 1* began its long journey to the edge of the solar system, it "looked back" at our home system of planets. In its view, Earth is a speck of light barely 0.1 pixels in size.

LEARNING CHECK

1. Refer to **Figure 3.22**. Predict why ancient civilizations created drawings of constellations.
2. Why do people continue to explore space and send space probes to the far reaches of our galaxy?
3. In your own words, compare the similarities and differences between ancient explorers and modern-day space explorers.

Making a DIFFERENCE

Stargazing is becoming more difficult because of light pollution. Light pollution is light that shines where it is not needed. A lit sign meant to be seen by people passing by produces light pollution if it also shines light into the sky where it isn't needed.

Shelby Mielhausen noticed that light pollution was increasing in Tobermory, Ontario, her home town. She studied the issue for a project in Grade 8. She wanted to know how light pollution affects the number of stars she could see. She counted visible stars in four communities and confirmed her prediction. Owen Sound, the largest community, had the most light pollution and the smallest star count. Shelby also studied outdoor lighting design and constructed a light shield that can help reduce light pollution.

What could you do to decrease light pollution in your neighbourhood?

When Nishant Balakrishnan was in Grade 9, he observed that every winter the hydro lines in the city freeze and communities lost power until the lines could be repaired. He thought about the way earthworms and leeches move vertically, and he wondered if he could design a robot that could climb hydro towers in a similar way.

Nishant named his design Leech Bot. The robot is controlled by a microprocessor. It uses springs, magnets and motors to creep across surfaces and flip over obstacles. "I've always tried to model solutions to problems based on what I see around me, because it's the basic things in nature that offer the most elegant and effective solutions," says Nishant. He believes that Leech Bot could be modified to do repair work on Earth as well as in space and on remote worlds such as Mars or the Moon.

What kind of robot can you imagine designing for use in space or on Earth?

Topic 3.5 Review

Key Concepts Summary
- We develop technologies that shape our lives.
- We are challenged to think and act locally, globally, and universally.
- We gain a deeper appreciation for ourselves and our home planet.

Review the Key Concepts

1. **K/U** Answer the question that is the title of this topic. Copy and complete the graphic organizer below in your notebook. Fill in four examples from the topic using key terms as well as your own words.

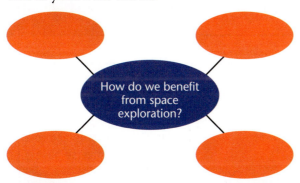

2. **C** Refer to **Figure 3.5**. In your notebook, draw a one-frame cartoon identifying one spin-off of space exploration that has a direct impact on your life.

3. **A** Below is a list of technologies that are direct spin-offs of space exploration. Identify a possible occupation that would benefit from each of these technologies and explain how it might benefit an individual in that job.
   a) Water purification technology used on the Apollo spacecraft
   b) Cordless power tools and appliances
   c) Freeze-dried food developed for the long *Apollo* missions
   d) A sensitive, hand-held infrared camera that observes the flames produced during the launch of the space shuttle
   e) Airocide, an air-purifier that kills 93.3 percent of airborne disease-causing organisms that pass through it, including the anthrax bacterium

4. **C** On January 11, 2007 China launched an anti-satellite missile and destroyed one of its aging weather satellites. The Anti-satellite (ASAT) Test produced as many as 35 000 pieces of debris about 1 cm in size. Nearly 1500 pieces were 10 cm and larger. Space scientists say there have already been three known cases in the last 15 years when satellites were disabled by collisions with space debris. In your opinion, should China have conducted this type of test? Justify your answer.

5. **C** NASA is preparing to send a small satellite into orbit that can be propelled by solar sails. When light from the Sun strikes the surface of the sail, solar energy is transferred to it. This energy provides a force that propels the satellite through space. Solar sails provide very low but inexhaustible thrust to move the satellite. In your opinion, should scientists use nuclear-powered or solar-powered sails as a propulsion system for interplanetary probes? Justify your answer.

6. **A** One day, humans could make a planet such as Mars suitable for life by transforming its environment through terraforming. If you were a scientist in charge of this massive project, how would you change the environment of Mars to make it suitable for life forms from Earth? Justify your answer.

Unit 3 Summary

Topic 3.1: What do we see when we look at the night sky?

Key Concepts
- We see stars that we organize into patterns.
- We see celestial objects of the universe.
- We see objects separated by immense distances.

Key Terms
constellation (page 171)
universe (page 172)
gravitational pull (page 172)
orbit (page 172)
solar system (page 172)
star (page 172)
galaxy (page 173)
astronomical unit (AU) (page 174)
light-year (page 175)

Big Ideas
- Celestial objects in the solar system and universe have specific properties that can be investigated and understood.

Topic 3.2: What are the Sun and the Moon and how are they linked to Earth?

Key Concepts
- The Sun is our nearest star.
- Interactions of Earth and the Sun make life possible.
- The Moon is our nearest neighbour in space.
- The Sun, Moon, and Earth interact to create eclipses.

Key Terms
magnetosphere (page 186)
aurora (page 186)
solar eclipse (page 190)
lunar eclipse (page 191)

Big Ideas
- Celestial objects in the solar system and universe have specific properties that can be investigated and understood.

Topic 3.3: What has space exploration taught us about our solar system?

Key Concepts
- The four inner Earth-like planets are small and rocky.
- The four outer gas-giant planets are large and ringed.
- Rocky chunks of various sizes make up the rest of the solar system.

Key Terms
inner planets (page 198)
outer planets (page 200)
asteroid (page 202)
meteoroid (page 203)
comet (page 203)

Big Ideas
- Celestial objects in the solar system and universe have specific properties that can be investigated and understood.

Topic 3.4: What role does Canada play in space exploration?

Key Concepts
- Canada contributes people and technology to explore space.
- Canada helps build the future of space exploration.

Big Ideas
- Celestial objects in the solar system and universe have specific properties that can be investigated and understood.
- Technologies developed for space exploration have practical applications on Earth.

Topic 3.5: How do we benefit from space exploration?

Key Concepts
- We develop technologies that shape our lives.
- We are challenged to think and act locally, globally, and universally.
- We gain a deeper appreciation for ourselves and our home planet.

Big Ideas
- Celestial objects in the solar system and universe have specific properties that can be investigated and understood.
- Technologies developed for space exploration have practical applications on Earth.

Unit 3 Projects

Inquiry Investigation: Space Thirst

Astronauts need water to drink, but water is heavy and costly to carry. Instead of bringing a supply for the whole mission, astronauts need to capture all sources of water (including waste water) and recycle it, the way cities do.

Initiate and Plan

1. For this investigation, use simulated urine made of glucose and food colouring. Your goal is to produce clean, clear, colourless water from the simulated urine.

2. Consider different methods that can be used to purify your water sample (for example, filtering or boiling). Design an investigation to test one of these methods.

3. Write a plan for your investigation. Include:
   - equipment and materials from the science lab
   - step-by-step method
   - safety precautions
   - a procedure for recording results
   - criteria for measuring and judging results

4. Have your teacher approve your plan.

Perform and Record

5. Conduct your investigation and record the results.

Analyze and Interpret

1. Describe the results of your investigation.

2. Were you satisfied with the quality of your final water sample? Explain why or why not.

3. Do you think your water sample is safe to drink? Why or why not?

4. How did knowing what was in your water sample influence your investigation?

5. Could your method be used to test a different type of waste water, for example, sweat or wash water?

6. Could your procedure be used on a space mission? Explain why or why not.

Communicate Your Findings

7. Present your results in a brief written or oral report. Include either a detailed outline of your procedure or a series of graphics.

> **Assessment Checklist**
> Review your project when you complete it.
> Did you...
> - ☑ determine which method of purifying water would work with your sample? **K/U**
> - ☑ create a plan using appropriate equipment and following all safety precautions? **T/I**
> - ☑ demonstrate that the goal of clean, clear, colourless water would be achieved as the result of your procedure? **T/I**
> - ☑ organize and record your results using an appropriate format? **C**
> - ☑ explain how your purification process would (or would not) work on a space mission? **A**

An Issue to Analyze: The Costs and Benefits of Space Travel

The transport of material into space via the space shuttle or existing rockets costs more than $22 000 for each kg sent! Researchers are investigating the possibility of using a space elevator, electromagnetic launch vehicles, or other technologies to do the job instead. What are the costs and benefits of new space travel technologies?

Issue

What are the costs and benefits of two different space travel technologies that could be used to carry equipment into space?

Initiate and Plan

1. As a class, find out what new technologies are being developed for space transport. Choose two you would like to investigate.
2. Make a list of the information you need to determine the costs and benefits of each space travel technology.
3. Record your information in chart form in your notebook.
4. Review your research plans with your teacher.

Perform and Record

5. Use the Internet or material provided by your teacher to find the information.
6. Record where you obtained your information and specific details about costs and benefits for the different technologies.

Analyze and Interpret

1. Evaluate the costs and potential benefits of each space travel technology.
2. Prepare a t-chart to show the costs and potential benefits of each technology.
3. Use your analysis to make a recommendation regarding the space travel technology that appears to have the greatest potential for development.

Communicate Your Findings

4. Prepare a brief written or oral report to describe the costs and benefits of two space travel technologies and recommend what you consider to be the best option.

Assessment Checklist

Review your project when you complete it. Did you...

- ✔ identify which two space travel technologies you would like to investigate? **K/U**
- ✔ use a variety of sources to gather the information you need to determine the costs and benefits of each space travel technology? **T/I**
- ✔ summarize your information in a t-chart to show the costs and potential benefits of each option for space travel? **C**
- ✔ make a recommendation regarding the space travel technology that appears to have the greatest potential for development? **A**

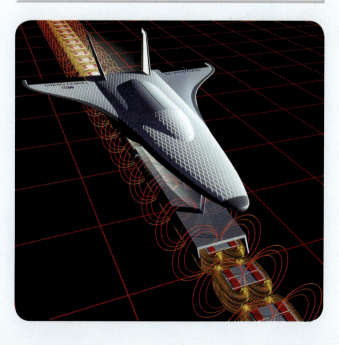

Unit 3 Review

Connect to the Big Ideas

You can answer these first two questions now or after you have completed the other Unit 3 Review questions.

1. Celestial objects in the solar system and universe have specific properties that can be investigated and understood. Canada's participation in space exploration can be traced back to 1839 with the establishment of the first observatory to study the Northern Lights. Prepare a timeline or a brief report on the role that Canada has played in the exploration of space. Include some details about Canada's role as a leader in space robotics and satellite communications.

2. Technologies developed for space exploration have practical applications on Earth. Use a cartoon, drawing, or story to describe how your life might have been different had humans not explored space.

Knowledge and Understanding K/U

3. In your notebook, draw and label a sketch that shows the relationship between stars and constellations.

4. Use a series of drawings to predict how the Little Dipper constellation would appear to move around Polaris during one evening.

5. Explain how the terms "Sun," "gravitational pull," "planets," and "solar system" are related.

6. Look back at **Figure 3.6**. Identify the major components of a galaxy such as this.

7. Identify the unit that scientists would use to measure the distance between Earth and each of the following celestial objects.
   a) Sun
   b) Uranus
   c) Betelgeuse (a star)
   d) Andromeda Galaxy

8. Use words, a picture, or a graphic organizer to explain the relationship among Earth's magnetosphere, the aurora borealis, and solar wind.

9. Use the diagram below to answer the questions that follow.

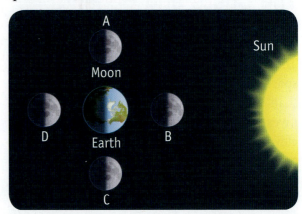

   a) In which position is the Moon at third quarter?
   b) In which position is the Moon at first quarter?
   c) For a lunar eclipse to occur, in which position in its orbit must the Moon be?
   d) For a solar eclipse to occur, in which position in its orbit must the Moon be?

10. Identify four features that the inner planets have in common with Earth.

11. Identify five features that the outer planets have in common with each other.

12. Name all the planets that have one or more moons.

13. On November 20, 2008, a large fireball raced across the skies of Alberta and Saskatchewan. More than 1000 pieces of the fireball were found in the Buzzard Coulee region of Alberta. Use this information to describe a meteoroid, a meteor, and a meteorite.

14. Identify the Canadian satellite that is "keeping an eye" on asteroids, and explain the significance of this satellite.

15. List two contributions that Canadian-developed technologies have made to the International Space Station (ISS).

16. In terms of space exploration, define the term "spinoff" and provide two examples of spinoffs that have affected your life.

Thinking and Investigation T/I

17. Based on the data in this graph, determine which planet is closer to Earth: Saturn or Jupiter. Show your work.

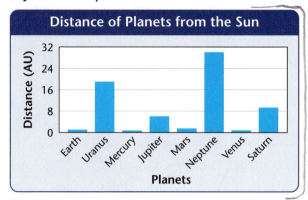

1 AU = 150 000 000 km

18. Stars can be classified by their surface temperatures. Based on the data in the table below, determine the colour of the following stars:

a) Betelgeuse (3100 K);
b) Arcturus (4500 K);
c) Antares (3000 K);
d) Rigel (11 000 K); and
e) the Sun (6000 K).

Classification of Stars by Colour and Temperature

Star Classification	Colour	Temperature (K)
O	blue	30 000–60 000
B	blue-white	11 000–30 000
A	white	7500–11 000
F	yellow-white	6000–7500
G	yellow	5000–6000
K	yellow-orange	3500–5000
M	red	less than 3500

19. Based on the diagram and the caption below, describe the relationship between the meteoroids associated with the Perseid meteor shower and the comet Swift-Tuttle.

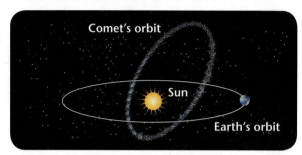

Comet Swift-Tuttle takes approximately 130 years to circle the Sun and was last observed in 1992. Earth passes through the debris left in space by the comet. Every summer, we see a dazzling display of meteoroids striking Earth's surface. This display is called the Perseid meteor shower.

Communication C

20. Refer to the heading on page 226: "We are Timekeepers." Suppose the information below about the Maya pyramid were added to that page. Write a heading for this information, and then explain its significance.

The Pyramid of Kukulkan (the feathered serpent) is a popular site for visitors today to Chichén Itzá, in the Yucatan region of Mexico. The pyramid has four stairways, each with 91 stairs. If you add the total number of stairs, plus the platform at the top, you get 365.

22. The Canadian Space Agency is working on a mission to bring soil samples back from Mars. Research the pros and cons of this mission. Then have a class discussion about sending humans to Mars.

Unit 3 Review

Application A

22. Based on the picture and caption below, create a diagram or a graphic organizer to summarize how the glass in a greenhouse is similar to Earth's atmosphere.

Many people use a greenhouse for growing plants. Solar energy (yellow arrow) reaches the greenhouse, passes through the glass, and is absorbed by the ground and plants inside. Some of this solar energy is given off by the ground and plants as heat (orange arrows). Most of this heat cannot escape through the glass (red arrows).

23. The space probe named *NEAR Shoemaker* was launched in February 1996. (NEAR stands for Near Earth Asteroid Rendezvous.) In February 2001, it landed on the rocky surface of the asteroid Eros, which is nearly 322 million kilometres from Earth. Identify some technological challenges that scientists must have faced attempting to land a space probe on the surface of this asteroid.

24. Astrobiology is the study of the origin, distribution, and future of life in the universe. Astrobiology includes the search for environments in our solar system and beyond our solar system that could support life. Based on the factors and resources needed to sustain life on Earth, predict three factors that astrobiologists would look for in their search for life on other planets. Explain your reasoning.

25. People from different cultures have different ways to explain and understand celestial objects. For instance, some have said that the northern lights (the *aurora borealis*) form a narrow, dangerous pathway that souls travel after the death of the physical body. Others have said that the shimmering lights could be reflected firelight from the edge of the world, sunlight reflected from ice caps, or reflected light from ice crystals in the sky. Use words, diagrams, or a graphic organizer to summarize how scientists today explain the northern lights.

26. Imagine that you are going to create a scale model of our solar system that compares the sizes of the planets in relation to each other and to the Sun.

   a) Identify the pages in Unit 3 that you would look on to find the information you need. (Hint: There are three pages that would be most helpful.)

   b) Name an object that you could use to represent the sizes of the following solar-system objects in your scale model. Explain why these objects are good choices.
   - an object to represent the size of Mercury
   - an object to represent the size of Earth
   - an object to represent the size of Jupiter
   - an object to represent the size of Uranus

27. Unit 3 began with a song written by Canadian artist, Buffy Sainte-Marie and a photograph from the movie, *E.T. The Extra-Terrestrial*, which is now used as the logo for the company, Amblin Entertainment.

   a) Suggest a different photograph that you think would be a good choice to start this unit.

   b) Suggest a different song that you think would be a good choice to start this unit.

   c) Explain how your choices of photograph and song work together well for this unit.

Literacy Test Prep

Read the selection below, and answer the questions that follow it.

The *Hubble Space Telescope* (*HST*) is a large space telescope that has revolutionized astronomy by providing extremely deep and clear views of the universe. It was launched in 1990 by the United States and has been in operation ever since. *MOST* (Microvariability and Oscillation of STars) is Canada's first space telescope. It was launched in 2003. The purpose of *MOST* is to study stars that are similar to the Sun. The table compares the *HST* and *MOST*.

A Comparison of the *HST* and *MOST*

	Hubble Space Telescope	MOST
Cost and year of launch	$10 billion; 1990	$10 million; 2003
Dimensions and mass	13.2 m × 4.2 m (about the size of a large school bus); 11 110 kg	65 × 65 × 30 cm (about the size of a suitcase); 60 kg
Instruments	7 science instruments; diameter of the primary optical mirror: 2.4 m	telescope mirror diameter: 15 cm
Altitude above Earth	600 km	820 km
Power source	solar energy	solar energy

Multiple Choice

In your notebook, record the best or most correct answer.

28. The short form for Canada's space telescope is
 a) CST
 b) HST
 c) MOST
 d) MST

29. Refer to the table on this page. One feature that both telescopes have in common is
 a) their power source
 b) their year of launch
 c) their size
 d) the number of instruments

30. The size of the HST is similar to that of
 a) a suitcase
 b) a large school bus
 c) a mirror
 d) a large van

31. The cost of building and launching *MOST* was
 a) $10 billion
 b) $13.2 million
 c) $10 milllion
 d) $60 million

32. *MOST*'s altitude above Earth is
 a) 600 km
 b) 820 m
 c) 220 km
 d) 820 km

Written Answer

33. In 2008, the tiny *MOST* space telescope received the Alouette Award for its outstanding contributions to space technology and research. Imagine that you interviewed one of the scientists that invented *MOST*. Write a one-paragraph newspaper article based on your interview. Include these elements in your article:
- a strong lead sentence
- a made-up name for the scientist
- a quote from the scientist describing how she or he felt after learning of the award. (Be creative!)

Unit 4 Electrical Applications

Big Ideas

- Electricity is a form of energy produced from a variety of non-renewable and renewable sources.
- The production and consumption of electrical energy has social, economic, and environmental implications.
- Static and current electricity have distinct properties that determine how they are used.

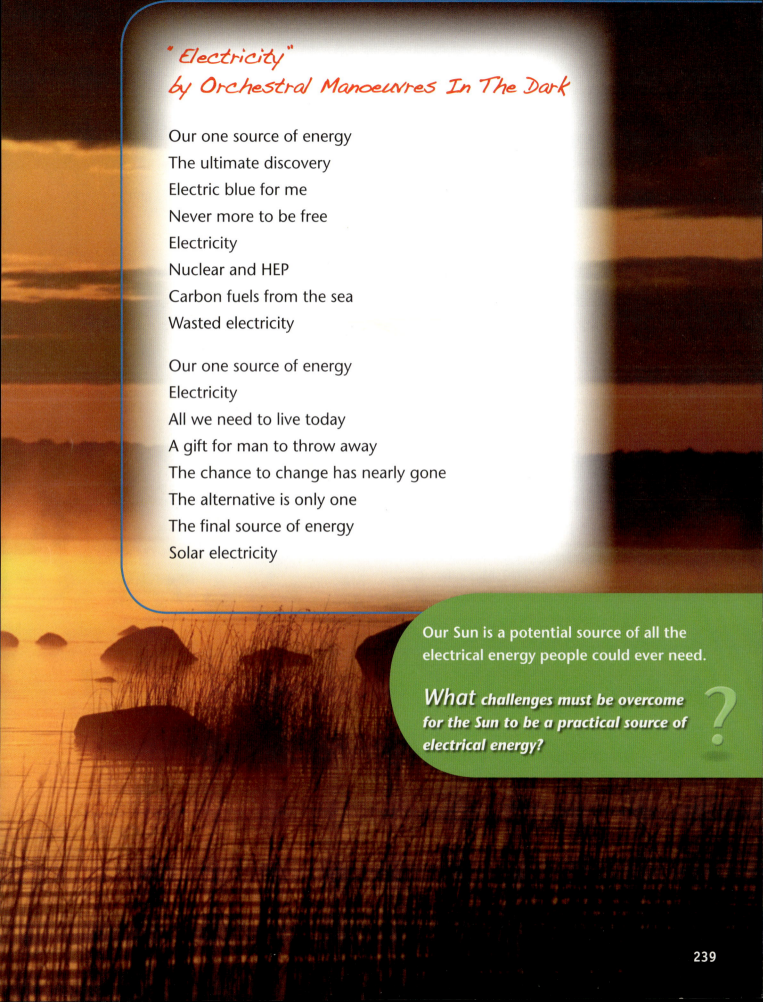

"Electricity" by Orchestral Manoeuvres In The Dark

Our one source of energy
The ultimate discovery
Electric blue for me
Never more to be free
Electricity
Nuclear and HEP
Carbon fuels from the sea
Wasted electricity

Our one source of energy
Electricity
All we need to live today
A gift for man to throw away
The chance to change has nearly gone
The alternative is only one
The final source of energy
Solar electricity

Our Sun is a potential source of all the electrical energy people could ever need.

What challenges must be overcome for the Sun to be a practical source of electrical energy?

Unit 4 At a Glance

In this unit you will learn that electricity is a form of energy that can be produced from a variety of non-renewable and renewable sources. You will also learn that the uses of static and current electricity are determined by their properties and that the production and consumption of electrical energy has social, economic, and environmental implications.

Topic 4.1: How do the sources used to generate electrical energy compare?
Key Concepts
- Different sources of energy can be converted into electrical energy.
- Renewable and non-renewable energy sources have advantages and disadvantages.

Topic 4.7: How can we conserve electrical energy at home?
Key Concepts
- Conserving energy at home requires an understanding of how energy is measured.
- People can conserve energy by making informed choices.

Electrical Applications

Topic 4.6: What features make an electrical circuit practical and safe?
Key Concepts
- Practical wiring for a building has many different parallel circuits.
- Circuit breakers and fuses prevent fires by opening a circuit with too much current.
- Higher-voltage circuits, larger cords and cables, and grounding help make home circuits safe.

Topic 4.5: What are series and parallel circuits and how are they different?
Key Concepts
- The current in a series circuit is the same at every point in the circuit.
- The current in each branch in a parallel circuit is less than the current through the source.
- The sum of the potential differences across each load in a series circuit equals the potential difference across the source.
- The potential difference across each branch in a parallel circuit is the same as the potential difference across the source.

Topic 4.2: What are charges and how do they behave?

Key Concepts
- Negative charges are electrons, and positive charges are protons.
- Opposite charges attract each other, and like charges repel each other.
- Negative charges can move through some materials but not others.

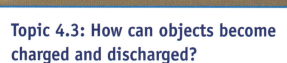

Topic 4.3: How can objects become charged and discharged?

Key Concepts
- Objects can become charged by contact and by induction.
- Charged objects can be discharged by sparking and by grounding.

Topic 4.4: How can people control and use the movement of charges?

Key Concepts
- A constant source of electrical energy can drive a steady current (flow of charges).
- An electric current carries energy from the source to an electrical device (a load) that converts it to a useful form.
- A source, load, and connecting wires can form a simple circuit.
- Meters can measure potential difference and current.
- Potential difference and resistance affect current.

Looking Ahead to the Unit 4 Project

At the end of this unit, you will do a project. The **Inquiry Investigation** involves planning a way to reduce energy consumption in a room of your home. The **Issue to Analyze** challenges you to find out which power companies in Ontario use "greener" energy sources. Read pages 328–329 With tips from your teacher, start your project planning folder now.

Unit 4

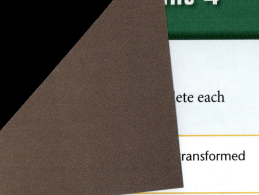

transformed

a) Electrical energy can be changed into other forms of energy. What is another word for "changed?"
b) What type of electrical energy is the build-up of an electrical charge on the surface of an object?
c) What type of electrical energy can be described as the movement of electrical charges?
d) In which type of circuit is there a single path for charges to flow?
e) In which type of circuit are there two or more paths for charges to flow?

2. Examine the picture below. Identify materials that are either insulators or conductors of electricity. Record your answers in a two-column chart with the headings "Insulating Materials" and "Conducting Materials."

3. Electrical energy can be converted into other forms of energy. These include mechanical energy, sound energy, light energy, and heat. Sometimes an electrical device converts electrical energy into more than one form of energy. Use the phrases in the box below to answer the questions about energy conversion in some common electrical devices. Write your answers in your notebook.

| mechanical energy | sound energy | light energy heat |

a) In an MP3 player, into what form of energy is electrical energy converted?
b) Into what forms of energy is electrical energy converted in a toaster oven?
c) Into what forms of energy is electrical energy converted in a TV set?
d) You turn on a blender to make a milkshake. Into what forms of energy is electrical energy converted?

Inquiry Check

4. **Plan** You predict that a rubber balloon will allow a static charge to build up on it, if the balloon is rubbed with another object. Design a test you could perform to show the balloon's ability to hold a static charge.

5. **Analyze** Use the circuit diagrams below to complete each sentence in your notebook.
   a) Which circuit, A or B, is a series circuit?
   b) Which circuit has two loads (electrical devices)?
   c) One of the circuits shows how electrical eenrgy flows in our homes. Identify this circuit.
   d) In which circuit will all loads (electrical devices) stop working if one of the loads burns out?

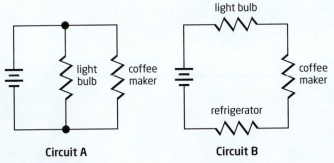

Circuit A   Circuit B

Numeracy and Literacy Check

7. **Analyze** The Ontario Energy Board sets the price of electricity in Ontario based on the time of day.

Electricity Use Pricing in Ontario (2008)

Day	Time	Use	Price Rate (cents per kW•h)
Weekends and holidays	All day	Non-peak	0.04
Summer weekdays (May 1st–Oct 31st)	7 A.M.–11 A.M.	Non-peak	0.07
	11 A.M.–5 P.M.	Peak	0.08
	5 P.M.–7 A.M.	Non-peak	0.05
Winter weekdays (Nov 1st–Apr 30th)	7 A.M.–11 A.M.	Peak	0.08
	11 A.M.–5 P.M.	Non-peak	0.07
	5 P.M.–8 P.M.	Peak	0.08
	8 P.M.–7 A.M.	Non-peak	0.05

a) When are the most expensive times for electricity use?
b) When are the least expensive times for electricity use?

8. **Writing** Write a one-minute school PA announcement encouraging students and staff to reduce their daily electricity use.

Topic 4.1

How do the sources used to generate electrical energy compare?

Key Concepts

- Different sources of energy can be converted into electrical energy.
- Renewable and non-renewable energy sources have advantages and disadvantages.

Key Skills

Literacy
Research

Key Terms

renewable energy source
non-renewable energy source

Starting Point Activity

Examine the two pictures shown here. Share your ideas about these questions.

1. How are the two pictures related?
2. For what tasks do you use electrical energy at home, at school, and in the world around you?
3. What impact does your use of electrical energy have on the environment?

Different sources of energy can be converted into electrical energy.

You use electrical energy many times every day. You probably have also seen power lines like the ones shown on the previous page. Power lines such as these carry electrical energy from great distances to your community. What is at the other end of those power lines? How is the electrical energy that travels along those power lines generated?

Most of the electrical energy in Canada is made by converting kinetic energy (the energy of motion) into electrical energy. This is done with a device called a generator. **Figure 4.1** shows a simple model of a generator used to make electrical energy.

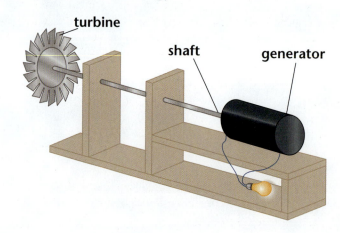

▶ **Figure 4.1** The key parts of a system used to produce electrical energy are a turbine, a shaft, and a generator. The shaft connects the turbine to a rotor inside the generator. As the turbine spins, it makes the shaft and rotor spin. The kinetic energy of the rotor is converted into electrical energy inside the generator.

Resources for Generating Electrical Energy

Any type of energy that can be used to turn a turbine can be used to generate electrical energy.

The vast majority of electrical energy used in Ontario (and Canada) is generated from three sources of energy: falling water, fossil fuels (coal, oil, natural gas), and uranium. **Figure 4.2** outlines the processes that convert these three sources to electrical energy.

Go to **scienceontario** to find out more

> **Literacy Focus**
>
> ### Activity 4.1
> **YOUR SOURCE OF ENERGY**
>
> Is the source of electrical energy for your community one of the three sources described in **Figure 4.2**? If so, which one? If not, what is the source of your electrical energy? If you don't know or aren't sure of the source of electrical energy for your community, check with your classmates and friends. If you're still not sure, how else can you find out?

A

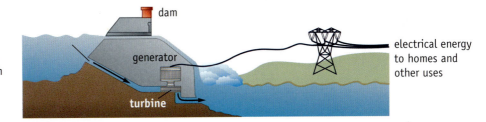

1. Water flowing through a dam spins giant turbines, which spin a generator to produce electrical energy.

Hydroelectric sources of energy convert the kinetic energy of moving water to electrical energy.

B

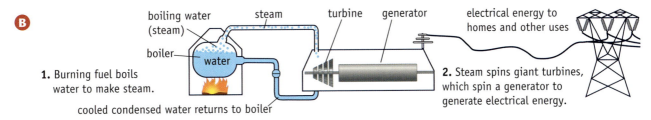

1. Burning fuel boils water to make steam.
2. Steam spins giant turbines, which spin a generator to generate electrical energy.

Thermoelectric sources of energy convert the chemical energy of burning fossil fuels (mostly coal) into heat that boils water into steam. The kinetic energy of the hot steam spins a turbine to generate electrical energy.

C

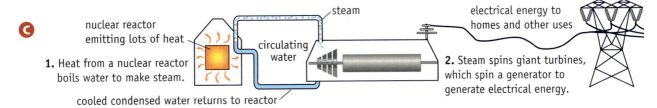

1. Heat from a nuclear reactor boils water to make steam.
2. Steam spins giant turbines, which spin a generator to generate electrical energy.

Nuclear sources of energy convert the energy released from the splitting of uranium atoms into heat that boils water into steam. The kinetic energy of the hot steam spins a turbine to generate electrical energy.

▲ **Figure 4.2** Compare the ways in which moving water, fossil fuels, and uranium are used to generate electrical energy.

LEARNING CHECK

1. Describe how most electrical energy is generated in Canada.
2. Refer to **Figure 4.1**. List the three key parts of a generator system, and briefly describe their functions.
3. Refer to **Figure 4.2**. Use a table to compare the similarities and differences among the use of moving water, the burning of fossil fuels, and nuclear reactions to generate electricity.
4. Describe how your life would be different if the electric generator had not been invented.

Energy sources have advantages and disadvantages.

renewable energy source: an energy source, such as moving water, that can be replaced or restocked within a human lifetime, or less

non-renewable energy source: an energy source, such as fossil fuels and uranium, that cannot be replaced or restocked within a human lifetime, or longer

You probably recall that sources of energy can be classified as renewable and non-renewable. A **renewable energy source** can be replaced or restocked within a short period of time. The water that is used to generate hydroelectric energy is an example of a renewable energy source. Wind, solar, biomass, tides, and geothermal (heat from below Earth's surface) are other examples of renewable energy sources. **Figure 4.3** shows two of these energy sources being converted into electrical energy.

A **non-renewable energy source** is one that cannot be replaced or restocked within a human lifetime. Fossil fuels and uranium are examples of non-renewable energy sources. Fossil fuels took millions of years to form on Earth, and millions of years will be needed to create new stocks of them. Uranium was formed in the explosions of stars before Earth was formed. Some of the uranium condensed with the dust that formed Earth billions of years ago. It can never be replaced. So when all available supplies of fossil fuels and uranium on Earth are used up, they are gone forever.

Whether it is renewable or not, every energy source has pros and cons—advantages and disadvantages.

▶ **Figure 4.3** Two energy sources used to generate electrical energy

A A generator is located directly behind the blades of each wind turbine. As the blades turn the shaft, electrical energy is generated.

B Solar cells convert solar energy directly into electrical energy.

LEARNING CHECK

1. Define the terms "renewable energy source" and "non-renewable energy source."
2. Explain why uranium is a non-renewable energy source and why water is a renewable energy source.
3. In your opinion, why is it important to assess every energy source in terms of its advantages and disadvantages?

INVESTIGATION LINK
Investigation 4A, on page 250

Activity 4.2
ASSESS THE SOURCES

Research Focus

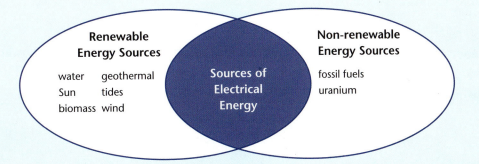

Environment
- What impact does the use of the energy source have on ecosystems?
- What impact does extracting or providing it have on ecosystems?

Society
- What impact does using this energy source have on the ability of people to live where and how they choose?
- How abundant are the supplies of this energy source?
- Is it possible or practical to use the energy everywhere or only in specific places?

Technology
- Is there technology to convert the energy source to electrical energy?
- Is the technology energy-efficient and cost-effective?
- Does the technology solve one problem by creating another?

What do you think is the best choice of energy source or combination of sources to generate electrical energy? To help you assess the sources and reach a decision, you will use a R.A.D.D. chart like the one below. The letters that make up R.A.D.D. stand for **R**esearch, **A**dvantages, **D**isadvantages, and **D**ecision.

Work in small groups. Your group will fill out a R.A.D.D. for one specific energy resource. Use the Venn diagram above and the questions that follow it to help make your decision. When you and the other groups are done, you will share your information with one another. Combine all of the information on one large chart, and use it to make a class decision. Decide which energy source or combination of sources will be best for Ontario. Give reasons to justify your decision.

Research the Energy Source	**Advantages** of the Energy Source	**Disadvantages** of the Energy Source	**Decision**
(List key characteristics of using the energy source for electrical energy. Include effects on the environment, on society, and the economy.)	(Identify the characteristics that make this energy source an attractive choice.)	(Identify the characteristics that make this energy source an unattractive choice.)	(Weigh the advantages and disadvantages for this one energy source, and decide if it represents a good choice or a bad choice.)

Investigation 4A

Skill Check

Initiating and Planning

Performing and Recording

✓ Analyzing and Interpreting

✓ Communicating

What You Need

Internet access

Leapin' 'Lectricity

What To Do

Read the information below. Use the website of Bullfrog Power and information at a local library to answer the "What Did You Find Out" questions that follow.

> To some North American Aboriginal people, such as the Sinixt Nation of British Columbia and the Seminole of Florida, frog is a symbol of survival, offering hope in times of fear and concern. The green colour of the frog also makes it a natural symbol of hope for the "green movement" and its promise of products and energy that are more environmentally sensitive and responsible.
>
> One company that has adopted the frog as a symbol is Bullfrog Power. Since 2005, residents and businesses in parts of Ontario, Alberta, and British Columbia can sign up with Bullfrog Power as their supplier of electrical energy. In return, the company promises energy that is generated from renewable energy sources such as wind and water.

What Did You Find Out?

1. Bullfrog Power refers to itself as a "100% green electricity provider."
   a) What do you think this means?
   b) Check the Bullfrog Power website to check your answer.
2. The company has the endorsement of Canada's EcoLogo program.
   a) What is the EcoLogo program?
   b) What types of power generation qualify for the EcoLogo?
3. Name three wind-farm sites that are providers for Bullfrog Power. Compare these sites in terms of the amount of electrical energy they can, or are predicted to, provide.

Frog Mountain, B.C.

Topic 4.1 Review

Key Concept Summary
- Different sources of energy can be converted into electrical energy.
- Renewable and non-renewable energy sources have advantages and disadvantages.

Review the Key Concepts

1. **K/U** Answer the question that is the title of this topic. Copy and complete the graphic organizer below in your notebook. Fill in four examples from the topic using key terms as well as your own words.

   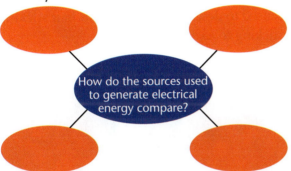

2. **C** In your notebook, draw and label the parts of a simple generator used to generate electrical energy. Include a brief explanation of how a generator works.

3. **K/U** a) Explain what a non-renewable energy source is.

   b) Explain what a renewable energy source is.

4. **K/U** Use a double bubble graphic organizer or a table to compare the similarities and differences between renewable energy sources and non-renewable energy sources.

5. **C** Use words, pictures, or a graphic organizer such as a flowchart to explain how falling water can be used to generate electrical energy.

6. **T/I** Refer to the pie graph.
   a) In your notebook, list the sources of electrical energy production in Canada from largest to smallest.

   b) Predict what you think the percentage of each source of electrical energy will be in 2015. Re-draw the pie graph to reflect these changes.

   c) Explain the reasons for your predictions.

   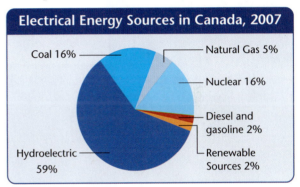

7. **A** When producers of "green power", such as Bullfrog Power, sell their electricity to electrical utility companies, the amount they are paid often does not cover all their costs to produce the electricity. Consumers who buy "green power" are paying a premium (extra fee) that is the difference between the standard price the producer gets for the sale of electricity and the cost to generate it.

   a) In your own words define the term "green power."

   b) Refer to **Figure 4.2**. Identify the different methods used to generate electrical energy that you think would qualify as "green power." Justify your answer.

   c) Would you be willing to pay a premium to buy electrical energy produced by green power? Write a blog explaining why you would or would not pay this premium.

Topic 4.2

What are charges and how do they behave?

Key Concepts

- Negative charges are electrons, and positive charges are protons.
- Opposite charges attract each other, and like charges repel each other.
- Negative charges can move through some materials but not others.

Key Skills

Inquiry
Literacy

Key Terms

negative charges
positive charges
electrically neutral
conductor
conductivity
insulator

In a clothes dryer, the clothes constantly rub together as the dryer drum turns. If no anti-cling product was used, some clothes will stick to each other when you take them out of the dryer. People commonly refer to this sticking effect as static cling. The same effect can happen when you put on or pull off a wool or polyester sweater and it rubs against your hair. You might even hear popping or crackling sounds as the fabric rubs against your hair. The term "static electricity" is used to describe effects such as hair sticking to fabric and clothes sticking to each other.

Starting Point Activity

Allergy Alert! If you are allergic to latex, do not touch the balloons.

1. Read the captions that go with the two pictures. Predict what will happen in each case. Give a reason for each prediction. Then test your predictions by doing the steps.

2. How did your results compare with your predictions?

3. What facts or experiences did you use to make your predictions?

- Rub a balloon on your hair.
- Move the balloon away, then bring it near your hair again.
- What happens?

- Rub the balloon on your hair again.
- Hold the balloon against a wall and let go.
- What happens?

Negative charges are electrons, and positive charges are protons.

> **Literacy Focus**
>
> ### Activity 4.3
> **REMEMBERING ATOMS**
>
> Sketch an atom. Add labels to show protons and electrons, and add a caption to describe their properties. Refer to Unit 2 of this textbook if you need to review the structure of an atom.

negative charges: the type of electrical charges that can be rubbed off a material

positive charges: the type of electrical charges that are left behind when negative charges are rubbed off a material

When you rub certain materials against each other, something comes off one of the materials and goes onto the other. As a result, the two materials attract each other. About 250 years ago, the American scientist Benjamin Franklin used the term **negative charges** to describe the "somethings" that were rubbed off materials. He also said that when negative charges were rubbed off a material, an excess of **positive charges** was left behind.

Today, we know that the negative charges are parts of the atom called electrons. The positive charges are protons, which are also parts of atoms. Protons cannot be rubbed off materials, but electrons can be. When electrons are rubbed off a material, it becomes positively charged. The material that gains the electrons becomes negatively charged. This process of charging materials by rubbing is called *charging by friction*.

Positively, Negatively, and Electrically Neutral

You just read that when electrons are rubbed off a material, it becomes positively charged. This means that the positive charges must have been there before the materials were rubbed. If the two materials had positive and negative charges before they were rubbed, why didn't they attract each other? The answer is: The materials had *equal numbers* of positive and negative charges before they were rubbed together. So all of the positive charges were balanced by all of the negative charges.

When objects have equal numbers of positive and negative charges, they are **electrically neutral**. It is only when the numbers of positive and negative charges are *unequal* and *unbalanced* that an object is electrically charged. **Figure 4.4** shows two materials before and after they were rubbed together to help you understand this idea.

electrically neutral: having equal numbers of positive charges and negative charges

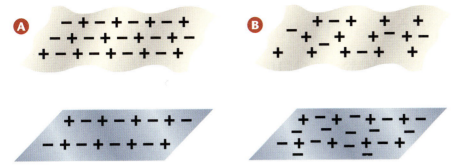

▶ **Figure 4.4** Diagram A shows a paper towel (top) and an acetate strip (bottom) that are electrically neutral. They have equal numbers of positive and negative charges. Diagram B shows the two materials after they are rubbed together. Now the paper towel has fewer negative charges, and the acetate strip has more negative charges. Therefore, the paper towel is positively charged and the acetate strip is negatively charged. Both materials still have the same number of positive charges before and after rubbing.

Explaining Why Rubbed Hair Is Attracted to a Balloon

You now have enough information to explain why your hair is attracted to a balloon after it has been rubbed on your hair. As shown in **Figure 4.5**, some negative charges (electrons) have been rubbed off your hair and are now on the balloon. The excess negative charges on the balloon attract the excess positive charges on your hair.

◀ **Figure 4.5** (A) Before rubbing a balloon on hair, both have equal numbers of positive and negative charges. They are electrically neutral. (B) When the balloon is rubbed on the hair, negative charges are transferred from the hair to the balloon. So now the hair has an excess of positive charges and the balloon has an excess of negative charges.

LEARNING CHECK

1. Explain the relationship among negative charges, positive charges, electrons, and protons.
2. Refer to **Figure 4.4**. Describe what happens when you rub two different materials together.
3. Make a sketch that shows the positive and negative charges on the following objects:
   a) a balloon that is neutrally charged
   b) a rubber duck that is negatively charged
   c) a wool sweater that is neutrally charged
   d) a wool sweater that is positively charged

ACTIVITY LINK
Activity 4.4, on page 256

INVESTIGATION LINK
Investigation 4B, on page 252

Opposite charges attract each other, and like charges repel each other.

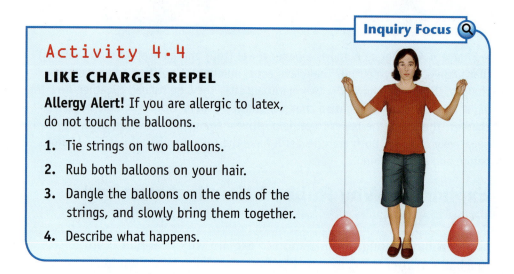

Activity 4.4
LIKE CHARGES REPEL

Allergy Alert! If you are allergic to latex, do not touch the balloons.

1. Tie strings on two balloons.
2. Rub both balloons on your hair.
3. Dangle the balloons on the ends of the strings, and slowly bring them together.
4. Describe what happens.

You have learned that objects that are positively charged attract objects that are negatively charged. If you brought two negatively charged balloons near each other, you saw that they repelled each other—pushed each other away. All of these ideas are summarized by the law of electric charge in the box below.

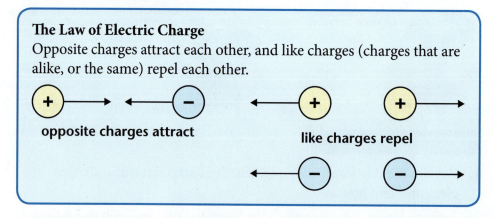

The Law of Electric Charge
Opposite charges attract each other, and like charges (charges that are alike, or the same) repel each other.

opposite charges attract   like charges repel

The law of electric charge applies to individual charges, such as electrons and protons. Every negative charge attracts every positive charge, every negative charge repels every other negative charge, and every positive charge repels every other positive charge. When you bring together objects that have an excess of charges, you see the overall result of the attractions and the repulsions.

Charged Objects and Neutral Objects

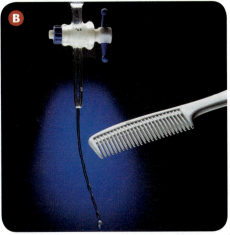

◀ **Figure 4.6** The balloon in photo A is charged, but the wall is not. The comb in photo B is charged, but the water is not.

Examine the photographs in **Figure 4.6**. You have probably seen charged balloons sticking to walls before. But have you ever seen a charged object such as a comb brought close to a stream of water? Notice how the water is actually attracted to the charged comb.

You can use the law of electric charge to explain why charged objects attract neutral objects. As you know, all neutral objects have an equal number of positive and negative charges. When you bring a charged object near a neutral object, the charges in the neutral object do not come off the object. Instead, they stretch out apart from each other. **Figure 4.7** shows this happening to a neutral wall when a charged balloon is close to it. The positive charges in the wall are pulled to the surface by the negative charges on the balloon. Then the positive charges in the wall are attracted to the negative charges on the balloon enough to make the balloon stick to the wall.

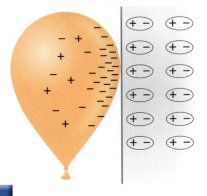

◀ **Figure 4.7** This diagram shows why a charged balloon sticks to an electrically neutral wall.

LEARNING CHECK

1. State the law of electric charge.
2. Make a labelled sketch, including charges, to show what happens when you bring two negatively charged balloons together.
3. Refer to **Figure 4.6B**. Make a labelled sketch, including charges, to explain why a stream of neutral water bends toward a positively charged comb.

ACTIVITY LINK
Activity 4.6, on page 260

INVESTIGATION LINK
Investigation 4B, on page 262

Negative charges can move through some materials but not others.

If an object is charged and the charges are repelling each other, why don't the charges just move away from each other? Actually, sometimes the charges *do* move. Whether or not they move depends on two things.

First, negative charges are the only kind of charge that can move in solid materials. (In other words, negative charges can move easily in solid materials, but positive charges can't.) However, both negative charges and positive charges are able to move in liquids.

Second, electrical charges can move only in certain types of materials. Any material in which electrical charges can move is called a **conductor**. How easily charges move through a material is called the material's **conductivity**. Any material in which electrical charges can't move easily is called an **insulator**. Refer to **Figure 4.8**.

Metals such as copper, iron, gold, and silver are conductors. In fact, most metals are conductors. Materials that are not metals, such as glass, plastic, wood, and Styrofoam®, are insulators. In fact, most non-metals (materials that are not metals) are insulators.

conductor: a material in which electrical charges can move easily

conductivity: an indication of how easily charges move within a material

insulator: a material in which electrical charges cannot move easily

A Electrical cords have a metal conductor, such as copper wires, in the centre. Charges can move easily along the conductor.

B The metal conductor must be covered by an insulator, such as plastic to prevent the charges from moving from one wire to the other. The insulator also prevents the charges from moving to other objects outside of the wire.

▲ **Figure 4.8** The electrical cords of most appliances and electrical tools are made of a conductor that is covered by an insulator.

LEARNING CHECK

1. Name the type of charge that can move only in a solid.
2. Use a t-chart to compare a conductor and an insulator.
3. Explain why electrical wires are covered by an insulator.

Activity 4.5

CONDUCTORS OR INSULATORS?

You will use a conductivity meter to test materials and find out if they are conductors or insulators. Many conductivity meters have two metal probes. When the meter is turned on, one probe is positive and the other is negative. When you touch the two probes to a material, the dial or digital display will tell you whether any charges are moving from one probe, through the material, and to the other probe. Numbers on the meter indicate the amount of movement of charges. Zero means that no charges are moving.

What You Need
conductivity meter
a variety of materials such as paper, glass, plastic, metal, water, salt solution, and sugar solution

What To Do
1. Make a t-chart, with the first column labelled *Material* and the second column labelled *Meter Reading*.
2. Turn on the conductivity meter. Choose a material to test its conductivity.
3. To test it, touch one probe to the material and—keeping that probe in contact—touch the other probe to the material. If the material you're testing is a liquid, put the tips of the probes in the liquid. Rinse and dry the probes before using the meter again.
4. Record the values on the meter in your t-chart for each material.
5. Make two lists. First list the materials that gave a value of zero, or very close to zero. Then list the materials that gave readings with values greater than zero.

What Did You Find Out?
1. Which materials are conductors and which are insulators? Explain how you know.
2. Which of the conductors has the greatest conductivity? The least?
3. Why might it be helpful to be able to rank the conductivity of different conductors? (Hint: Think about uses for conductors.)

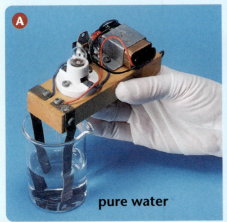
pure water

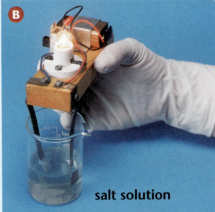
salt solution

The light bulb of the conductivity meter in A is not lit, so the liquid is a poor conductor. The light bulb in B is lit, so the liquid is a conductor.

Activity 4.6

Inquiry Focus

RUBBING AND STATIC ELECTRICITY

When some objects are rubbed together, static electricity is produced. How do these objects interact with each other and with objects that have not been rubbed?

What You Need

ebonite rod   fur   small pieces of paper

What To Do

1. Make a table like the one shown here to record your data.

   Table of Observations: Predicted interactions between an ebonite rod, fur, and pieces of paper when the rod is rubbed with fur

Object	Ebonite Rod	Fur	Pieces of Paper
ebonite rod		Your prediction: Your observations:	Your prediction: Your observations:
fur			Your prediction: Your observations:

2. Predict what will happen when you rub the ebonite rod with the fur, and then bring the rod and fur near each other and near small pieces of paper. Record your predictions in the table.

3. Place several bits of paper on your desk.

4. Rub the ebonite rod with the fur. Then separate them.

5. Slowly move the rod close to the fur. Record your observations of the interaction between the rod and the fur.

6. Move the rod close to the pieces of paper. Record your observations of the interaction between the rod and the paper.

7. Rub the ebonite rod with the fur again. Move the fur close to the pieces of paper. Record your observations of the interaction between the fur and the paper.

What Did You Find Out?

1. How did your predictions compare with your observations?

2. Imagine any situation in which you bring two charged objects (one positively charged and one negatively charged) near each other. Predict what would happen in such a situation. Record your prediction in the form of a statement.

3. Imagine any situation in which you bring a charged object near an uncharged (electrically neutral) object. Predict what would happen in such a situation. Record your prediction in the form of a statement.

Topic 4.2 Review

Key Concept Summary

- Negative charges are electrons, and positive charges are protons.
- Opposite charges attract each other, and like charges repel each other.
- Negative charges can move through some materials but not others.

Review the Key Concepts

1. **K/U** Answer the question that is the title of this topic. Copy and complete the graphic organizer below in your notebook. Fill in four examples from the topic using key terms as well as your own words.

   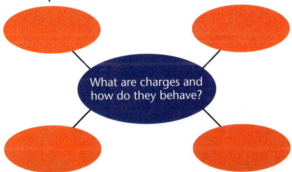

2. **K/U** Use a Venn diagram to compare the similarities and differences between negative charges and positive charges. Include how these charges are related to the parts of an atom.

3. **K/U** Define "insulators" and "conductors", and provide some common examples of each.

4. **C** In your notebook, draw and label a diagram of an electrically neutral balloon. Write a caption for your diagram, describing an electrically neutral object.

5. **C** Refer to **Figure 4.8B**. In your notebook, draw and label a sketch showing the cross section of an electrical cord. Indicate the part of the cord that is a conductor and the part that is an insulator.

6. **C** Using words or diagrams, explain what happens to the positive and negative electric charges when you rub two different materials.

7. **T/I** Refer to "The Law of Electric Charge" box on page 256. Copy the table into your notebook. (Do not write in your textbook.)
   a) Complete Row 3 of the table.
   b) Complete Row 4 of the table.

Summarizing the Law of Electric Charge

Row 1			
Row 2	− −	+ −	+ +
Row 3	The two objects repel each other.		
Row 4			Objects with like charges repel each other.

8. **A** Air purifiers rid the air of impurities and airborne irritants. Air purifiers can filter out smoke, mould, pollen, pet dander, and some viruses that can damage your lungs and immune system. One type of air purifier is called an electrostatic purifier. Electrostatic purifiers draw air into a home by means of a fan. The particles in the air pass by a series of electrical wires. The particles become charged when they pass by these wires. The air and the charged particles then pass over several metal plates. The metal plates carry the opposite charge. Based on this information and your understanding of the law of electric charge, draw a diagram showing how you think an electrostatic air purifier removes impurities from the air entering a home.

Investigation 4B

Charging Materials

Skill Check

Initiating and Planning

✓ Performing and Recording

✓ Analyzing and Interpreting

Communicating

In this Investigation, you will test several different materials for their ability to become charged. As well, you will learn another way to charge certain materials.

What To Do

1. Fold over a 5 mm "handle" on a 10 cm piece of invisible tape, and stick it to your desk. Do the same with a second piece of tape, and stick it beside the first.

2. Hold the two pieces of tape by their handles. Then quickly pull the pieces of tape off the desk, and hold the two pieces close together. Record your results.

Safety

- Handle the glass rod carefully.
- Do not use a chipped or broken glass rod.

What You Need

support stand

iron ring

invisible tape

plastic comb

acetate strip

glass rod

ebonite rod

polyethylene

fur

wool

3. Make a table like the one below to record your results.

Table of Observations: Interaction of objects with charged invisible tape

Object	Tape 1	Tape 2
glass rod rubbed with polyethylene		
ebonite rod rubbed with fur		
plastic comb rubbed with wool		
finger		

4. Repeat step 2, but place the second piece of tape *on top* of the first piece of tape.

5. Quickly pull both pieces of tape off the desk, and then pull them apart.

6. Stick the two pieces of tape on the iron ring, so that they hang a short distance apart.

7. Rub the glass rod with polyethylene. Bring the rod close to one of the pieces of tape. Record your results.

8. Bring the rod close to the other piece of tape. Record your results.

9. Repeat steps 7 and 8 with the ebonite rod rubbed with fur and the plastic comb rubbed with wool.

10. Bring your finger close to one of the pieces of tape. Record your results.

11. Bring your finger close to the other piece of tape. Record your results.

What Did You Find Out?

1. Use the law of electric charge to describe the charges on the tape in step 2.

2. Use the law of electric charge to describe the charges (same or opposite) on:
    a) the two pieces of tape and the glass rod rubbed with polyethylene.
    b) the two pieces of tape and the ebonite rod rubbed with fur
    c) the two pieces of tape and the plastic comb rubbed with wool
    d) the two pieces of tape and your finger

3. Did any of the materials attract both pieces of tape? Explain this result.

4. Did any of the materials repel both pieces of tape? Explain this result.

Topic 4.3

How can objects become charged and discharged?

Key Concepts

- Objects can become charged by contact and by induction.
- Charged objects can be discharged by sparking and by grounding.

Key Skills

Inquiry

Key Terms

charging by contact
electroscope
charging by induction
discharged
grounding

It's winter. You walk across a carpet while wearing wool socks or rubber-soled shoes. With each step you take, charges are building up on your body. You reach for a metal doorknob. Snap! You get a shock. If the room is dark, you might even see a spark jump between your finger and the doorknob. The object you reach for doesn't even have to be metal. If the amount of charge that your body has built up is great enough, you can cause a spark to jump between your finger and another person's skin.

Lightning looks like a giant spark. In fact, that's exactly what it is! Lightning just has millions of times more energy than the spark you experience when you touch a doorknob. Scientists still aren't sure how charges in clouds become separated. However, scientists do know that the bottoms of the clouds in a thunderstorm are negatively charged, and the tops of the clouds are positively charged. This separation of charges in the clouds leads to the stupendous spark or clusters of sparks that we call lightning.

Starting Point Activity

1. Hang a pith ball from a clamp that is attached to a retort stand. (In the science of electricity, a pith ball is a very light-weight ball that is covered with foil or metallic paint.)

2. Rub an ebonite rod with fur.

3. Very slowly, bring the charged rod close to the pith ball without touching it. Observe any movements of the pith ball before the rod touches it.

4. Let the rod touch the pith ball, and observe its movement. (Don't let anything touch the pith ball before the next step or you'll ruin the effect.)

5. Rub a glass rod with silk. Very slowly, bring the charged glass rod close to the pith ball. Observe the movements of the pith ball before and after the glass rod touches it.

6. Explain what you think happened to cause the movements of the pith ball in this activity.

Objects can become charged by contact and by induction.

There are three ways that you can charge objects. One way that you learned about in Topic 4.2 is by rubbing different materials together (charging by friction). The second way involves objects touching. The third way involves objects being close to each other but not touching.

Charging by Contact: Objects Touch

If you touch an uncharged object with a charged object, some of the charge will move onto the uncharged object. This process is called **charging by contact**. You cannot tell whether an object is charged just by looking at it. Therefore, it is convenient to use a device that changes in shape when it is charged. This device is called an electroscope.

The **electroscope** shown in **Figure 4.9** has a metal ball connected to one end of a metal rod. The other end of the metal rod has two lightweight, flexible, metal strips called metal leaves. When you touch a charged object to the metal ball, the ball is charged by contact. The parts of the electroscope are conductors. So the charges spread out over the metal ball, down the metal rod, and onto the metal leaves. When the leaves are charged, they repel each other and move apart. Therefore, when you see the leaves move apart, you know that they are charged. You can also use an electroscope to test an object to find out if the object is charged. If you touch the ball of the electroscope with an uncharged object, the leaves will not move.

charging by contact: causing a neutral object to become charged by touching it with a charged object

electroscope: a device that enables you to test an object and find out if it is charged

A When an electroscope is not charged, the leaves hang down.

B When you touch the electroscope with a negatively charged rod, the charge spreads out over all the metal parts. This gives the metal leaves like charges, so they repel each other.

C If the rod is positively charged, negative charges in the metal parts will be attracted to the rod and will move into it. As a result, the electroscope will be positively charged.

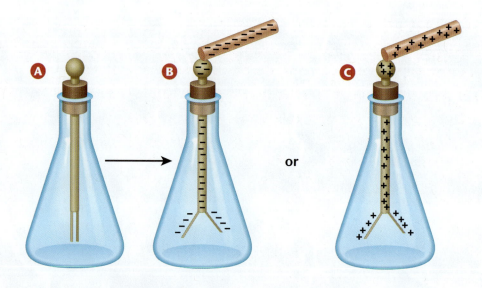

▲ **Figure 4.9** Charging an electroscope by contact. Only excess charges are shown. (There are many more positive and negative charges present.)

Charging by Induction: Objects Don't Touch

Figure 4.10 shows another way to charge an electroscope. This way involves bringing a charged rod near the metal ball but not touching it.

If, for example, the rod is negatively charged, the negative charges on the rod repel negative charges in the metal ball. The negative charges in the metal ball then move away from the rod and down to the leaves. Since the leaves become negatively charged, they repel each other. Positive charges are left behind on the metal ball. Now, without moving the rod, if you touch the ball of the electroscope with your finger, the positive charges on the ball will attract the negative charges on your finger. Because your body is a conductor, the negative charges will move onto the ball. When you remove your hand and move the rod away, the negative charges from your hand will remain on the electroscope, so it will now be charged. Charging an object in this way is called **charging by induction**.

charging by induction: causing a neutral object to become charged by bringing a charged object near to, but not touching, the object

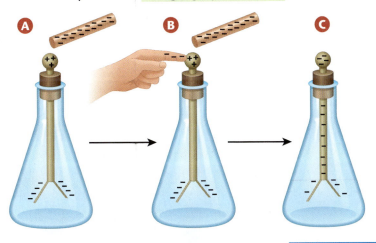

◀ **Figure 4.10** To charge an electroscope by induction, bring a charged rod near the ball. As shown in (A), if the rod is negatively charged, the electrons from the ball will be repelled and go down to the leaves. Now touch the ball with your finger, as shown in (B). Negative charges from your finger will be attracted by the positive charges on the ball of the electroscope. Some negative charges will move onto the ball. Remove your hand first and then the rod. As shown in (C), the electroscope now has an excess of negative charges even though the charged rod never touched it. (Only the excess charges are shown.) If the rod had been positively charged, all the motion of the negative charges would have occurred in the opposite direction.

Activity 4.7
Inquiry Focus

PREDICT THE RESULT

Predict what will happen if you follow the steps below. Sketch the electroscope leaves in each case. Will the leaves look different? If so, how?

1. You touch an uncharged electroscope with a negatively charged rod.
2. You touch an uncharged electroscope with a rod that carries a larger negative charge than the rod in step 1.

LEARNING CHECK

1. Describe how to charge an object by a) contact and b) induction.
2. Draw and label sketches to explain how to use an electroscope to predict if an object is charged.

Charged objects can be discharged by sparking and by grounding.

What is happening to the charged objects when a spark between them occurs? Sometimes, the attraction between the negative charges and positive charges becomes so great that charges actually jump across the gap between the objects. If all of the excess charge moved off the objects, the objects are said to be **discharged**. However, in a spark, usually only some of the charges leave the object. Figure 4.11 shows how charges that build up on a body and in a cloud can be discharged with a spark.

discharged: an object loses its excess charge

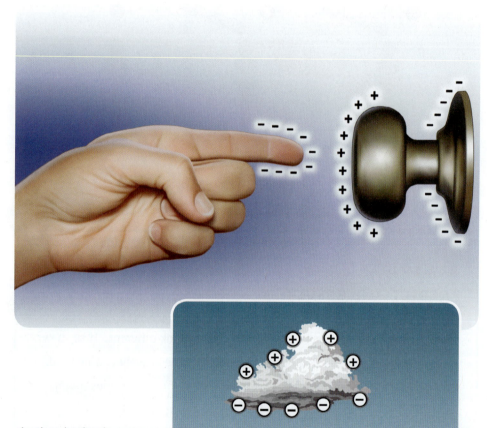

▶ **Figure 4.11A** When your shoes rub on a carpet, they rub negative charges off the carpet. Your body is a conductor, so the charges spread over the surface of your body. When your hand is near the doorknob, it induces a charge on the doorknob. If the attraction between the charges on your hand and the doorknob are great enough, the negative charges from your hand jump across the gap to the doorknob and you experience a shock.

▶ **Figure 4.11B** The turbulence in thunderclouds causes many collisions between water droplets and ice crystals. This somehow causes charges to separate so that the bottoms of the clouds become negatively charged and the tops of the clouds become positively charged. The negative bottom of the clouds induces a charge on the ground and objects on the ground. When the attractions between the negative charges in the clouds and positive charges on the ground are great enough, charges jump between the clouds and the ground.

How Grounding Discharges an Object

Lightning can be deadly, and a spark near flammable materials such as gasoline can start a fire. There must be a better way to discharge a charged object. Fortunately, there is. It is called grounding. You can use a conductor to carry charges away from an object safely, and guide them to a place where they will do no harm. That place is Earth. Earth is a giant conductor. Adding charges to Earth is like pouring a cup of water into the ocean. The change is not noticeable. Connecting a conductor to Earth with another conductor is called **grounding**. Figure 4.12 shows two ways that grounding protects people by providing a safe path for charges to follow.

grounding: connecting a conductor to Earth's surface so that charges can flow safely to the ground

◀ **Figure 4.12A** When gasoline flows from the hose of a tank truck, it rubs against the nozzle. This causes the metal of the truck to become charged. The rubber tires are insulators, so the charges cannot flow across them. A spark could ignite the gasoline. Therefore, the driver grounds the truck by leaning a metal rod against the truck so it can carry the charges away safely. Some trucks have dangling chains that are always touching the ground. This is another way to ground the truck.

▲ **Figure 4.12B** Lightning usually strikes the highest point in the area near a thunderstorm. Therefore, people attach metal lightning rods to the roofs of buildings. Metal conductors run from the lightning rod down to the ground and often go deep into the ground. When lightning strikes, it will most likely hit the lightning rod. The rod and conductors will carry the charges down to the ground, preventing the building from catching on fire.

LEARNING CHECK

1. Use a labelled sketch to show what happens when a spark occurs.
2. Describe two ways in which an object can become discharged.
3. How does grounding a metal object prevent sparks from occurring?

ACTIVITY LINK
Activity 4.9 on page 271

INVESTIGATION LINK
Investigation 4C on page 272

Inquiry Focus

Activity 4.8
CHARGING AN ELECTROSCOPE

In this activity, you will charge an electroscope by contact.

What You Need

electroscope
ebonite rod
fur

What To Do

1. Rub the ebonite rod with the fur.
2. Bring the rod near the ball of the electroscope. Observe any movement of the leaves. Record your observations.
3. Move the rod away from the ball of the electroscope, and observe any movement of the leaves. Record your observations.
4. Repeat steps 1 and 2, but touch the ball of the electroscope with the charged ebonite rod. Observe any motion of the leaves. Record your observations.
5. Move the rod away from the ball of the electroscope, and observe any motion of the leaves. Record your observations.

What Did You Find Out?

1. Explain why the leaves of the electroscope move as they do in each of the last four steps.
2. Make sketches for the electroscope for steps 2 through 5. Assume that the ebonite rod was negatively charged. Add charges to your sketches that show why the leaves of the electroscope moved as they did.

Inquiry Focus

Activity 4.9
GROUNDING AN ELECTROSCOPE

Because your body is a conductor, you can ground an object by touching it. Charges will flow along your body to your feet and into the ground. If the amount of charges flowing over your body is small, you will not notice anything. Find out what happens when you ground an electroscope under different conditions.

What You Need
electroscope
ebonite rod
fur

What To Do
1. Rub the ebonite rod with the fur.
2. Touch the ball of the electroscope with the charged rod. Observe the position of the electroscope leaves. Record your observations.
3. Touch the ball of the electroscope with your hand. Observe the final position of the leaves. Record your observations.
4. Rub the ebonite rod with the fur again. Hold the rod near the ball of the electroscope, but do not let it touch the ball. Observe the position of the leaves. Record your observations.
5. While the rod is near the ball of the electroscope, touch the ball with your hand. Remove your hand and then the rod. Observe the position of the leaves. Record your observations.

What Did You Find Out?
1. Make sketches of the positions of the leaves of the electroscope after the completion of each of the last four steps. Add charges to your sketches to show why the leaves were in the positions that you showed in your sketches.
2. Describe what happened when you touched the charged electroscope in step 2. Provide a possible explanation for your observations.
3. Describe the position of the leaves after you completed step 4. Provide a possible explanation for your observations.

Investigation 4C

Materials for Lightning Rods

Skill Check

Initiating and Planning

✓ Performing and Recording

✓ Analyzing and Interpreting

Communicating

In this activity, you will build an electroscope out of different materials to determine which of the materials would be best for making a lightning rod.

What To Do

1. Make a table like the one below to record your results.

Response of Foil Strips

Test Material	Observation
Metal	
Plastic	
Carbon	
Cardboard	
Wood	
Air (no material)	

What You Need

Styrofoam® cup

two 5 cm aluminum foil strips

invisible tape

2 cm × 10 cm test materials, including metal, plastic, carbon, cardboard, wood

ebonite rod

fur

2. Turn the cup upside down on your desk.
3. Tape two 5 cm strips of aluminum foil together at the same end so they hang like the leaves of an electroscope. Then tape the foil strips to one end of the 10 cm metal test material.
4. Lay the 10 cm metal test material on top of the cup, as shown in the picture to the left. Tape it to the cup if necessary.
5. Rub the ebonite rod with fur.
6. Touch the charged rod to the free end of the metal test material, as shown in the picture to the left.
7. Observe what happens to the aluminum foil strips. Record your observations in your table.
8. Touch the metal test material with your finger to discharge it.
9. Repeat steps 4 to 8 with each of the test materials listed in the table.
10. To test the effect of air, place the charged ebonite rod 10 cm away from the metal test material, without touching it, as shown in the diagram.

step 4

step 6

10 cm

step 10

What Did You Find Out?

1. Rank the materials from the best conductors (the ones that charged the foil strips the most) to the best insulators (the ones that did not charge the foil strips at all).
2. What material would make the best lightning rod? Explain.

Topic 4.3 Review

Key Concept Summary
- Objects can become charged by contact and by induction.
- Charged objects can be discharged by sparking and by grounding.

Review the Key Concepts

1. **K/U** Answer the question that is the title of this topic. Copy and complete the graphic organizer below in your notebook. Fill in four examples from the topic using key terms as well as your own words.

2. **K/U** Use a Venn diagram to compare the similarities and differences between charging an object by contact and charging an object by induction.

3. **K/U** Use a Venn diagram to compare the similarities and differences between discharging an object by sparking and discharging an object by grounding.

4. **C** Use a series of labelled diagrams to explain how you can charge an insulator such as a pith ball by contact. Make sure you write a clear caption for the diagrams as part of your explanation.

5. **A** Have you ever battled "static cling" when removing clothing from a clothes dryer on a dry day? Static cling in clothing is the result of different materials rubbing against each other in the clothes dryer. As a result, some of your clothing sticks together and you may feel, and maybe even see, small sparks. Use words, diagrams, or a graphic organizer to explain what is happening in this situation.

6. **A** Refer to **Figure 4.12A**. There have been serious accidents associated with fires spontaneously igniting when people were filling portable gas cans in the backs of pickup trucks equipped with plastic liners. In your own words, explain why one of the precautions you should follow is: "Remove the portable gas can from the pickup truck and place it on the ground a safe distance from the vehicle."

7. **C** Refer to **Figure 4.12B**. Use words, diagrams, or a cause-and-effect map to explain how a lightning rod protects a building from a lightning strike.

8. **A** The diagram below shows a neutrally charged comb and neutrally charged hair.
   a) Explain why both the hair and the comb shown in the diagram are neutrally charged.
   b) Imagine that the comb has been used to comb the hair. The comb is now negatively charged. Make a sketch to show the charges on the comb.
   c) Make another sketch to show the charges on the hair.

   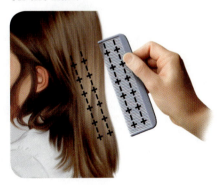

Topic 4.4 How can people control and use the movement of charges?

Key Concepts

- A constant source of electrical energy can drive a steady current (flow of charges).
- An electric current carries energy from the source to an electrical device (a load) that converts it to a useful form.
- A source, load, and connecting wires can form a simple circuit.
- Meters can measure potential difference and current.
- Potential difference and resistance affect current.

Key Skills

Inquiry
Literacy

Key Terms

source
potential difference
current
ampere
load
resistance
ohm
electrical circuit
voltmeter
ammeter

Western societies such as Canada depend greatly on electrical energy to run our homes, our communities, and the whole country. In three hours, you and the other 33 million Canadians use the same amount of electrical energy that is released in just one major electrical storm. That's a lot of energy. Unfortunately, there is no way to control the energy of a thunderstorm so that it can be used in lights and other common electrical devices. The electrical energy that you depend on to run a lamp, stove, or any other electrical device must be controlled and distributed to the device continuously. This is true whether you "plug into" electrical energy from a wall socket or "snap into" electrical energy from batteries.

Starting Point Activity

1. Obtain a 1.5 V battery, a flashlight bulb, and two insulated wires with alligator clips on the ends.
2. Decide how to connect the battery and light bulb in a way that will make the light bulb light up. (**Caution:** If the wires begin to get hot, disconnect them right away.)
3. Sketch the connection arrangement that made the light bulb light up.
4. Compare your sketch with those of other students. Describe how the successful arrangements are similar.

A constant source of electrical energy can drive a steady current (flow of charges).

source: device that supplies electrical energy

potential difference: change in the energy of a unit of charge after passing through a source or a load

The device that supplies electrical energy to operate any electrical equipment is called the **source**. Your source might be an electrical outlet or a battery. You would describe your source by a quantity called its **potential difference**. The symbol for potential difference is V.

Charges gain energy when they pass through a source. Potential difference describes how much their energy changes as they pass through a source. The potential difference across a source is the difference in the energy of a unit of charge entering one end of the source and the energy of a unit of charge leaving the other end of the source. **Figure 4.13** shows the potential difference across two common sources of electrical energy.

Some people also use the word "voltage" to mean potential difference. In fact, the SI unit of potential difference is the volt, and the symbol for the volt is V. This can get confusing, so be careful when you are reading and working with potential difference. Notice that the symbol for the quantity, potential difference, is in italics (V) and the symbol for the unit, the volt, is not in italics (V). They look similar, but they are not the same thing. The quantity of potential difference is measured in units of volts. In other words, V is measured in V.

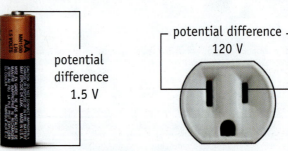

▶ **Figure 4.13** A normal electrical outlet in your home or classroom has a potential difference of 120 V. A typical battery such as an AA battery or AAA battery provides a potential difference of 1.5 V.

How Potential Difference across a Battery Works

Whenever charges are separated, there is a potential difference between the positive and negative charges. For example, when you rubbed different materials together and separated positive and negative charges, you generated a potential difference between the charged materials. Instead of rubbing, a battery uses energy from chemical changes to separate charges. **Figure 4.14** shows a model to help you understand the way this works. In the model, a miniature worker is carrying negative charges up a ladder and placing them on a shelf, leaving positive charges on the bottom. By separating charges in this way, the worker is generating a potential difference between the ends of the battery.

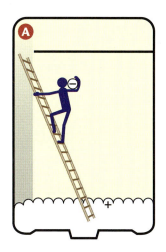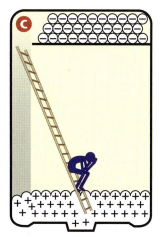

Figure 4.14 This model shows how charges are separated in a battery.

A The first charge is easy to carry up the ladder because only one pair of charges is being separated.

B After a few charges have been separated, all of the positive charges at the bottom are attracting the negative charge that the worker is carrying. Therefore, it takes more energy to carry each additional charge up the ladder.

C Eventually, the attraction between the positive and negative charges gets so strong that the worker cannot carry any more negative charges up the ladder. The potential difference across the battery represents the amount of energy it took to carry the last unit of charge up the ladder. For example, if this is a 1.5 V battery, it took 1.5 units of energy to carry that last unit of charge up the ladder.

Activity 4.10

Inquiry Focus

BATTERY SIZE

Five of these batteries have a potential difference of 1.5 V. One is a 3 V battery, one is a 6 V battery, and one is a 9 V battery. Can you match the voltages to the batteries?

What To Do

1. With a partner, choose a potential difference for each battery from the following possibilities: 1.5 V, 1.5 V, 1.5 V, 1.5 V, 1.5 V, 3 V, 6 V, 9 V.

2. Your teacher will give you a list of the correct values for the potential difference of each battery, so you can check your choices.

   a) Is the size of a battery related to its potential difference? Explain.

   b) What property do you think is affected by the size of a battery?

LEARNING CHECK

1. Describe what an electrical source is, and give two examples.
2. Define the terms "potential difference (V)" and "volt (V)."
3. What kind of energy does a battery use to separate charges?
4. Explain how the model in **Figure 4.14** describes the way that charges are separated in a battery.

An electric current carries energy from the source to an electrical device (a load) that converts it to a useful form.

When you want to use the energy from a source to make a device work, you must connect the source to the device. The connection is usually made with metal wire conductors that are covered by an insulator.

Current: The Flow of Charges

The energy from the source causes charges to move through the wires, carrying energy to the device. The moving charges are called an electric **current**. The symbol for current is *I*.

current: moving charges

Charges cannot build up in a conductor. The amount of current flowing into one end of the wire is the same as the amount of current flowing out the other end. In fact, the amount of current flowing past every point in the wire is the same. This idea is shown in **Figure 4.15**. You describe the amount of current flowing through a wire in units called **amperes**. The symbol for amperes is **A**.

ampere: unit of current

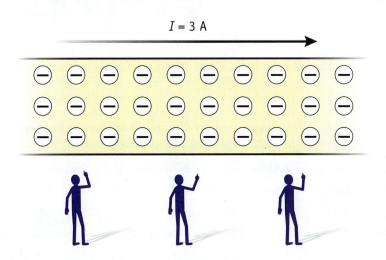

▶ **Figure 4.15** If it was possible to count charges as they flowed past any point in a wire, you would find that the count is the same at every point in the wire. The equation *I* = 3 A means that the current (*I*) is three amperes (3 A). In other words, three units of charge are passing each point in the wire each second.

The Load: An Energy Converter

Any device that converts electrical energy into a different form of energy is called a **load**. For example, a light bulb is a load. A light bulb converts electrical energy into light energy. A radio is also a load. A radio converts electrical energy into sound energy. A printer is also a load. A printer converts electrical energy into motion energy (mechanical energy). A load always converts electrical energy into another useful form of energy.

load: device that converts electrical energy into another form of energy

The Load and Resistance

A load resists the flow of current. This means that a load hinders the flow of the charges passing through it. **Figure 4.16** shows how resistance makes a filament-type light bulb light up.

The quantity that describes this hindering is called **resistance**. The symbol for resistance is **R**. And the unit used to measure resistance is the **ohm**. The symbol for an ohm is the Greek letter omega, Ω.

The filament in a light bulb is a good example of the resistance to the flow of charges. Use **Figure 4.16** to help you understand the role of resistance in making a filament-type light bulb light up.

As charges pass through a load, they lose energy. This happens because the electrical energy has been converted into another form of energy such as heat or light. Recall that the increase in the energy of a unit of charge passing through a source is the potential difference. You can also use potential difference to describe the amount of electrical energy that is lost by each unit of charge as it passes through a load. Both the increase and the decrease of the energy of a unit charge are described as a potential difference.

> **resistance:** describes the amount that current is hindered by load
>
> **ohm:** unit of resistance

▼ **Figure 4.16** A filament in a light bulb is a very thin wire. Many charges are trying to move from a much larger wire into it. As the charges move into the small wire, they collide with each other so hard that the filament gets very hot. This heat makes the filament glow.

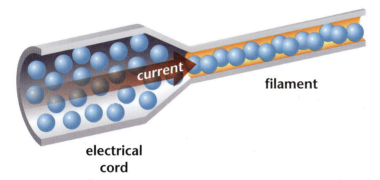

LEARNING CHECK

1. Identify what is required to connect a source of electricity to a load.
2. Use a main idea web or a spider map to show the relationships among the following terms: current, ampere, load, resistance, ohm. Add other terms if they help make your graphic organizer easier to understand.
3. Use the terms "source," "current," and "load" to describe how the heating element on an electric stove probably works. Refer to **Figure 4.16** to help you.

A source, load, and connecting wires form a simple circuit.

Now that you have learned about sources and loads, you can connect them and let the current flow. When a source, load, and conductor are connected in a way that can allow current to flow, it is called an **electrical circuit**. As you read on page 278, charges cannot build up in a conductor. All charges that leave the source must return to the source. Therefore, a circuit must form a closed loop. Examine the circuit picture in **Figure 4.17**. It uses a diagram like the ones on pages 278 and 279. **Figure 4.18** shows you how the circuit might look if you assembled it.

electrical circuit: at least a source, a load, and wires that allow current to flow

Remember, the "worker" represents a model of the energy that is provided by chemical changes in the battery.

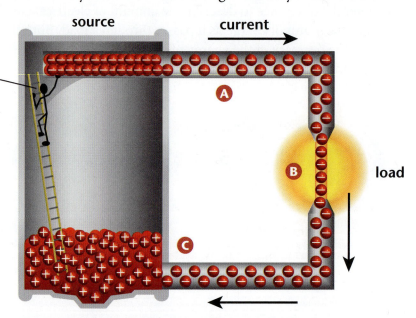

▲ **Figure 4.17** How charges move from a source and through a load in a working circuit

A The wires that are attached to the ends of the battery (source) already have charges in them that can move. The negatively charged end of the battery repels the negative charges in the wires. The positively charged end of the battery attracts the negative charges in the wire. As a result, the negative charges move along the conducting wires. Some negative charges that were inside the battery also start to move into the wire.

B As the negative charges pass through the load, they transfer some of their energy to the load. They then leave the load and return to the battery.

C When negative charges enter the battery, they combine with positive charges and make them neutral (no charge). The process results in a smaller number of negative charges at the negative end and a smaller number positive charges at the positive end of the battery. Now the "worker" inside the battery can carry more negative charges up the ladder and keep the number of separated positive and negative charges the same at all times.

▲ **Figure 4.18** This picture shows how the circuit in Figure 4.17 would look if you built it. When you look at a circuit like this one, try to imagine the negative charges moving in the battery, wires, and load.

A Switch: Controlling the Flow of Current

The circuit on page 280 is like the circuit in a flashlight. However, in a circuit connected this way, the light bulb would always be on. In a typical flashlight, you have a switch to turn the light on and off. To see how a switch works in a circuit, study **Figure 4.19**.

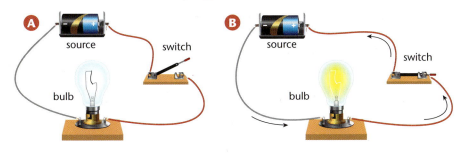

◀ **Figure 4.19** How a switch controls current in a circuit

A This circuit includes an open switch. With the switch open, the circuit is not a closed loop. Because negative charges cannot build up at any point in a circuit, the current cannot flow while the switch is open.

B When you close the switch, you complete the circuit. Current can flow, and the light bulb goes on.

Using Circuit Diagrams To Represent the Parts of a Circuit

Simple symbols are used to make it easier to draw circuits. **Table 1** lists the symbols for the basic parts of a circuit. The quantities used to describe the components and their units of measurement are also included. **Figure 4.20** shows how the symbols are used to draw a circuit with a battery, wires, a light bulb, and a switch.

Table 4.1 Symbols for Circuit Diagrams

Component of Circuit	Component Symbol	Quantity	Unit of Measurement
Source (battery)	—⊣|⊦—	Potential difference (V)	Volt (V)
Conducting wire	———	Current (I)	Ampere (A)
Load (resistance)	—⌇⌇⌇—		Ohm (Ω)
Switch: open closed	—•⁄•— —••—		

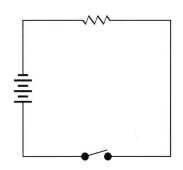

▲ **Figure 4.20** A circuit diagram uses the symbols for circuit components to make it simpler to communicate the parts of a circuit.

LEARNING CHECK

1. Use a flowchart to show charges moving from a source through a load in a working circuit.
2. Describe the role of a switch in an electrical circuit.
3. Refer to **Table 4.1**. Draw a circuit diagram for the circuit that you made in the Starting Point Activity for Topic 4.4.

Meters can measure potential difference and current.

voltmeter: instrument that measures the potential difference between two points in a circuit

The instrument you use to measure the potential difference across a battery or across a load is called a **voltmeter**. Because you always measure the difference in energy between two points in the circuit, the voltmeter must be connected to these two points. **Figure 4.21** shows you how to connect a voltmeter to a circuit to measure the potential difference across a load.

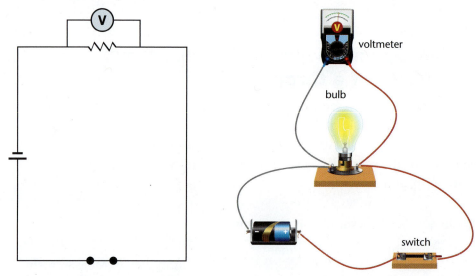

▶ **Figure 4.21** One side of the voltmeter is connected to one side of the load. The other side of the voltmeter is connected to the other side of the load. The readout tells you the value of the potential difference across the load. The potential difference across the load is the amount of energy that a unit of charge will lose while passing through the load.

ammeter: instrument that measures current in a circuit

The instrument that you use to measure the current passing through a circuit is called an **ammeter**. Current flows through every point in a circuit, so you must connect the ammeter into the circuit so the current flows through it. Study **Figure 4.22** to see how to connect the ammeter. Notice that the symbol for an ammeter is a circle with the letter "A" in the centre.

ACTIVITY LINK
Activity 4.11, on page 285

INVESTIGATION LINK
Investigation 4D, on page 286.

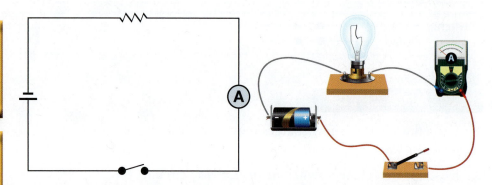

▲ **Figure 4.22** An ammeter is connected into the circuit so all of the current flows through the ammeter. In the picture, the switch is open, so no current is flowing. If you close the switch, the reading on the dial will tell you how much current is flowing through the circuit.

Potential difference and resistance affect current.

What would happen to the current in a circuit if you increased the potential difference of the source but kept the same resistance of the load? For example, in **Figure 4.23**, the first circuit has a 1.5 V battery and the second circuit has a 9.0 V battery. The resistance of the load stays the same. The higher potential difference of the source in the second diagram would give the charges more energy as they passed through it. If the charges have more energy and if the resistance to the flow of charges has not changed, the charges will flow more easily. The current will be larger.

INVESTIGATION LINK
Investigation 4E, on page 288
Investigation 4F, on page 290

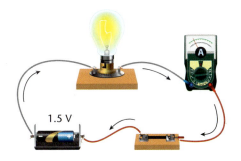 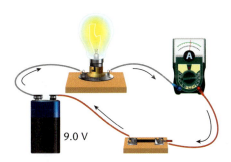

◀ **Figure 4.23** If you increase the potential difference of the source in a circuit but keep the resistance the same, the current will increase.

Now consider the opposite situation. What happen if you keep the potential difference of the source the same but increase the resistance in a circuit? You could just add another load or replace the load with another one that had a larger resistance. In **Figure 4.24**, the potential difference of each source is 1.5 V. In Circuit A, the resistance of the load is 100 Ω. In Circuit B, the resistance of the load is 500 Ω. Because the potential difference of the sources is the same, charges leaving the sources in the two circuits have the same amount of energy. But the resistance to the flow of charges in the second circuit is greater. So the current in the second circuit is smaller than the current in the first circuit.

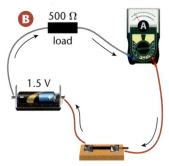

▲ **Figure 4.24** If you increase the resistance in a circuit and keep the potential difference the same, the current will decrease.

LEARNING CHECK

1. Compare a voltmeter with an ammeter in terms of what each measures and how each should be connected in a circuit.
2. Describe what happens to current if you increase the potential difference of the source but keep resistance the same.
3. Describe what happens to current if you increase resistance but keep the potential difference of the source the same.

Making a DIFFERENCE

When Vishvek Babbar was 13 years old, he was walking with an elderly relative during a visit to India when there was a power outage. He noticed that some people around him were having trouble walking safely in the dark. The incident motivated Vishvek to design an inexpensive electric cane to help elderly and disabled people walk in dark and crowded places.

The cane cost about $20.00 to build. It includes a light, an alarm that sounds when the cane hits an obstacle, and a light and sound system that activates if the user and the cane fall down. It runs on a battery. Vishvek thinks science students should always keep their eyes open for new ideas. "Good ideas never strike when you want them to," he says.

Have you ever witnessed a problem that could be solved with the help of a simple electrical device?

In Grade 9, Ghufran Siddiqui wanted to do a science project that could reduce waste and produce renewable energy. The Sarnia student chose to study biogas, a clean, renewable source of energy.

Ghufran built a working model of a biodigester to produce biogas. A biodigester uses bacteria to extract methane gas from plant waste. The chemical energy stored in methane can be converted into electrical energy. Meanwhile, the plant waste is changed to a rich soil to grow new plants. Ghufran used his model to calculate how large a full-scale biodigester would need to be to provide power to an average Canadian household during the summer months.

Since science projects can take time to work out and complete, Ghufran advises students who want to do projects to choose topics they are interested in. "The key is to do something you're curious about," he says.

What ideas about electrical applications could make a difference in your life and in the world around you?

Inquiry Focus

Activity 4.11

VOLTMETERS AND AMMETERS IN CIRCUITS

What To Do

Use the diagram below to check your understanding of the use of ammeters and voltmeters. Redraw the circuit, and add the ammeter or voltmeter as described in each of the following cases.

1. Redraw the circuit with the ammeter beside the source.
2. Redraw the circuit with the ammeter between load 1 and load 2.
3. Redraw the circuit with the voltmeter connected so it will measure the potential difference across the source.
4. Redraw the circuit with the voltmeter connected so it will measure the potential difference across load 1.
5. Redraw the circuit with the voltmeter connected so it will measure the potential difference across load 2.

What Did You Find Out?

1. Examine the circuits you have drawn. Use them to make these predictions.
   a) Make a prediction about the ammeter readings when the switch is open.
   b) Assume that the switch is closed. How do you predict the ammeter readings will compare for the two circuits described in steps 1 and 2?
   c) How do you think the voltmeter reading across load 1 will compare with the voltmeter reading across the battery?
   d) How do you think the voltmeter reading across load 2 will compare with the voltmeter reading across the battery?

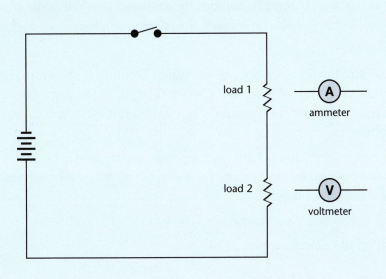

Investigation 4D

Skill Check

Initiating and Planning

✓ Performing and Recording

✓ Analyzing and Interpreting

✓ Communicating

Safety

- Before turning on any circuit, have your teacher check it to make sure that it is connected properly.

What You Need

ammeter

voltmeter

power supply

switch

2 identical light bulbs with bases

7 wire leads with alligator clips

Using Ammeters and Voltmeters

In this investigation, you will practise connecting and reading ammeters and voltmeters. You will also use a power supply, which is a source that provides a potential difference just like batteries do. However, you can choose the potential difference that you need to use.

What To Do

1. Make sure the power supply is off. Build Circuit A as shown in the diagram below. Circuit A includes the ammeter marked "ammeter for Circuit A," but it does not include the other ammeter.

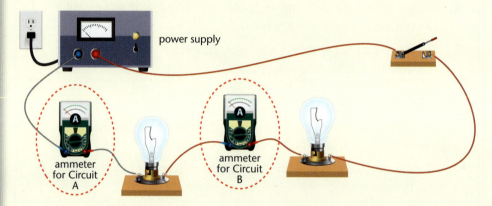

2. Leave the switch open. Turn on the power supply, and set it to 3.0 V. Read the ammeter, and record the value on the ammeter.

3. Close the switch. Read and record the value on the ammeter.

4. Turn off the power supply.

5. Make sure the power supply is off. Build Circuit B as shown in the diagram above. Circuit B includes the ammeter marked "ammeter for Circuit B," but it does not include the other ammeter.

6. Repeat steps 2, 3, and 4.

7. Make sure the power supply is off. Build Circuit C as shown in the diagram at the top of page 287. Circuit C includes the voltmeter marked "voltmeter for Circuit C," but it does not include the other voltmeters.

8. Leave the switch open. Turn on the power supply, and set it to 3.0 V. Read the voltmeter, and record the value of the potential difference across the power supply.

9. Close the switch. Read and record the value on the voltmeter.

10. Turn off the power supply.

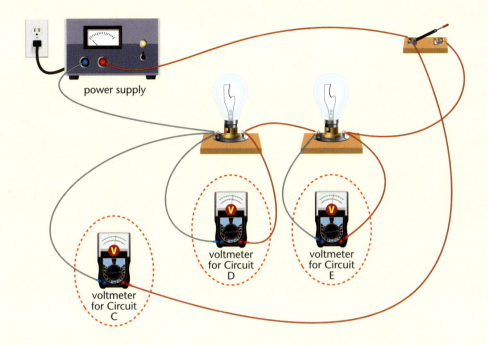

11. Make sure the power supply is off. Build Circuit D as shown in the diagram above. Circuit D includes the voltmeter marked "voltmeter for Circuit D," but it does not include the other voltmeters.

12. Repeat steps 8, 9, and 10.

13. Make sure the power supply is off. Build Circuit E as shown in the diagram above. Circuit E includes the voltmeter marked "voltmeter for Circuit E," but it does not include the other voltmeters.

14. Repeat steps 8, 9, and 10.

15. Take apart your circuit, and return the equipment to your teacher.

What Did You Find Out?

1. Look at your results for Circuits A and B. How do the values of the current compare for the two different positions of the ammeter?

2. Look at your results for Circuits C, D, and E. How does the potential difference you measured for each of the circuits compare when the switch was open? Suggest an explanation for your observations.

3. Look at your results for Circuits C, D, and E. How does the potential difference you measured for each of the circuits compare when the switch was closed? Make a general statement about the potential difference across an individual load compared with the potential difference across the source (power supply).

Investigation 4E

Skill Check

Initiating and Planning

✓ **Performing and Recording**

✓ **Analyzing and Interpreting**

✓ **Communicating**

Safety

What You Need

power supply

ammeter

switch

3 identical light bulbs with bases

6 wire leads with alligator clips

Observing the Effects of Resistance on Current

What happens to the current if you keep the potential difference of the source *the same* but *increase* the resistance of the load? You can't change the resistance of a light bulb. But you can increase the total resistance in a circuit by adding more light bulbs, one after the other. In this investigation, you will observe the current and the brightness of the light bulbs as you increase the number of light bulbs in the circuit. Before you start, make a table to record your data.

What To Do

1. With the switch open and the power supply off, build Circuit A.

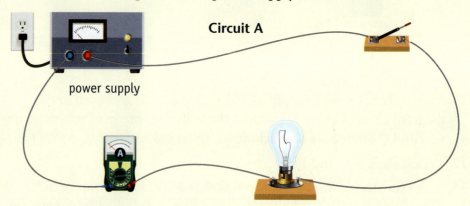

2. Close the switch. Turn on the power supply, and set it to 3.0 V.
3. Read and record the value on the ammeter.
4. Observe and record the brightness of the light bulb.
5. Turn off the power supply.
6. With the switch open and the power supply off, build Circuit B.

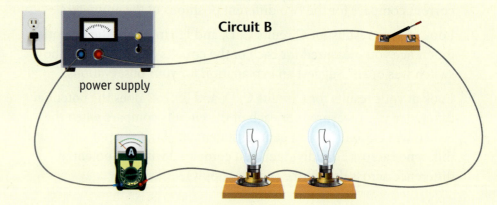

7. Repeat steps 2 and 3.
8. Compare the brightness of the two light bulbs with the brightness of the single light bulb you observed in Circuit A. Record the brightness.
9. Turn off the power supply.
10. With the switch open and the power supply off, build Circuit C.

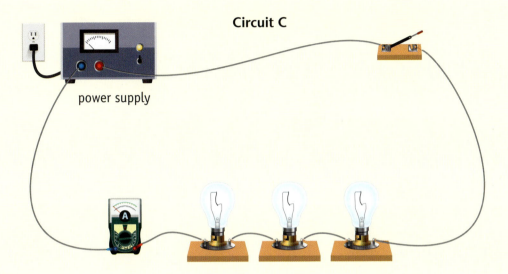

Circuit C

11. Repeat steps 2 and 3.
12. Compare the brightness of the three light bulbs with the brightness of the two light bulbs you observed in Circuit B. Record the brightness.
13. Take apart your circuit, and return the equipment to your teacher.

What Did You Find Out?

1. What happened to the current when you increased the resistance by adding more light bulbs, one after the other?
2. What happened to the brightness of the light bulbs as you added more light bulbs, one after the other?
3. Describe the relationship between current and resistance in a circuit when you keep the potential difference the same. For example, write a sentence that answers this question: "What happens to the current in a circuit when the resistance changes and the potential difference stays the same?"

Investigation 4F

Skill Check

✓ Initiating and Planning
✓ Performing and Recording
✓ Analyzing and Interpreting
✓ Communicating

Safety

What You Need

power supply
ammeter
switch
light bulb with base
4 wire leads with alligator clips

Potential Difference and Current

What will happen to the current if you keep the resistance of the load the same while you change the potential difference?

What To Do

1. Make a table like this, but with two extra rows for 4.5 V and 6.0 V.

Potential Difference	Current
1.5 V	
3 V	

2. Examine the diagram of the circuit. With the switch open and the power supply off, build this circuit.

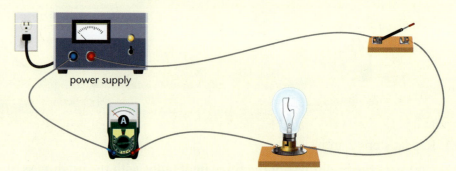

3. Close the switch. Turn on the power supply, and set it to 1.5 V.
4. Read the value on the ammeter and record it in your table.
5. Increase the potential difference of the power supply to 3.0 V. Read and record the value on the ammeter.
6. Repeat step 5 with a value of 4.5 V and then a value of 6.0 V.
7. Take apart your circuit and return the equipment to your teacher.
8. Draw a line graph of your data. Show current on the horizontal axis and potential difference on the vertical axis.

What Did You Find Out?

1. What happened to the current when you increased the potential difference of the source?
2. What did the line on your graph look like? Describe the appearance of your graph.
3. Describe the relationship between current and potential difference in a circuit with a load that stays the same.

Topic 4.4 Review

Key Concept Summary

- A constant source of electrical energy can drive a steady current (flow of charges).
- An electric current carries energy from the source to an electrical device (a load) that converts it to a useful form.
- A source, load, and connecting wires can form a simple circuit.
- Meters can measure potential difference and current.
- Potential difference and resistance affect current.

Review the Key Concepts

1. **K/U** Answer the question that is the title of this topic. Copy and complete the graphic organizer below in your notebook. Fill in four examples from the topic using key terms as well as your own words.

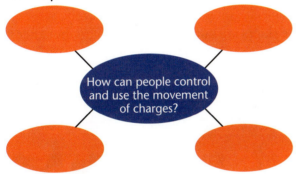

2. **K/U** Refer to **Figure 4.16**, and think about the filament. Use words, diagrams, or a graphic organizer to explain why an electric toaster is considered to be a "load."

3. **A** Imagine a fast-flowing river. If you were to describe the river's current, you might state the number of litres of water that flow past a certain point every minute. Compare the current in a river with the electrical current flowing in a conductor. Use words or pictures to describe a comparison of your own that illustrates what electrical current is.

4. **K/U** Refer to **Figure 4.18**. Use words, diagrams, or a graphic organizer to explain what happens to electric charges as they pass through a load.

5. **T/I** The graphs below show the relationship among potential difference of the source, resistance of the load, and current in a closed electric circuit. Based on the data in the graphs, explain how changing the potential difference of the source and changing the resistance of the load affect the current flowing through this circuit.

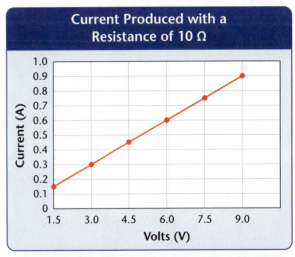

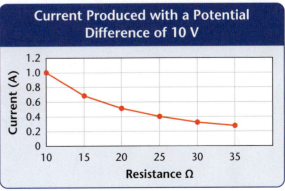

Topic 4.5

What are series and parallel circuits and how are they different?

Key Concepts

- The current in a series circuit is the same at every point in the circuit.
- The current in each branch in a parallel circuit is less than the current through the source.
- The sum of the potential differences across each load in a series circuit equals the potential difference across the source.
- The potential difference across each branch in a parallel circuit is the same as the potential difference across the source.

Key Terms

series circuit
parallel circuit

The skiers in Picture A are all following the same path. They are going up the ski lift one at a time. They are following each other down the ski run, one after the other. You could say that they are skiing in series. The word "series" refers to objects following along, one after the other, along a single path. With your finger, trace the path the skiers are taking to make sure that you see only one path they can follow.

The skiers in Picture B are not following the same path. They have a choice of three different runs to take while skiing down the hill. You could say that they are skiing in parallel. The word "parallel" refers to objects going side-by-side and in the same direction. All of the skiers are taking the same ski lift, but after they reach the top, they branch out onto the parallel runs.

Picture A

Starting Point Activity

Examine Picture A and Picture B closely. Use an organizer such as this one to compare all the features of the two pictures that are the same and all the features that are different.

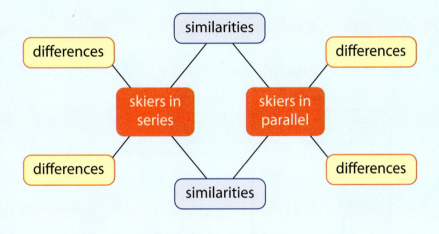

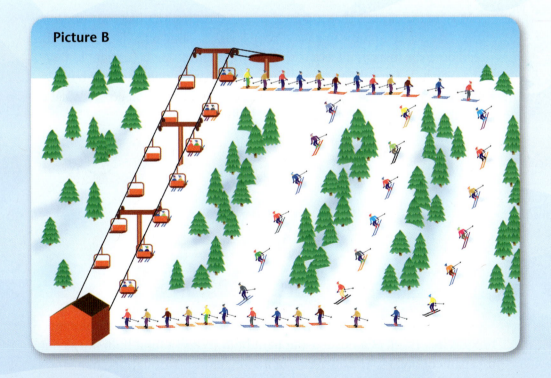

Picture B

The current in a series circuit is the same at every point in the circuit.

series circuit: a circuit that has only one path for current to follow

All of the circuits that you have worked with so far are series circuits. A **series circuit** has only *one path* for the current to follow. **Figure 4.25** reviews the circuit on page 280. You now know that it is a series circuit.

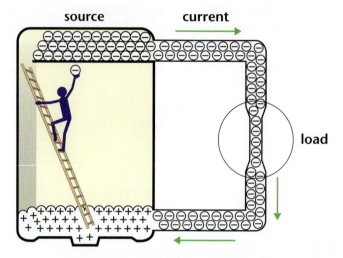

▶ **Figure 4.25** The current is the same at every point along a series circuit. Although a load resists the flow of current and the charges lose energy as they pass through, the current is still moving at the same rate after it leaves a load as it was when it entered the load.

Most of the circuits that you have analyzed in this unit had only one load. However, a series circuit can have several loads, as shown in **Figure 4.26**. You would describe this circuit as a 6 V battery with three loads (or light bulbs) and a switch that are all in series. When the switch is open, the light bulbs will all be off. If you close the switch, all of the light bulbs will go on. What do you think will happen if one of the light bulbs burns out?

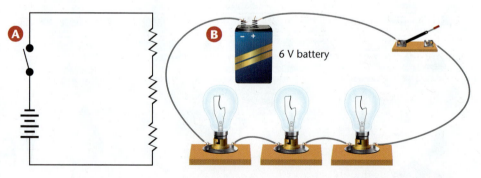

▶ **Figure 4.26** The circuit diagram (A) shows a battery, a switch, and three loads that have resistance. The circuit (B) shows a large 6 V battery, a switch, and three light bulbs. Study A and B until you can see that both types of diagrams represent the same circuit.

In a series circuit, all of the loads must be on and working at the same time. Why? Think of a burned-out light bulb. When a light bulb burns out, the filament breaks. The current can no longer flow through the filament. As a result, the current cannot flow anywhere in the circuit. The same thing would happen if you removed a light bulb from a series circuit. In a series circuit, all of the light bulbs, or other kinds of loads, must be on and working for charges to flow through the circuit.

The current in each branch in a parallel circuit is less than the current through the source.

Examine the circuit in **Figure 4.27**. With your finger, trace the paths of the negative charges leaving the source. Notice how similar these pathways are to the ski hill on page 293. The source is like the ski lift. Instead of lifting skiers to the top of the run, the source is raising negative charges to a higher level of energy. The charges are like the skiers. From the top of the hill, the skiers have a choice of three different runs to ski down. The negative charges have three different paths that they can follow to return to the positive end of the source. A circuit that has two or more paths for the current to follow is called a **parallel circuit**.

parallel circuit: a circuit that has two or more paths for current to follow

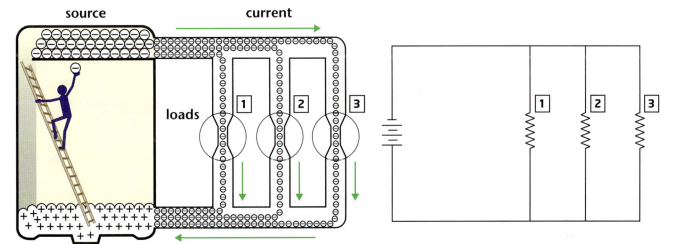

▲ **Figure 4.27** These two diagrams show the same parallel circuit. Compare them closely to be sure you recognize why they are the same.

Now think about what happens to the amount of current as it leaves the source and reaches a point where the current separates. Some of the current goes down toward the first load and the rest continues on. Then the current divides again and follows paths 2 and 3. The amount of current following each of the paths is less than the amount of current that left the source. Because the current separates before passing through the loads, there is *less current* passing through the loads than there is leaving or entering the source.

LEARNING CHECK

1. Refer to **Figures 4.26** and **4.27**. Draw and label a diagram of
    a) a series circuit with a source, a switch, and two lights
    b) a parallel circuit with a source, a switch, and two lights
2. Explain what you think would happen if you had 10 garden lights wired in series and one of the bulbs burned out.

The sum of the potential differences across each load in a series circuit equals the potential difference across the source.

Assume that you measure the potential difference between points A and B in the series circuit shown in **Figure 4.28**. If the meter reads 6 V, what does this really mean? It means two things. First, it means that the potential difference across the battery is 6 V. Second, it means that the potential difference across all three loads together (L_1, L_2, and L_3) is 6 V. It does not matter which path you follow from point A to point B. The potential difference between A and B must be the same.

As you know, when a charge passes through any load, it loses energy. So there is a potential difference across each of the three loads in **Figure 4.28**. If you measured the potential difference across each of the loads and added them together, the sum would be 6 V.

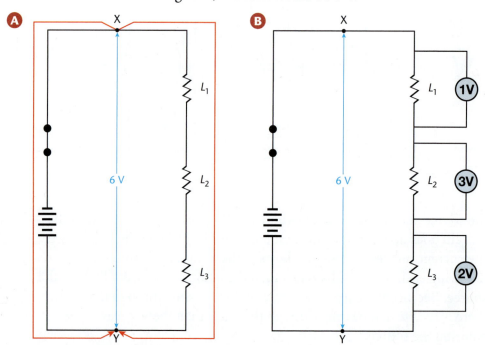

▲ **Figure 4.28**

A If you connect a voltmeter at points X and Y in the circuit, you are measuring both the potential difference across the battery and across all three of the loads.

B If, instead, you connect a voltmeter across each load separately, you could get readings such as 1 V, 3 V, and 2 V. The sum of the values on each meter would then be 1 V + 3 V + 2 V = 6 V. So the potential difference across each of the loads is lower than the potential difference across the source. But the sum of the potential differences across all the loads is equal to the potential difference across the source.

INVESTIGATION LINK
Investigation 4G, on page 298

The potential difference across each branch in a parallel circuit is the same as the potential difference across the source.

Examine the parallel circuit shown in **Figure 4.29**. If you measure the potential difference between points A and B, what are you really measuring? Notice that there are four different paths to get from A to B. The red path goes across the source. Each of the green paths goes across one of the loads. If the reading on the meter is 6 V, this means that the potential difference across the source and across each of the loads is 6 V. In a parallel circuit, the potential difference across the source and each of the branches is the same.

> **INVESTIGATION LINK**
> Investigation 4H, on page 300

◀ **Figure 4.29** If you placed a voltmeter across the source and across each of the loads separately, you would get the following results. $V_S = 6$ V, $V_1 = 6$ V, $V_2 = 6$ V, and $V_3 = 6$ V.

LEARNING CHECK

1. In Circuit 1, the battery provides a potential difference of 12 V. If the reading on voltmeter V_1 is 4 V, what is the reading on voltmeter V_2?

2. In Circuit 1, the reading on ammeter A_1 is 2 A. What is the reading on ammeter A_2?

3. In Circuit 2, the battery provides a potential difference of 12 V. What are the readings on voltmeters V_1, V_2, and V_3?

4. In Circuit 2, ammeter A_1 reads 3 A, ammeter A_2 reads 2 A, and ammeter A_3 reads 4 A. What is the reading on ammeter A_4?

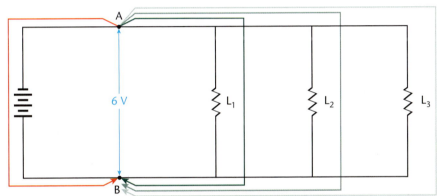

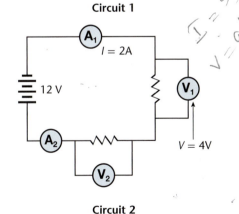

Investigation 4G

Skill Check

Initiating and Planning

✓ Performing and Recording

✓ Analyzing and Interpreting

✓ Communicating

Safety

- Disconnect any wire that gets too hot.
- When using a power supply, always follow your teacher's instructions.
- Always unplug or turn off a power supply before working on a circuit.

What You Need

power supply

switch

ammeter

3 flashlight bulbs with bases

3 voltmeters (or use one and move it for each measurement)

12 wire leads with alligator clips

Observing Characteristics of Series Circuits

In this investigation, you will assemble series circuits and observe their characteristics. You will compare the current in the circuits and potential difference across the loads (flashlight bulbs) while you vary the number of loads.

What To Do

1. Make a table like this one. Give it a suitable title.

Circuit	Current	Potential Difference			Brightness of Bulbs		
		Bulb 1	Bulb 2	Bulb 3	Bulb 1	Bulb 2	Bulb 3
A			✕	✕		✕	✕
B				✕			✕
C							

2. You are going to build three circuits. These circuits will be the same, except they will have different numbers of loads and voltmeters.
   - Circuit A will have one load (one bulb) and one voltmeter.
   - Circuit B will have two loads (two bulbs) and two voltmeters.
   - Circuit C will have three loads (three bulbs) and three voltmeters.

   Refer to the diagram on page 298 for the three circuits you will build.

3. Draw a circuit diagram for Circuit A (one bulb). With the power supply off and the switch open, build this circuit.

4. Turn on the power supply and set the potential difference to 6 V. Close the switch.

5. Observe and record the current on the ammeter and the potential difference across the bulb.

6. Observe the brightness of the bulb. Make a note in the table that will help you remember the brightness so you can compare the brightness of the bulbs in the next two circuits that you build.

7. Turn off the power supply and open the switch. Add a second bulb, in series with the first, and a second voltmeter. This is Circuit B. Draw a circuit diagram of Circuit B.

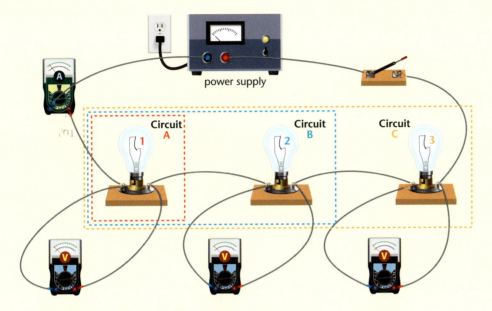

8. Repeat steps 4 to 6 with the new circuit. When describing the brightness of the two bulbs, note whether they are brighter, the same, or less bright than the bulb in Circuit A.

9. Turn off the power supply, and open the switch. Add a third bulb, in series with the first two bulbs, and a third voltmeter. This is Circuit C. Draw a circuit diagram of Circuit C.

10. Repeat steps 4 to 6 with Circuit C.

11. Turn off the power supply and open the switch. Remove one of the bulbs from its base, but leave the base in the circuit. Turn on the power supply and close the switch. Note what happens.

What Did You Find Out?

1. When you added more loads (bulbs) to your circuit, did the current reading go up, stay the same, or go down?

2. **a)** What happened to the brightness of the bulbs as you added more bulbs to the circuit?
   b) What is the relationship between the current through the bulbs and their brightness?

3. When you added more loads (bulbs) to your circuit, did the potential difference across each individual bulb go up, stay the same, or go down?

4. In step 11, you removed one bulb from its base and turned the power supply back on and closed the switch. What happened? Explain why it happened.

5. Write a statement describing how practical it would be to put many loads in a series circuit.

Investigation 4H

Skill Check

Initiating and Planning

✓ Performing and Recording

✓ Analyzing and Interpreting

✓ Communicating

Safety

- Disconnect any wire that gets too hot.
- When using a power supply, carefully follow your teacher's instructions.
- Always unplug or turn off a power supply before working on a circuit.

What You Need

power supply

switch

ammeter

3 flashlight bulbs with bases

3 voltmeters (If there are not enough, use one and move it for each measurement.)

13 leads with alligator clips

Observing Characteristics of Parallel Circuits

In this investigation, you will start with a simple parallel circuit with one load (bulb), and then you will add two more loads, one at a time, in parallel with the bulb or bulbs already present. You will observe and measure the current near the source and the potential difference across each load.

Examine the circuit shown here. Notice that loads 2 and 3 are not yet connected to the first load and the power supply. This is the arrangement for your first measurements. You will add the second and third loads in parallel and make measurements after adding each one to the circuit.

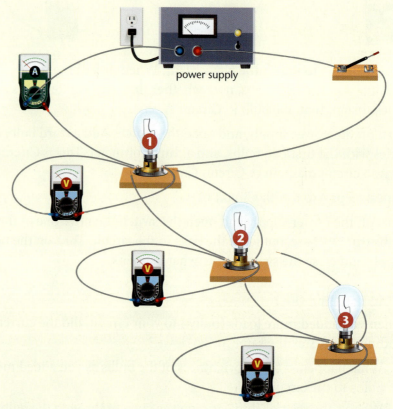

What To Do

1. Make a table like this one. Give it a suitable title.

Circuit	Current	Potential Difference			Brightness of Bulbs		
		Load 1	Load 2	Load 3	Load 1	Load 2	Load 3
A			╳	╳		╳	╳
B				╳			╳
C							

300 MHR • UNIT 4 ELECTRICAL APPLICATIONS

2. Examine the circuit shown in the diagram. With the power supply off and the switch open, build this circuit.
3. Turn on the power supply and set the potential difference to 6.0 V. Close the switch.
4. Observe and record the current reading on the ammeter and the potential difference across the bulb.
5. Observe the brightness of the bulb. Make a note in the table that will help you remember the brightness so you can compare the brightness of the bulbs in the next two circuits that you build.
6. Turn off the power supply and open the switch. Add a second bulb and voltmeter to your circuit so they are in parallel with the first. You can do this by connecting the two ends of the leads from bulb 2 to the sides of bulb 1.
7. Repeat steps 3 to 5 with the new circuit. When describing the brightness of the two bulbs, note whether they are brighter, the same, or less bright than the bulb in the first circuit.
8. Turn off the power supply and open the switch. Add a third bulb and voltmeter to your circuit in parallel with the first two. Once again, study the diagram and connect the ends of the leads from bulb 3 to the sides of bulb 2.
9. Repeat steps 3 to 5 with the third circuit.
10. Turn off the power supply and open the switch. Remove one of the bulbs from its base but leave the base in the circuit. Turn on the power supply and close the switch. Note what happens.

What Did You Find Out?

1. When you added more loads (bulbs) in parallel to your circuit, did the current beside the power supply go up, stay the same, or go down?
2. Is the amount of current that is passing through the power supply the same, higher, or lower than the current that is passing through an individual bulb? Explain.
3. What happened to the brightness of the bulbs as you added more bulbs to the circuit?
4. When you added more loads (bulbs) to your circuit, did the potential difference across each individual bulb go up, stay the same, or go down?
5. In step 10, you removed one bulb from its base and then turned the power supply back on and closed the switch. Explain what happened.
6. Write a statement describing how practical it would be to connect many loads in a parallel circuit.

Strange Tales Of Science

SPARKS OF GENIUS

POWER SERIES

PREMIUM ISSUE

There are some images that immediately "spark" our curiosity, and this is one of them. Who is the man reading in this image, seemingly without a care in the world? Shouldn't he at least be a little "shocked" by the electrical currents that are electrifying the air around him? The man's name is Nikola Tesla. Tesla was a genius when it came to electricity, but it turns out that not getting shocked was one of the least weird things about him!

Nikola Tesla (1856–1943) has over 500 inventions to his credit, most of which revolve around electricity. His inventions rank with those of Thomas Edison, inventor of the light bulb, who was Tesla's main rival and possible arch-enemy.

So... What do you think?

1. Find out why Tesla was perfectly safe while posing for this photograph!
2. Tesla would only stay in a hotel room with a number that could be divided by three. Find out three other really weird things about Tesla's life.
3. Tesla's contributions to the world of electricity may have even surpassed those of his rival Thomas Edison. Find out three things Tesla invented.
4. The machines generating the sparks in this photo are called Tesla coils. What are Tesla coils and how do they generate electricity?

Topic 4.5 Review

Key Concept Summary

- The current in a series circuit is the same at every point in the circuit.
- The current in each branch in a parallel circuit is less than the current through the source.
- The sum of the potential differences across each load in a series circuit equals the potential difference across the source.
- The potential difference across each branch in a parallel circuit is the same as the potential difference across the source.

Review the Key Concepts

1. **K/U** Answer the question that is the title of this topic. Copy and complete the graphic organizer below in your notebook. Fill in four examples from the topic using key terms as well as your own words.

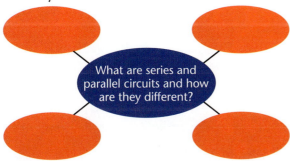

2. **C** Use a double bubble organizer to compare the properties of parallel circuits and series circuits in terms of current and potential difference.

3. **K/U** Look at the circuit diagram below.
   a) Describe this circuit and identify all of its components.
   b) Identify the meter labelled "M" in this circuit, describe what it measures, and determine the reading you would expect to see on this meter.
   c) Identify the meter labelled "N" in this circuit, describe what it measures, and determine the reading you would expect to see on this meter.

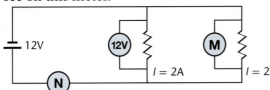

4. **K/U** Look at the circuit diagram below.
   a) Describe the circuit and identify all of its components.
   b) Identify the meter labelled "X" in this circuit, describe what it measures, and determine the reading you would expect to see on this meter.
   c) Identify the meter labelled "Y" in this circuit, describe what it measures, and determine the reading you would expect to see on this meter.

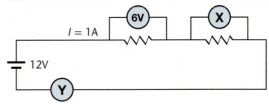

5. **T/I** Assume the loads in the diagram above are light bulbs. Predict what would happen to the brightness of the bulbs if you added a third bulb to this circuit. Justify your answer.

6. **K/U** Examine pictures A and B below. Decide whether each picture shows a series circuit or a parallel circuit. Explain the evidence you used to make your decision.

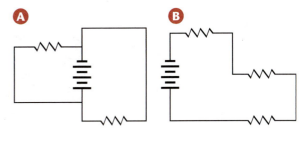

Topic 4.6

What features make an electrical circuit practical and safe?

Key Concepts

- Practical wiring for a building has many different parallel circuits.
- Circuit breakers and fuses prevent fires by opening a circuit with too much current.
- Higher-voltage circuits, larger cords and cables, and grounding help make home circuits safe.

Key Skills

Numeracy

Key Terms

circuit breaker
fuse

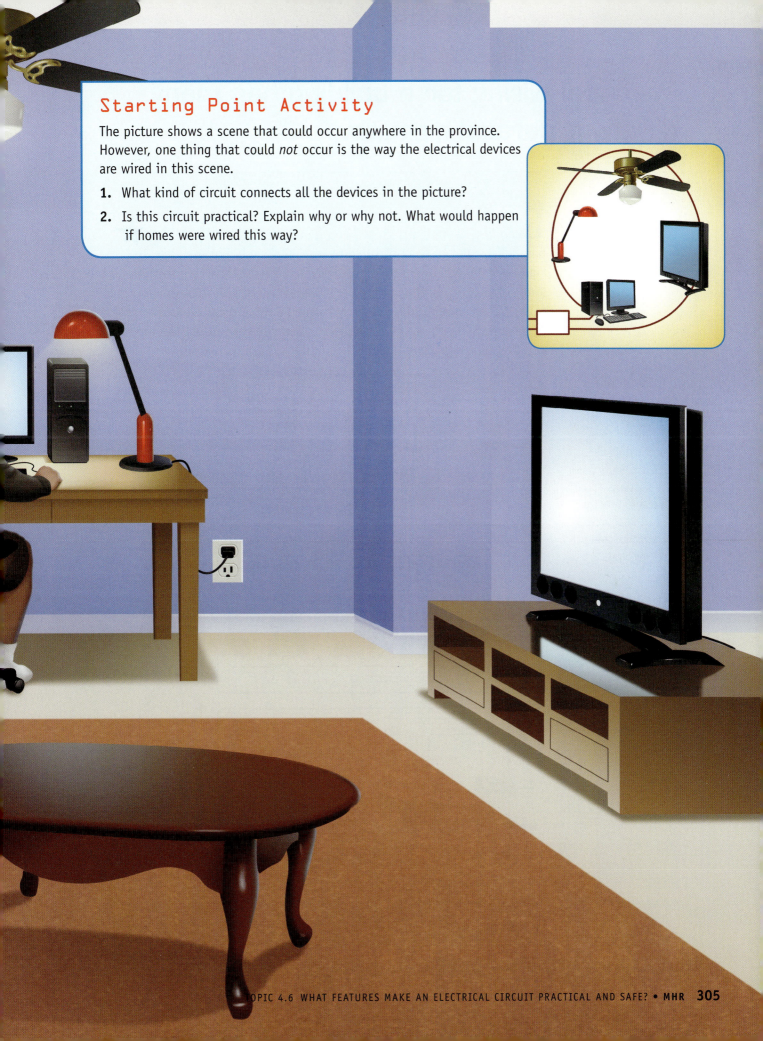

Starting Point Activity

The picture shows a scene that could occur anywhere in the province. However, one thing that could *not* occur is the way the electrical devices are wired in this scene.

1. What kind of circuit connects all the devices in the picture?
2. Is this circuit practical? Explain why or why not. What would happen if homes were wired this way?

Practical wiring for a building has many different parallel circuits.

Examine the parallel circuit shown in **Figure 4.30**. As you know, current can flow around any closed conducting loop. So each device in a parallel circuit can be controlled by its own switch. For instance, in **Figure 4.30A**, when you turn on the switch for the toaster, it forms a closed loop so current can flow. Only the toaster is on. If you then turn on the radio, as in **Figure 4.30B**, you have formed a second closed loop that lets current flow. **Figure 4.30C** shows that when you turn on the ceiling lamp and the microwave oven, all of the appliances are turned on.

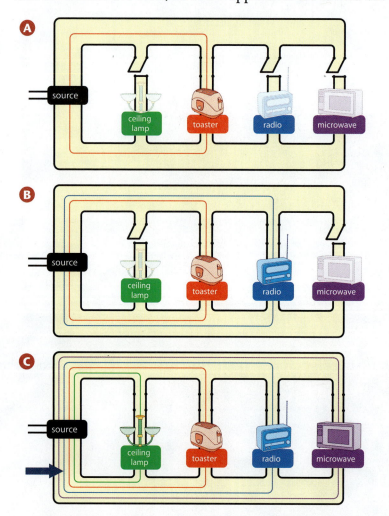

▶ **Figure 4.30** In this parallel circuit, each of the coloured lines represents the current flowing to a specific electrical device. Notice that all colours pass through the sections of the wire that are directly connected to the source. A large current like this can make a wire very hot.

Take a closer look at **Figure 4.30C**. Look at the place where the arrow in the picture is pointing. Notice that, when all appliances are on, all the current that is going to each appliance is passing through the conductor near the source. When large amounts of current flow through a wire, the wire can get very hot.

Now imagine that all the electrical devices in an entire home were connected to the same parallel circuit. The current flowing to each device also would be flowing through the wire conductors connected to the source. This large amount of current would make those wires extremely hot and would certainly start a fire. A parallel circuit with too many electrical devices connected to it is not practical, because it is not safe.

A house, apartment building, school, or any building must have many electrical outlets, because many electrical devices are used in it. So it is safer and more practical to install many separate parallel circuits in the building. Of course, all of the current flowing in all circuits in a building must be flowing in the conductors that lead into the building from a power company. Therefore, very large electrical cables are used to carry electrical energy from a power company to a building. Large cables are designed to carry a large current without becoming too hot. Inside the building, many different parallel circuits are connected to the large cables. This idea is shown in **Figure 4.31**.

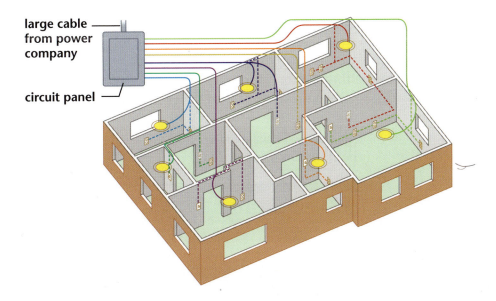

◀ **Figure 4.31** The box labelled "circuit panel" is the place in the building where large electrical cables connect the building to the electrical energy flowing from the power company. From the circuit panel, many smaller parallel circuits are wired throughout the building. Each colour in this diagram represents one small parallel circuit.

LEARNING CHECK

1. Refer to **Figure 4.30**. Describe how a switch controls the flow of electrical current through individual appliances.
2. Explain why a parallel circuit with too many electrical devices connected to it is not practical.
3. Explain what a circuit panel is.
4. Explain why a building must have many electrical outlets.

Circuit breakers and fuses prevent fires by opening a circuit with too much current.

The current that flows through a wire conductor can become very high, even if the parallel circuit has just a few outlets connected together. For example, people sometimes use a gadget like the one in **Figure 4.32**. It converts a double outlet into an outlet that has more places to plug in additional pieces of equipment. People do this for convenience. But connecting too many electrical devices to one outlet is a safety hazard, because it increases the amount of current in the circuit.

Circuit Breakers

A safety device called a **circuit breaker** will prevent any circuit from carrying too much current and starting a fire. Refer to **Figure 4.33**. A circuit breaker has a strip made of two metals. When the metals get too hot, they bend and cause a switch to open. This open switch prevents current from flowing through the entire circuit. When a circuit breaker opens the switch, you can close it by going to the breaker panel and pushing the switch back in place. However, you need to turn off some of the electrical devices that caused the large current to flow, or the circuit breaker will break (open) the circuit again.

▲ **Figure 4.32** An outlet designed to connect two electrical devices to the source is now connecting many devices. If every device is turned on at the same time, they could draw a large amount of current.

▶ **Figure 4.33** Every switch in a circuit panel controls one parallel circuit. Circuit breakers are always located beside the source. When they break the circuit, no current flows to any of the pathways. The circuit breaker in the diagram will open the circuit when the current goes above 15 A.

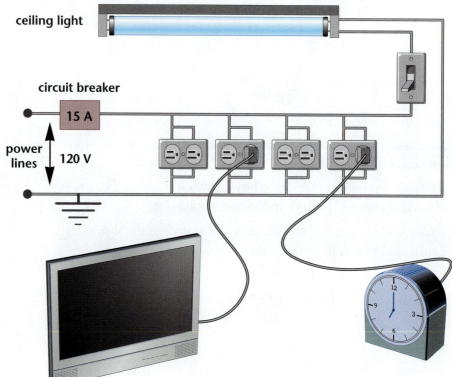

Fuses

Many years ago, fuses were used in homes instead of circuit breakers. Some very old buildings still have a fuse box with fuses instead of circuit breakers. A **fuse**, shown in **Figure 4.34**, has a small wire that will melt and break apart when the current gets too high. This has the same effect as opening a breaker switch. Both fuses and breaker switches stop the current from flowing when the wire becomes too hot. However, when a fuse "blows out," it has to be replaced with a new one.

circuit breaker: a safety device that opens a circuit if the current gets too high; it can be reset

fuse: an older safety device that opens a circuit if the current gets too high. it must be replaced.

◀ **Figure 4.34** Before circuit breakers were invented, everyone used fuses similar to these.

A Each fuse has a label on it that shows the amount of current that would cause the fuse wire to melt.

B Fuses are still used today in many applications. For example, electrical systems in cars use fuses like this one.

LEARNING CHECK

1. Refer to **Figure 4.32**. Explain why connecting too many electrical devices to one outlet is a safety hazard.
2. Use a Venn diagram to compare a circuit breaker with a fuse.
3. Use a flowchart or other graphic organizer to outline the steps you would follow to reset a circuit breaker in your home.

Numeracy Focus

Activity 4.12

MAKE AND BREAK THE CIRCUIT

Most circuit breakers in homes are designed to open if the current becomes greater than 15 A. The chart contains common home appliances and the current they use. List all the combinations of devices that could be on and working at the same time. Then list all the combinations of devices that would make the circuit breaker open the switch.

Device	Approximate Current (A)
coffee maker	10
microwave oven	6.25
clothes iron	15
laptop computer	0.4
toaster	6.5
toaster oven	10
refrigerator	6
ceiling fan	1.5
dishwasher	20
clock radio	0.1

Higher-voltage circuits, larger cords and cables, and grounding help make home circuits safe.

Some electrical appliances use so much current that they must have their own circuit. The most common example in the home is the electric stove. However, an electric water heater, an air conditioner, or an electric clothes dryer might also have its own circuit. Even if one of these appliances was connected to its own 120 V circuit, it would require so much current that it would still make the conductor dangerously hot. Therefore, electricians create a circuit that provides a potential difference of 240 V instead of 120 V.

By doubling the potential difference from 120 V to 240 V, a circuit uses half as much current and still provides the same amount of energy. To prevent people from plugging an appliance into the wrong outlet, outlets that provide 240 V are different from those that provide 120 V. **Figure 4.35** shows what a 240 V outlet looks like and the kind of plug that fits it.

▶ **Figure 4.35** The plug for an electric stove usually is very large and has three flat prongs and one curved prong. The cord is also very large. The large size of the cord and prongs allows them to safely carry a large amount of current without becoming too hot.

Safety with Larger Cables

You might have seen electrical wiring that is inside the walls of a home. These cables are much larger than a typical electrical cord on an appliance. Their size allows them to carry more current than is usually needed for a single appliance, without becoming too hot. The cables leading from the power company into your home are even larger.

Safety with Grounding

Grounding of some wires is also a safety feature. Of the two wires for every parallel circuit, one is grounded at the source. Recall that grounding means that the wire is attached to some type of conductor buried in the ground. Any excess current will go to the ground. Accidentally touching a grounded wire would not cause a shock. The wire in a circuit that is not grounded is called the hot wire. You would get a dangerous shock if you touched it.

Safety with Outlets

Figure 4.36 shows three types of outlets commonly found in homes and other buildings. You may have seen two-hole outlets like the one in **Figure 4.36A** in very old buildings. Three-hole outlets like the one in **Figure 4.36B** are required by law in newer homes. The third hole is for a different type of grounding that is not connected to either of the two wires in the electrical circuit. The third prong on the plug is connected to metal parts of the lamp or appliance. When plugged in, these metal parts are grounded. If, for any reason, the hot wire inside the appliance became frayed or damaged and touched a metal part of the appliance, all of the metal would be "hot." If you touched any metal part of the appliance, you would get a serious shock. The third prong prevents this type of shock from occurring.

Figure 4.36C is a ground fault interrupter or GFI. It is a special safety device that is installed in bathrooms and other locations near water faucets. If water splashes on an appliance like a hair dryer or radio and on you, it can create a conductor that includes your body. Current leaves the circuit and passes through you and goes to the ground that you are standing on. Such a shock can be fatal. The GFI measures the current leaving one end of the circuit and current entering the other. If these currents are not the same, the GFI immediately opens the circuit and stops any more current from flowing. A normal circuit breaker would also open the circuit when current gets too high, but it would not act soon enough. You could still get a fatal shock. The GFI responds more quickly than a circuit breaker or fuse. GFIs save lives.

◀ **Figure 4.36** Types of outlets found in many homes

A Two-hole outlets like this are found in very old buildings. Newer buildings are designed with three-hole outlets.

B In three-hole outlets, the third hole is for a grounding wire that safely channels current back into the ground.

C This type of outlet is called a ground fault interrupter, or GFI outlet. It is very sensitive and is commonly found in bathrooms or other locations that are within 2 m of water. A GFI outlet opens a circuit if there is any difference between current leaving one hole and entering the other.

LEARNING CHECK

1. Refer to **Figure 4.35**. Explain why a 240 V outlet has a different shape than a 120 V outlet.
2. Use a t-chart with the headings "Method" and "Where It Is Used" to summarize the three methods used to make home circuits safe.
3. Refer to **Figure 4.36**. Explain, in terms of your personal safety, why it is important that all of the electrical outlets in a bathroom are protected by a ground fault interrupter (GFI).

Activity 4.13
DELIVERING ELECTRICAL ENERGY TO YOUR HOME

Exposure to high voltages can stop your heart. Large currents also can heat body tissues very quickly, causing severe burns.

Perhaps you have seen a fenced area like the one in the photo, with warning signs about the danger. This type of site is called an electrical substation. The electrical current that flows into your home passes through a number of substations on its way from where it is first generated at a power plant. Along the way, the voltage is increased and decreased several times.

Suppose a power company wanted to transmit 1.0 MW (megawatts) of power over 100 km of power lines. Power represents the amount of energy that is transmitted or changed to another form of energy every second. What would be the advantage of transmitting the power at 500 000 V instead of 20 000 V?

A power company would want ways to minimize the loss of power between the generating plant and its customers. Since a current passing through a wire causes heating, the most significant losses would occur due to heating of the transmission lines.

To find the amount of power lost to heat, or the energy lost to heat every second, you can use a formula: $P = I^2R$. (P is the symbol for power, I is the symbol for current, and R is the symbol for resistance.) For typical copper conductors that are used in high voltage lines, the resistance of 100 km of wire is about 100 Ω.

Numeracy Focus

What To Do

1. Use the formula $I = \frac{P}{V}$ to find the current in the transmission lines, when 1.0 MW of power is transmitted at 500 000 V. Note that 1.0 MW is 1 000 000 W or 1.0×10^6 W. The value 500 000 V also can be written as 5.0×10^5 V.

2. Use your answer for step 1 and the formula $P = I^2R$ to find the amount of power that is lost to heat in 100 km of transmission lines, when 1.0 MW of power are transmitted at 500 000 V.

3. Find the percent of power lost to heat when 1.0 MW of power is transmitted at 500 000 V. To do this, use the formula

   $$\text{percent power lost} = \frac{\text{power lost to heat} \times 100\%}{\text{total power transmitted}}.$$

4. Find the current in the transmission lines when 1.0 MW of power are transmitted at 20 000 V.

5. Use your answer for step 4 and the formula $P = I^2R$ to find the amount of power that is lost to heat in 100 km of transmission lines, when 1.0 MW of power are transmitted at 20 000 V.

6. Find the percent of power lost to heat when 1.0 MW of power are transmitted at 20 000 V.

What Did You Find Out?

1. Write a paragraph that explains why power companies transmit power at very high voltages when transmitting power over long distances.

Topic 4.6 Review

Key Concept Summary

- Practical wiring for a building has many different parallel circuits.
- Circuit breakers and fuses prevent fires by opening a circuit with too much current.
- Higher-voltage circuits, larger cords and cables, and grounding help make home circuits safe.

Review the Key Concepts

1. Answer the question that is the title of this topic. Copy and complete the graphic organizer below in your notebook. Fill in four examples from the topic using key terms as well as your own words.

   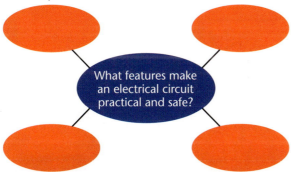

2. **K/U** Use words or diagrams to show how a) a circuit breaker and b) a fuse work to prevent electrical fires.

3. **T/I** Look at the graph on this page.
   a) In your notebook, draw a circuit diagram that includes a source with a potential difference of 120 V and that has two loads—the toaster and the iron—wired in a parallel circuit.
   b) Calculate the total amount of current in this circuit if both the toaster and the iron were turned on. Show your work. (Hint: When you have loads in parallel, you can find the total current in the circuit by adding up the current drawn by each device.)
   c) Predict what would happen if this circuit were protected by a 15 A circuit breaker.

4. **T/I** Refer to the graph on this page.
   a) In your notebook, draw a circuit diagram that includes a source with a potential difference of 120 V and that has four loads—the radio, television, and two lights—wired in a parallel circuit.
   b) Calculate the total amount of current in this circuit if all four appliances were turned on. Show your work.
   c) Predict what would happen if this circuit were protected by a 15 A circuit breaker.

   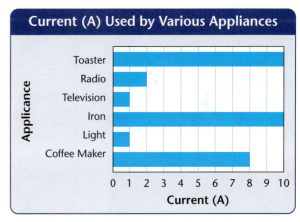

5. **A** You are helping your neighbour Freda construct a wooden fence around her garden. Freda is using an electric drill to screw the boards to the frame. You notice the drill has a damaged plug and is missing the ground wire prong. You ask Freda why she removed the ground wire prong. She answers that an old extension cord had only a two-prong outlet, so she cut off the ground wire prong. Explain to Freda why this was not a good idea.

Topic 4.7: How can we conserve electrical energy at home?

Key Concepts

- Conserving energy at home requires an understanding of how energy is measured.
- People can conserve energy by making informed choices.

Key Skills

Inquiry
Literacy

Key Terms

EnerGuide label
ENERGY STAR® label

At any given moment, on any given day, you are linked to each of the 12.7 million or so people with whom you share the province of Ontario. This linkage exists because you and your fellow Ontarians draw on the same overall "pool" of the province's supply of electrical energy. It is up to all of us—as individuals and as a provincial community—to use our share of electrical energy wisely and sustainably.

ONTARIO DEMAND
13, 377 MW
10:00 a.m. EDT – May. 3, 2009

TODAY'S PROJECTED PEAK AT
10:00 PM

GENERATOR AVAILABILITY AT PEAK
19, 077 MW

Starting Point Activity

1. Why should we be concerned about using electrical energy in a sustainable manner? If you need to review the meaning of "sustainable," turn to page 65.
2. What are some examples of sustainable uses of electrical energy?
3. What are some examples of unsustainable use? How could these examples be made more sustainable?

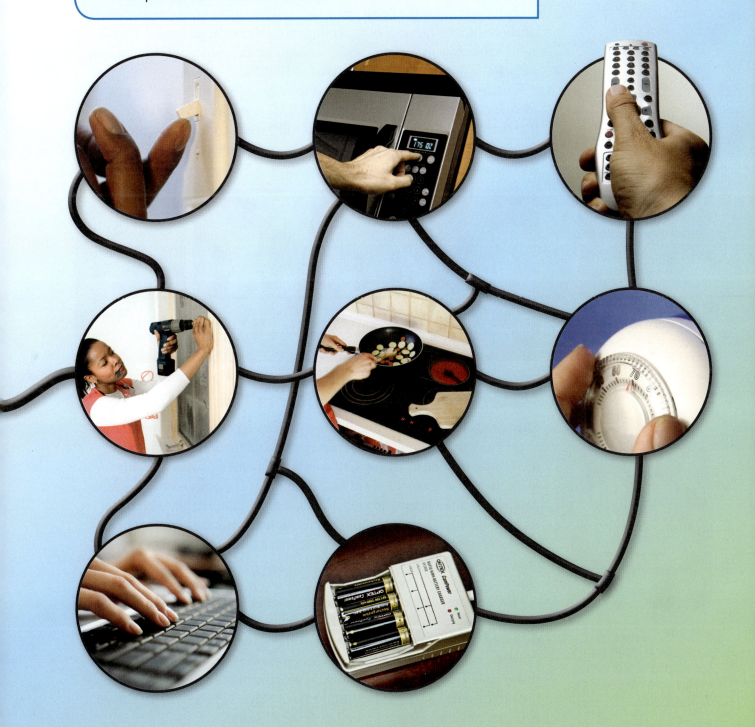

Conserving energy at home requires an understanding of how energy is measured.

If you look at an energy bill, you will see that the electrical energy you use at home is measured in units called kilowatt hours (kWh). You probably recognize the "watt" part of that term. Most home appliances are labelled with the number of watts they use. For example, the old types of light bulbs came in varieties such as one hundred watts (100 W) and sixty watts (60 W). An iron might be rated at 1000 W.

A kilowatt has the prefix *kilo-*, which means "one thousand." So a kilowatt is one thousand watts, or 1000 W. If you use an appliance that is rated at 1000 W for 1 hour, you will have used 1 kWh of energy. Figure 4.37 shows some examples to help you get an idea of the amount of energy in 1 kilowatt hour (1 kWh).

▶ **Figure 4.37** Examples of tasks that involve 1 kWh of energy

A You have used 1 kWh of energy if you have used a 100 W light bulb for 10 h.

B You have used 1 kWh of energy if you have used a 1000 W iron for 1 h.

C You have used approximately 1 kWh of energy if you have jogged for 1 h.

D You have used 1 kWh of energy if you have taken a hot shower for about 3 min.

Meters for Measuring Home Energy Use

Most houses and apartment buildings have meters that are connected to the cable that brings electrical energy into the buildings. Old-style meters like the one in Figure 4.38 continuously measure the amount of electrical energy that is used in the building. The power-supply company sends a person to read and record the numbers on each meter on a regular basis. At the main office, computers calculate the cost of the energy used that month. For example, if a building used 345 kWh of energy and the company charges 10.9¢ per kWh, the company would charge 345 kWh × 10.9¢/kWh = 3760.5¢ ($37.61) for the energy.

By the year 2010, technicians from power-supply companies will have replaced the old-style meters with new smart meters. A smart meter like the one shown in Figure 4.39 measures energy use in a way that is different from the old-style meters. Smart meters measure the amount of energy that is used every hour in the home or apartment building. The data are then transmitted automatically to the head office. No one will have to go to every house or business and read each individual meter.

▲ **Figure 4.38** The old-style type of meter records the amount of electrical energy used continuously, 24 hours a day, 365 days a year.

Time-of-Use Prices

Use of the smart meter lets power-supply companies charge different prices for electrical energy that is used at different times of the day. Prices will be highest when the most electrical energy is being used. This is referred to as on-peak use. Prices will be lowest when the least electrical energy is being used. This is referred to as off-peak use. Periods between on-peak use and off-peak use are called mid-peak hours. The graph in **Figure 4.40** compares on-peak, mid-peak, and off-peak periods during winter. The schedule is different for winter energy use and for summer energy use, because people use electrical energy at different times and in different ways. For example, many people use air conditioning in the summer but not in the winter.

▲ **Figure 4.39** The new smart meters measure energy use each hour. The meters encourage "smart" behaviour on the part of Ontarians by giving us the means to think about how and when we use electrical energy.

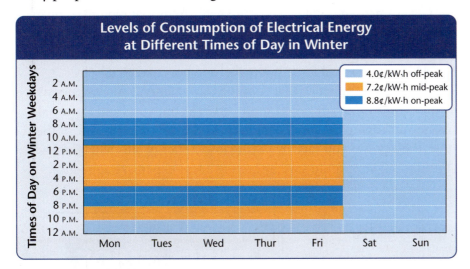

◄ **Figure 4.40** The key on the upper right of the graph shows possible costs for energy during the different time periods.

LEARNING CHECK

1. Identify the unit used to measure the amount of electrical energy used by most appliances.
2. Define the term "kilowatt" by using an example.
3. Predict how many kilowatt hours you might use at home in a day.

Inquiry Focus

Activity 4.14
BEST TIME TO USE

1. Predict how your use of electrical energy in summer would be different compared with winter. Sketch a graph similar to Figure 4.40 to show your predictions.
2. Explain why it makes sense to have different prices and different time periods during these two seasons.
3. Give three examples of how time-of-use data would help you change the way your family uses energy.

People can conserve energy by making informed choices.

In a typical home, certain appliances tend to use more electrical energy than others. These include the refrigerator, the washing machine, the clothes dryer, and the electric stove. However, some appliances of the same type are more energy-efficient than others. For example, front-loading washing machines use less energy than top-loading washing machines.

The Government of Canada has set up regulations that require companies to put a label on all new electrical appliances to show how much energy the appliances use in a typical year. That label is called the **EnerGuide label**. **Figure 4.41** shows a typical EnerGuide label and how to read it.

Sometimes, the appliance that uses the least energy is more expensive to buy than some others. It might sound strange, but you would probably save money by buying the more expensive appliance. Over the life of the machine, the amount you save by paying lower energy bills will be greater than the amount you would save by buying a less-expensive appliance.

EnerGuide label: a label that gives details about the amount of energy that an appliance uses in one year of normal use

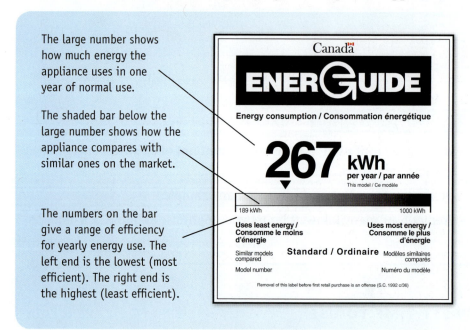

▲ **Figure 4.41** How to interpret an EnerGuide label

▲ **Figure 4.42** Products with this label use 10 to 50 percent less energy compared with a standard product in the same category.

ENERGY STAR®: a label that identifies a product as meeting or exceeding certain standards for energy efficiency

The **ENERGY STAR® label** makes it even easier to identify the most efficient appliances. The Government of Canada has set minimum standards of efficiency for electrical appliances and equipment that will save energy. Appliances or equipment that meet or exceed these standards can have the ENERGY STAR® label that is shown in **Figure 4.42**.

Fight the Phantom (Load)

Many electrical devices are on even when you think they are switched off. They are in stand-by mode. For example, if you have a remote control to turn on a television, the television must be able to sense the signal, and this requires energy. The electrical energy that is used by a device when it is turned off is called a phantom load. Clock displays, such as those on microwaves and coffee makers, and external power adapters also require phantom loads. If you touch an external adapter, you will observe that it is quite warm. This is evidence of a phantom load.

Studies carried out by Thunder Bay Hydro estimate that phantom loads account for about 900 kWh of energy use each year! **Figure 4.43** shows one way this total can be reduced. Of course, people can also unplug devices when they are not in use. The cost of this inconvenience is a few seconds. On the other hand, the energy savings can be in the tens or even hundreds of dollars.

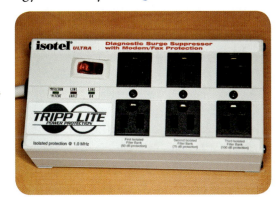

◄ **Figure 4.43** Many surge protectors have a light that tells you if they are on. They also have a switch. When you turn off the switch, power is cut off to all equipment that is plugged into the surge protector. Turning off a switch is much easier than unplugging several pieces of equipment.

LEARNING CHECK

1. Compare the information on an EnerGuide label with the information on an ENERGY STAR® label.
2. Explain how the information in **Figure 4.43** can help your family save electrical energy and money.

Literacy Focus

Activity 4.15

READING ENERGUIDE LABELS

Your teacher will give you several EnerGuide labels. Answer these questions about each one.

1. What is the estimated consumption of energy per year of the least efficient appliance that is similar to this one?
2. What is the estimated consumption of energy per year of the most efficient appliance that is similar to this one?
3. What is the estimated consumption of energy per year for this appliance?
4. Is the energy consumption of this model closer to the most efficient model or closer to the least efficient model?
5. Would you recommend buying this appliance? Why or why not?

Go to **scienceontario** to find out more

Case Study Investigation: People Power

1979 DAY 24

Late summer, 1979
Day 24 of Dmitra Constantino's (me!) exercise captivity
Can you believe Dad is still on this energy conservation kick? Okay, here's the skinny—24 days ago (I counted each and every one of them!) he hooked the TV up to a bike! He went on and on about how the bike links to the generator to make electricity. I told him it's bad enough that we don't have a colour TV when every single one of my friends does, but now I have to exercise to watch the Dukes of Hazzard? TV is all about relaxation, not exercise! How can he be so cruel? He says it is quite easy and we all need to do our part during the energy crisis. I asked him if there is an energy crisis, why our family has to be in crisis too? He just frowned. Dad says he lost his sense of humour when he had kids but I suspect he never had one! Next he'll probably hook up my record player so I have to jog in place to listen to my new Jaws soundtrack album. (That movie was so awesome, to the max!) Okay, gotta go bike some more. Catch ya on the flip side!

The Science Behind the Story

While the 1970s energy crisis was initially linked to oil shortages, people around the world were urged to use all forms of energy more efficiently. Everyone jumped on the energy efficiency band wagon: rich and poor, young and old, politicians and citizens alike. People began to think about using alternative energy sources to generate electrical energy. Many people also developed an interest in using our own bodies to generate electricity. Using pedal power to run an electrical device, such as a TV or washing machine, is an example of how people can produce power.

Pause and Reflect

1. How did people try to improve energy efficiency in the 1970s?

Can "people power" help reduce our use of electricity today?

Fast forward to the 21st century. In British Columbia, Professor Max Donelan is working on a device that captures the energy we generate every day—by walking! His device, the Bionic Energy Harvester, is secured around the knee. While it looks a lot like a common knee brace, this device generates energy in an unexpected way. Each time we take a step, our leg muscles slow our knees just before our feet touch the ground. The Bionic Energy Harvester recovers the energy our leg muscles absorb during this braking process. How much energy can be generated this way? A mere minute of walking can power a cell phone for 30 minutes!

The Bionic Energy Harvester, invented by Max Donelan of British Columbia

Pause and Reflect

2. What is the Bionic Energy Harvester?
3. How does it work?

Inquire Further

4. What caused the energy crisis in the 1970s? Do you think we are in one now? Why or why not?
5. Find out more about a pedal powered machine or vehicle that interests you. Learn its history and determine how it works. Present your findings in a poster or other suitable medium.
6. Design your own pedal powered machine. Your pedal power can generate electricity or it can mechanically move a device (i.e. pedaling can mechanically move the agitator in a washing machine). Create a sketch of your device and write a paragraph explaining how it works.

Making a DIFFERENCE

Katie Pietrzakowski was brushing her hair when she came up with the idea for her award-winning science fair project. She observed how different particles were attracted to her brush and wondered if these forces of attraction could be used to clean recycled grey water.

Grey water is household waste water from sinks, showers, washing machines, and dishwashers. It can be collected and re-used for lawn irrigation. Re-using grey water helps to conserve water safely and appropriately. Using a system she designed, Katie found that introducing an electric field to grey water could reduce the particulates suspended in the water.

Katie took her project, "Shock the Grey," to the 2006 Canada Wide Science Fair in Saguenay, Québec. She won a bronze medal and a scholarship to the University of Western Ontario. Katie is now a high school student in Sault Ste. Marie and hopes to become a teacher.

What other uses for grey water can you think of?

Fluorescent tubes are popular alternatives to conventional light bulbs. The tubes, however, contain mercury, which is toxic. Many tubes are not recycled and end up in landfills. Hamilton student Patrick Bowman was in Grade 7 when he studied the issue for his science fair project. He learned that up to 1 350 000 mg of mercury from fluorescent tubes enter his city's landfill each year.

To determine the effects these tubes might have on the environment, Patrick made two model landfills using compost. He put broken tubes into one landfill and left the other uncontaminated. He added rainwater to the landfills to model leachate, a liquid produced when precipitation and landfill waste mix. Patrick found that the uncontaminated landfill and leachate supported plant growth and micro-organism life better than the contaminated landfill and leachate. His project "Shedding the Lights from Landfill Sites" won several awards at the 2006 Canada Wide Science Fair.

How could you help raise your community's awareness of the hazards linked with throwing away fluorescent tubes?

Topic 4.7 Review

Key Concept Summary

- Conserving energy at home requires an understanding of how energy is measured.
- People can conserve energy by making informed choices.

Review the Key Concepts

1. **K/U** Answer the question that is the title of this topic. Copy and complete the graphic organizer below in your notebook. Fill in four examples from the topic using key terms as well as your own words.

   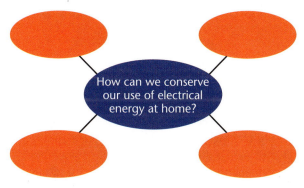

2. **K/U** Refer to **Figure 4.41**.
   a) Summarize, in your own words, the information provided on an EnerGuide label.
   b) Explain how you could use EnerGuide ratings when shopping for a new appliance.

3. **K/U** Use a main idea web to identify six things that you can do to conserve electrical energy in your home.

4. **T/I** While shopping with your family to buy a new refrigerator, you find two models that have the desired features. Model A consumes 400 kWh per year, and Model B consumes 460 kWh per year. Model A costs $20.00 more than Model B. Assume the price of electricity is $0.08/kWh. In your notebook, calculate the annual cost to run each refrigerator. Show your work. Which model will cost less in the long term, given that the average life span of a refrigerator is 17 years?

5. **T/I** Assume that the phantom load for an average television set is 100 kWh per year and that the cost of electrical energy is $0.08 per kWh. In 2003, there were an estimated 21 million (21 000 000) television sets in Canada. Calculate how much electrical energy was wasted in 2003 by the phantom loads of television sets alone, and how much this energy cost Canadians. Show your work.

6. **C** Write a blog explaining why replacing old appliances with modern ones could save you money and reduce your impact on the environment.

7. **A** During Earth Hour on Saturday, March 28, 2009, between 8:00 P.M. and 9:00 P.M., people all over the world were encouraged to use less electricity. More than 900 megawatts (MW) of electrical energy were saved in Ontario.
   a) How might the organizers of future Earth Hours encourage more people to turn out their lights for one hour on the day selected for Earth Hour?
   b) How might the organizers of future Earth Hours inspire people to make long-lasting lifestyle changes that will reduce their use of electrical energy?

SCIENCE AT WORK

CANADIANS IN SCIENCE

Tiffany Hando enjoys detailed work that requires a lot of precision. After completing a high school co-op placement in the electrical department of a mill, Tiffany knew she wanted to work with electricity. She decided to study electrical instrumentation. "I like to work with my hands and use computers," says Tiffany. "I knew I would enjoy instrumentation because it is a little bit of everything I am interested in." She is earning an instrumentation engineering technician diploma. Electrical instrumentation technicians install, maintain, and fix different electronic instruments used to measure and control the function of equipment.

▲ Tiffany Hando is a student at Confederation College in Thunder Bay. She is enrolled in a two-year program to earn an instrumentation engineering technician diploma.

What advice do you have for high school students interested in a career in electrical instrumentation?

Tiffany says there are a lot of opportunities available to students interested in the field, but they should be prepared to work hard. After she earns her instrumentation engineering technician diploma, Tiffany plans to attend college for a third year to earn a diploma in electrical engineering technology. Students can also complete a fourth year at university to obtain an engineering degree.

What challenges do people in your field face?

People in the field must keep up to date on regulations and advances in technology. "It is a constant learning process," she says.

What is most rewarding about working in your field?

Tiffany finds it rewarding when she can relate what she is working on to what she learned in school. She also enjoys working with people who have been in the field for a long time. "It is really great when the worker trusts you enough and believes you have enough knowledge to work on equipment or a process by yourself. That feels great. It is definitely rewarding!"

◄ Electrical instrumentation technicians can work in many types of areas. For example, they can work at plants, mills, and electrical generating stations. They can also work for biomedical equipment manufacturers or telecommunications businesses.

Electricity at Work

The study of electricity contributes to these careers, as well as many more!

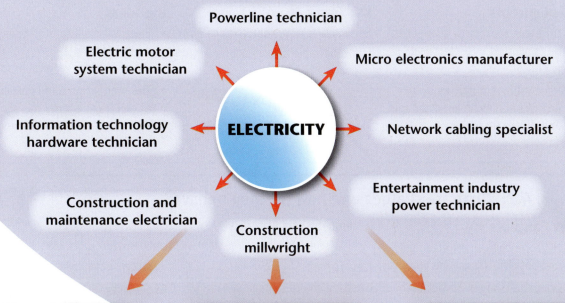

▲ Construction and maintenance electricians set up, test, maintain, and fix electrical equipment, fixtures, wiring, and other systems in homes, offices, and industrial buildings. They make sure electrical systems are safe in renovations or new construction projects.

▲ Construction millwrights install and fix machines during the construction of new plants and other facilities. They may be involved in the maintenance of machines and equipment.

▲ Entertainment industry power technicians plan, build, set up, maintain, and take apart power distribution systems. They work in film, television, live theatre, trade shows, and musical events.

Over To You

1. In which types of industries do electronic instrumentation technicians work?
2. Research a career involving electricity that interests you. If you wish, you may choose a career from the list above. What are the essential skills needed for this career? What would you need to do to pursue this career?

Go to **scienceontario** to find out more

Unit 4 Summary

Topic 4.1: How do the sources used to generate electrical energy compare?

Key Concepts
- Different sources of energy can be converted into electrical energy.
- Renewable and non-renewable energy sources have advantages and disadvantages.

Key Terms
renewable energy source (page 248)
non-renewable energy source (page 248)

Big Ideas
- Electricity is a form of energy produced from a variety of non-renewable and renewable sources.
- The production and consumption of electrical energy has social, economic, and environmental implications.

Topic 4.2: What are charges and how do they behave?

Key Concepts
- Negative charges are electrons, and positive charges are protons.
- Opposite charges attract each other, and like charges repel each other.
- Negative charges can move through some materials but not others.

Key Terms
negative charges (page 254)
positive charges (page 254)
electrically neutral (page 254)
conductor (page 258)
conductivity (page 258)
insulator (page 258)

Big Ideas
- Static and current electricity have distinct properties that determine how they are used.

Topic 4.3: How can objects become charged and discharged?

Key Concepts
- Objects can become charged by contact and by induction.
- Charged objects can be discharged by sparking and by grounding.

Key Terms
charging by contact (page 266)
electroscope (page 266)
charging by induction (page 267)
discharged (page 268)
grounding (page 269)

Big Ideas
- Static and current electricity have distinct properties that determine how they are used.

Topic 4.4: How can people control and use the movement of charges?

Key Concepts
- A constant source of electrical energy can drive a steady current (flow of charges).
- An electric current carries energy from the source to an electrical device (a load) that converts it to a useful form.
- A source, load, and connecting wires can form a simple circuit.
- Potential difference and resistance affect current.
- Meters can measure potential difference and current.

Key Terms
source (page 276)
potential difference (page 276)
current (page 278)
amperes (page 278)
load (278)
resistance (page 279)
ohm (page 279)
electrical circuit (page 280)
voltmeter (page 282)
ammeter (page 282)

Big Ideas
- Static and current electricity have distinct properties that determine how they are used.

326 MHR • UNIT 4 ELECTRICAL APPLICATIONS

Topic 4.5: What are series and parallel circuits and how are they different?

Key Concepts
- The current in a series circuit is the same at every point in the circuit.
- The current in each branch in a parallel circuit is less than the current through the source.
- The sum of the potential differences across each load in a series circuit equals the potential difference across the source.
- The potential difference across each branch in a parallel circuit is the same as the potential difference across the source.

Key Terms
series circuit (page 294)
parallel circuit (page 295)

Big Ideas
- Static and current electricity have distinct properties that determine how they are used.

Topic 4.6: What features make an electrical circuit practical and safe?

Key Concepts
- Practical wiring for a building has many different parallel circuits.
- Circuit breakers and fuses prevent fires by opening a circuit with too much current.
- Higher-voltage circuits, larger cords and cables, and grounding help make home circuits safe.

Key Terms
circuit breaker (page 308)
fuse (page 309)

Big Ideas
- Static and current electricity have distinct properties that determine how they are used.
- The production and consumption of electrical energy has social, economic, and environmental implications.

Topic 4.7: How can we conserve electrical energy at home?

Key Concepts
- Conserving energy at home requires an understanding of how energy is measured.
- People can conserve energy by making informed choices.

Key Terms
EnerGuide label (page 318)
ENERGY STAR® label (page 318)

Big Ideas
- Electricity is a form of energy produced from a variety of non-renewable and renewable sources.
- The production and consumption of electrical energy has social, economic, and environmental implications.

Unit 4 Project

Inquiry Investigation: Energy Savings

In this project, you will identify the room in your home that uses the most electrical energy. Then you will devise a plan to reduce that usage.

> **Inquiry Question**
> How can you reduce the amount of electrical energy used in one room of your home?

Initiate and Plan

1. List the appliances and lighting fixtures that require electricity in three rooms in your home.
2. Summarize the information in a table. Include a column that estimates power consumption of each item.
3. Identify the room that uses the most electricity. Determine whether each appliance is plugged into an outlet or into a power bar.
4. Research the potential impact of dimmer switches and power bars on use of electricity.

Perform and Record

5. Draw a circuit diagram to show the wiring of the room's lighting fixtures and appliances.
6. Revise your circuit diagram to include changes based on your findings on the use of dimmer switches and power bars.

Analyze and Interpret

1. Revise the table you prepared in Step 2 above to reflect the changes you've suggested.
2. Summarize the change in use of electrical energy that you are proposing.

Communicate your Findings

3. Present your original and revised circuit diagrams along with a brief report.
4. Explain how your plan would reduce the amount of energy used in the room.

Assessment Checklist

Review your project. Did you…

- [x] list all the appliances and lighting fixtures that need electricity in the rooms? **K/U**
- [x] accurately identify, with supporting evidence, the room that uses the most electricity? **T/I**
- [x] draw two circuit diagrams: one to show the wiring of the room's lighting fixtures and appliances, and a revision to include dimmer switches and power bars? **C**
- [x] present your circuit diagrams and a brief written or oral report? **C**
- [x] explain what your results suggest about the amount of energy that can be saved using your plan? **A**

An Issue to Analyze: Choosing Energy Sources in Ontario

You will choose two electrical power companies and do research to decide which company uses "greener" energy sources for the electrical energy that it provides to Ontario home-owners.

Issue

How should home-owners evaluate and choose a company to supply electrical power?

Initiate and Plan

1. Choose two companies that provide electrical energy in Ontario from different sources of energy.

Perform and Record

2. Make two tables to compare the companies.
3. Use these headings in the first table:
   - Number of Power Plants
   - Total Power Output
   - Source(s) of Energy
   - Reliability
   - Cost per kilowatt hour
   - Number of Customers
4. In the second table, use the headings below to compare the environmental effects of each company's source(s) of energy:
   - Renewable or Non-renewable Energy Source
   - Effects of Energy source on the Environment
   - Planned Changes for Improvement

Analyze and Interpret

1. Use the information in the tables to evaluate the environmental impact of each company. Take into account the size of the company as well as the impact of its energy source(s).
2. Decide which power company is better for the environment.
3. Decide which power company is better for the home-owner.
4. If you were a home-owner, which company would you choose? Explain your choice.

Communicate your Findings

5. Prepare a written or oral report to communicate your recommendation and your reasoning.

Assessment Checklist

Review your project. Did you...

- [✔] choose two companies that provide electrical energy in Ontario from different sources of energy? **T/I**
- [✔] gather information from a variety of sources to research each company and its impact on the environment? **T/I**
- [✔] use tables to summarize your information and compare the companies? **C**
- [✔] clearly state which power company you would recommend, based on your research? **A**
- [✔] prepare a written or oral report to communicate your findings, with evidence to support your decision? **C**

Unit 4 Review

Connect to the Big Ideas

1. Electricity is a form of energy produced from a variety of non-renewable and renewable sources. In your notebook, draw a chart or table similar to the one shown below. Complete this chart by filling in the missing information. Please do not write in your textbook!

Energy Source	General Advantages	General Disadvantages	Three Examples of Each Source
Renewable energy sources			
Non-renewable energy sources			

2. The production and consumption of electrical energy has social, economic, and environmental implications. Your local government plans to build a new electrical generating plant near your community. A decision has not yet been made about the type of generating plant that will be built. Assume that your community is able to use any renewable and non-renewable sources for generating electrical energy. Write a letter to the Ontario Minister of Energy and Infrastructure. (Infrastructure refers to basic systems that are needed to support a community, such as water treatment stations and power plants.) Identify which type of electrical generating plant you would like to see built in your community. Support your position by including information on the advantages and disadvantages of this particular energy source and why you think it would be the best choice.

3. Static and current electricity have distinct properties that determine how they are used. Use a t-chart to compare the properties of static electricity with the properties of current electricity.

Knowledge and Understanding K/U

4. Look back at Figure 4.2. Use a Venn diagram to compare the similarities and differences among hydroelectric, thermoelectric, and nuclear sources of energy used to produce electrical energy.

5. Explain why fossil fuels and uranium are considered to be non-renewable sources of energy.

6. Explain why wind, water, and the Sun are considered to be renewable sources of energy.

7. Use a t-chart to compare the similarities and differences between positive and negative charges. Include how these charges are related to the different parts of an atom.

8. Use a main idea web to summarize the law of electric charge. Include diagrams in your main idea web.

9. Describe the similarities and differences between a conductor and an insulator. Include examples of each in your comparison.

10. Describe the similarities and differences between charging by contact and charging by induction.

11. Use a labelled drawing to describe how lightning is generated.

12. Look back at Figure 4.16. Use this diagram to explain why a hair dryer gets hot when charges flow through it.

13. Draw a circuit diagram showing a circuit that contains a light bulb, two wires, and a source.

14. In your notebook, draw and label two diagrams. In the first diagram, show an electric circuit that includes a source, a switch, and three loads in series. In the second diagram, show an electric circuit that includes a source, a switch, and three loads in parallel.

15. You build a series circuit that consists of a source that is connected to two loads with different resistances. Describe the properties of the current and potential difference in this circuit.

16. You build a parallel circuit that consists of a source that is connected to two loads with different resistances. Describe the properties of the current and potential difference in this circuit.

17. Explain how unplugging your "instant-on" television when you are not watching it will save energy and money.

18. Rice puffs are very light and are electrically neutral. If a negatively charged ebonite rod is placed in a bowl of rice puffs, the puffs cling to the rod. However, a short time later, the puffs fly off in all directions. Explain why this happens.

Thinking and Investigation T/I

19. A friend complains that she sometimes experiences a small shock after using a telephone. She wonders whether the telephone has an electrical fault. You know that your friend often places the telephone handset on one of her shoulders while she talks. The telephone company tests its equipment and reports that neither the telephone nor the electrical circuit powering the telephone has an electrical fault. Predict the likely reason why your friend is receiving the shocks. Outline a simple investigation to test your prediction.

20. Suppose you have a part-time job assembling electronic components at a factory. The electronic components arrive at your metal work table after sliding through a plastic delivery tube. Some of the components you assemble do not work properly, and your manager is concerned. You think you know what could be damaging the electronic components. Write an e-mail message to your manager describing what you think is causing the damage to the components and predicting a way to solve this problem.

21. The data in the table below compare sources used to generate electrical energy in Ontario and Alberta.

  a) Make a bar graph based on the data.
  b) Identify the major differences in the sources used by the two provinces.
  c) Predict possible reasons why these differences exist.

Sources of Electrical Energy in Ontario and Alberta

Source	Ontario (percentage)	Alberta (percentage)
Nuclear	52	—
Hydroelectric	21	7
Coal	18	49
Natural gas	8	38
Wind and other	1	6

22. In the past, strings of decorative lights were connected in circuits like the one below.

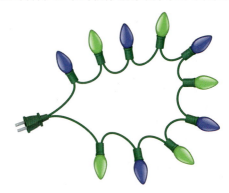

  a) Predict what would happen in this circuit if one bulb burned out.
  b) If one bulb burned out and you had another to replace it, describe how you would find out where the problem in the circuit occurred.
  c) Draw a circuit for another string of lights that are connected in a more practical way.

Unit 4 Review

Communication C

23. Use a cartoon, a drawing, or a story to explain how a generator produces electrical energy.

24. Do research to determine how a technological device called an electrostatic precipitator uses static electricity to control pollution. In your notebook, draw and label a diagram of an electrostatic precipitator and write a caption that briefly explains how this device works.

25. Draw a diagram to show the methods that you can use to charge an object.

26. In your notebook, draw a cartoon, create a diagram, or write a blog explaining the dangers associated with plugging too many appliances into a single circuit.

27. At the start of this unit, you read the lyrics for a song called "Electricity." Write the lyrics for your own song with the same title. The song can be about anything related to the concepts and skills you have learned in this unit.

Application A

28. The six objects in the picture below are all electrical loads.

   a) List the objects in the order in which you think they would use energy. Start with the object that would use the most energy, and end with the object that would use the least.

   b) Look at the top two items in your list. What form of energy do they convert electrical energy into? Do you think this is a coincidence? Explain your answer.

29. Describe the factors that contribute to the environmental costs of burning fossil fuels to generate electrical energy.

30. Most metals are good conductors of electricity. However, the metal wire used in the heating element in a toaster is not a good conductor. Look back at **Figure 4.16**. What properties are needed for a wire in a toaster, compared with copper wire?

31. Suppose you want to connect speakers in a bedroom to the multi-room audio system located in a living room. Look back at **Figure 4.28** and **Figure 4.29**. Identify the type of circuit that would ensure that the speakers receive the strongest possible signal. Explain why this circuit would be the best.

32. An environmental website claims that a large coal-burning plant emits more radioactive materials than a nuclear plant that has the same generating capacity. Do research to investigate this claim. Determine if the claim is accurate or inaccurate and support your position by quoting information from at least two sources.

33. Examine the circuits below.

   a) Use symbols to draw circuit diagrams for these circuits.

   b) Label each of your circuit diagrams to indicate if the light bulbs should be on or off. Explain how you know.

Literacy Test Prep

Read the selection below, and answer the questions that follow it.

> On June 15, 1752, Benjamin Franklin launched a kite into the dark clouds of a developing storm. He correctly assumed that the thunderclouds would have a static charge before there was a lightning strike. His goal was to collect the electricity from these storm clouds. Had lightning actually struck his kite, the precautions that Franklin had put in place would not have been enough to prevent his being electrocuted.
>
> Franklin's apparatus consisted of a kite attached to a long hemp string tied to an iron key. This string was damp from the storm and therefore would conduct the electricity. Franklin held onto the kite by a dry silk string that was attached to the key. Franklin and the silk string were under cover so that they stayed dry. Franklin understood that electricity would not travel easily along the dry silk string. A further safety precaution was a metal wire also attached to the key that led to a Leyden jar. (A Leyden jar is a device that can store charges.)
>
> After flying the kite for a few minutes, Franklin brought his knuckles close to the iron key and a spark jumped from the key to his knuckles. This spark was identical to those produced by friction. Benjamin Franklin had demonstrated that lightning was caused by a build-up of charges in the storm clouds.

Multiple Choice

In your notebook, record the best or most correct answer.

34. Benjamin Franklin believed that
    a) thunderclouds would have current electricity before a lightning strike
    b) thunderclouds would have static electricity before a lightning strike
    c) thunderclouds made electricity when they rubbed together
    d) electricity could be harvested from thunderclouds

35. Before flying his kite, Franklin took some safety precautions that included
    a) wearing rubber boots and staying under cover
    b) attaching a lightning rod and a Leyden jar to his apparatus
    c) holding onto the kite by a dry silk string and staying under cover
    d) tying a long hemp string with an iron key to the kite

36. A Leyden jar is a device that can
    a) store charges
    b) store rainwater
    c) hold a key
    d) produce electricity

37. Franklin attached the Leyden jar to his kite with a(n)
    a) long hemp string
    b) dry silk string
    c) metal wire
    d) iron key

38. Franklin flew his kite for
    a) a few seconds
    b) a few minutes
    c) a few hours
    d) a few days

39. After flying his kite, Franklin observed
    a) a spark jumping from his knuckles to the key
    b) a spark jumping from the key to his knuckles
    c) a spark jumping from the key to the kite
    d) a spark jumping from the kite to the key

Written Answer

40. Summarize this selection. Include the main idea and two relevant points that support it.

Guide to the Toolkits and Appendices

TOOLKITS

Science Skills Toolkit 1 Analyzing Issues—Science, Technology, Society, and the Environment 335

Science Skills Toolkit 2 Scientific Inquiry 339

Science Skills Toolkit 3 Technological Problem Solving 344

Science Skills Toolkit 4 Estimating and Measuring 346

Science Skills Toolkit 5 Precision and Accuracy 352

Science Skills Toolkit 6 Scientific Drawing 354

Science Skills Toolkit 7 Using Models and Analogies in Science 356

Science Skills Toolkit 8 How to Do a Research-Based Project 358

Science Skills Toolkit 9 Using Electric Circuit Symbols and Meters 362

Science Skills Toolkit 10 Creating Data Tables 365

Numeracy Skills Toolkit 1 Scientific Notation 366

Numeracy Skills Toolkit 2 Significant Digits and Rounding 367

Numeracy Skills Toolkit 3 The Metric System 368

Numeracy Skills Toolkit 4 Organizing and Communicating Scientific Results with Graphs 370

Numeracy Skills Toolkit 5 The GRASP Problem-Solving Method 376

Literacy Skills Toolkit 1 Preparing for Reading 377

Literacy Skills Toolkit 2 Reading Effectively 381

Literacy Skills Toolkit 3 Reading Graphic Text 385

Literacy Skills Toolkit 4 Word Study 388

Literacy Skills Toolkit 5 Organizing Your Learning: Using Graphic Organizers 390

APPENDIX

Properties of Common Substances 396

Science Skills Toolkit 1

Analyzing Issues—Science, Technology, Society, and the Environment

Can you think of an issue that involves science, technology, society, and the environment? An **issue** is a topic that can be seen from more than one point of view. How about the use of salt to de-ice roads in the winter? Roads are safer in winter when they are clear of ice and snow.

In a conversation with a friend, however, you find out that road salt may damage the environment. How might you use science and technology to solve this problem?

Suppose your town council is in the process of deciding whether to expand its road salting program. How will you analyze this issue and determine what action to take? The concept map on this page shows a process to help you focus your thinking and stay on track.

A Process for Analyzing Issues

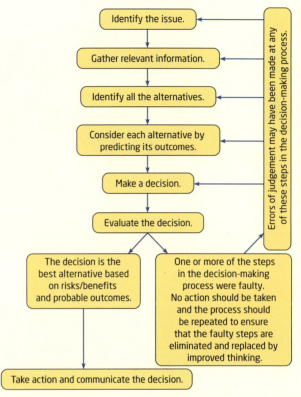

Identifying the Issue

Soon after talking with your friend about road salting, you go to your friend's house. You find your friend sitting in front of the computer, writing a letter to the town council. In it, your friend is asking that the salting program not be expanded to your area.

Gathering Information

Once you have identified the issue, you will need to find out more information.

The Internet and other sources, such as books or experts, are great places to find information about an issue. One thing that is important to do when gathering information is to look for bias.

Bias is a personal and possibly unreasonable judgement of an issue. For example, a person who makes his or her living putting salt on the roads may have a bias that salt does not harm the environment. It is important to check the source of information to determine whether it is unbiased. Refer to **Science Skills Toolkit 8** for more information about how to research information.

Another important part of gathering information is taking notes so that you can analyze what you have learned. You may read about different viewpoints or solutions and advantages and disadvantages for each one. It is helpful to be able to organize your notes in the form of a graphic organizer such as a concept map, a flowchart, or a Venn diagram. You will find information on using graphic organizers in **Literacy Skills Toolkit 5**.

Identifying and Considering Alternatives

Your research may lead you to ask new questions about alternative solutions and how successful they might be. For example, you might think about how a combination of salt and sand would work to keep roads clear of ice. Would this be a safer environmental alternative? Answering these questions often leads to more research or possibly doing your own scientific inquiry.

Making a Decision

When you have all of the information that your research can provide, you will need to weigh the pros and cons of each option and make a decision. Sometimes it helps to organize your thoughts in a PMI chart that lists the pros and cons of an issue, or a t-chart that compares two possible solutions. You will find more information on using these charts in **Literacy Skills Toolkit 5**. It might even be helpful to rate how important you feel each point (pro or con) is.

PMI Chart for Salt Use

Plus	Minus	Interesting
• very effective	• may contaminate drinking water	•
• relatively inexpensive		•
•	•	
•	•	

t-chart Comparing Salt and Sand

Salt	Sand
• more effective than sand	• not as harmful to organisms as salt
•	•
•	•

Your decision will still involve some very human and personal elements. People have strong feelings about the social and environmental issues that affect them. Depending on their point of view, other people may feel differently than you do about an issue. Something that seems obvious to you might not be so obvious to them, and vice versa. Even the unbiased scientific evidence you found during your research might not change people's minds. If you are going to encourage a group to make what you consider a good decision, you have to find ways to persuade the group to think as you do.

Evaluating the Decision

After you have made a decision, it is important to evaluate your decision. Is the decision the best alternative considering the risks and benefits? Have you thought about the possible consequences of the decision and how you might respond to them? If you determine that your decision-making process was faulty—if, for example, you based your decision on information that you later learned was false—you should begin again. If you find that you are comfortable with your decision, the next step is to take action.

Taking Action

Issues rarely have easy answers. People who are affected have differing, valid points of view. It is easier for you to act as an individual, but if you can persuade a group to act, you will have greater influence. In the issue discussed here, you might write a letter to your town council. As a compromise, you might suggest a combination of salt and sand on the roads. Your research can provide you with appropriate statistics. As a group, you could attend a town council meeting or sign a petition to make your views known.

Over time, you can assess the effects of your actions: Are there fewer accidents if less salt is used on roads? Does less salt end up in the water than when more salt alone is used?

Sometimes taking action involves changing the way you do things. After you have presented your findings to the town council, one of your friends makes you stop and think. "I have noticed you putting a lot of salt out on your sidewalk," your friend says. "You could use a bit of time and muscle power to chip away the ice, but that is not the choice you make."

> **Instant Practice—Analyzing Issues**
>
> We live in an energy-intensive society. One of the most common sources of the energy we use is fossil fuels. Complete the following exercise in a group of four.
>
> 1. Start by dividing your group into two pairs.
> 2. One pair will research and record the advantages of using fossil fuels and how this use has affected members of our society in a positive way.
> 3. The second pair will research and record the disadvantages of using fossil fuels and their negative impacts on society.
> 4. The pairs will then regroup, and both sides can present their findings. Record key points on a PMI chart or a t-chart for comparison.
> 5. Determine which pair has the more convincing evidence for its point of view concerning the use of fossil fuels.
> 6. As a group, research alternative energy sources, including advantages and disadvantages of each. Determine the best alternative, based on the information you found in steps 2 and 3 above.

You realize your friend is right—it is not only up to the town council or any other group to act responsibly; it is also up to you and your friends. How easy is it for you to give up an easy way of doing a task in order to make an environmentally responsible decision?

Science Skills Toolkit 2

Scientific Inquiry

Scientific inquiry is a process that involves many steps, including making observations, asking questions, performing investigations, and drawing conclusions. These steps may not happen in the same order in each inquiry. However, one model of the scientific inquiry process is shown here:

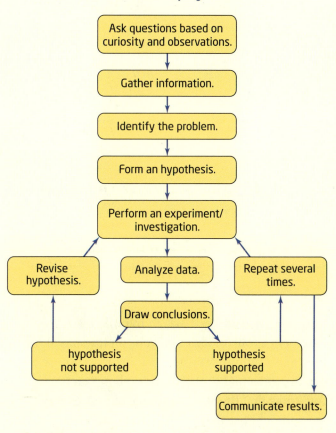

The Scientific Inquiry Process

You may have an idea about what happened to the puddle, but you need evidence that supports your idea. In order to test your idea, you need to carry out a scientific inquiry.

Gathering Information and Identifying the Problem

First, you might observe what happens to some other puddles. You would watch them closely until they disappeared and record what you observed.

Making Observations and Asking Questions

The rain has stopped, and the Sun is out. You notice that a puddle of water has disappeared from the sidewalk.

SCIENCE SKILLS TOOLKIT 2 • MHR **339**

One observation you might make is "The puddle is almost all gone." That would be a **qualitative observation**, an observation in which numbers are not used. A little later, you might also say, "It took five hours for the puddle to disappear completely." You have made a **quantitative observation**, an observation that uses numbers.

Although the two puddles were the same size, one disappeared (evaporated) much more quickly than the other one did. Your quantitative observations tell you that one evaporated in 4 h, whereas the other one took 5 h. Your qualitative observations tell you that the one that evaporated more quickly was in the sunlight. The one that evaporated more slowly was in the shade. You now have identified one problem to solve: Does water always evaporate more quickly in the sunlight than in the shade?

> **Instant Practice—Making Qualitative and Quantitative Observations**
>
> Copy the observations below in your notebook. Beside each observation, write "Qual" if you think it is a qualitative observation and "Quan" if you think it is a quantitative observation.
>
> 1. a. The bowling ball is heavier than the basketball.
>   b. The red ball weighs 5 g more than the blue ball.
> 2. a. The temperature increased by several degrees.
>   b. The temperature increased by 2°C.
> 3. a. The water was lukewarm.
>   b. The water was cooler than the oil.
> 4. a. The owl ate 3 mice.
>   b. The owl was larger than the nighthawk.
> 5. a. The second light bulb was the brightest.
>   b. The 60 W bulb was brighter than the 40 W bulb.
> 6. a. The colour of the surface water in the lake was green.
>   b. The lake contained 15 species of fish.

Stating an Hypothesis

Now you are ready to make an **hypothesis**, a statement about an idea that you can test, based on your observations. Your test will involve comparing two things to find the relationship between them. You know that the Sun is a source of heat energy, so you might use that knowledge to make this hypothesis: If a puddle of water is in the sunlight, then the water will evaporate faster than if the puddle is in the shade.

Instant Practice—Stating an Hypothesis

Write an hypothesis for each of the following situations. You may wish to use an "If…then…" format. For example: *If* temperature affects bacterial growth, *then* bacterial culture plates at a higher temperature will have more bacterial colonies than those at a lower temperature.

1. The relationship between temperature and the state of water
2. The relationship between types of atmospheric gases and global warming
3. The amount of time batteries last in different devices
4. The effect of the colour of flowers on honeybee visitations

Performing an Investigation

As you know, there are several steps involved in performing a scientific investigation, including identifying variables, designing a fair test, and organizing and analyzing data.

Identifying Variables

As you prepare to make your observations, you can make a **prediction**, a forecast about what you expect to observe. In this case, you might predict that puddles A, B, and C will dry up more quickly than puddles X, Y, and Z. A prediction will help you to decide whether your hypothesis is correct. In the case of the puddles, if puddles A, B, and C do not dry up more quickly than puddles X, Y, and Z, you'll know that your hypothesis was likely incorrect.

The breeze is one factor that could affect evaporation. The Sun is another factor that could affect evaporation. Scientists think about every possible factor that could affect tests they conduct. These factors are called **variables**.
It is important to test only one variable at a time.

You need to control your variables. This means that you change only one variable at a time. The variable that you change is called the **independent variable** (also called the manipulated variable). In this case, the independent variable is the condition under which you observe the puddle (one variable would be sunlight; another would be wind).

According to your hypothesis, sunlight will change the time it takes for the puddle to evaporate. The time in this case is called the **dependent variable** (also called the responding variable).

Often, experiments have a **control**. This is a test that you carry out with no variables, so that you can observe whether your independent variable does indeed cause a change. Look at the illustration below to see some examples of controls and variables.

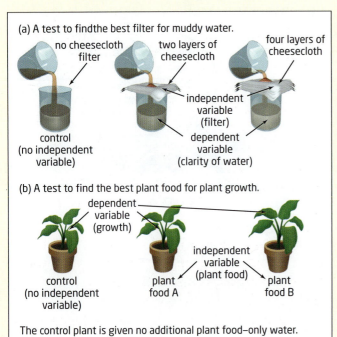

(a) A test to find the best filter for muddy water.
- no cheesecloth filter
- two layers of cheesecloth
- four layers of cheesecloth
- control (no independent variable)
- independent variable (filter)
- dependent variable (clarity of water)

(b) A test to find the best plant food for plant growth.
- dependent variable (growth)
- control (no independent variable)
- plant food A
- plant food B
- independent variable (plant food)

The control plant is given no additional plant food—only water.

Instant Practice—Identifying Variables

For each of the following questions, state your control, your independent variable, and your dependent variable.

1. Does light travel the same way through different substances?
2. Does adding compost to soil promote vegetable growth?
3. How effective are various kinds of mosquito repellent?

Controlling Variables for a Fair Test A controlled experiment tests only one variable at a time, while keeping all other variables constant. If you consider more than one variable in a test, you are not conducting a **fair test** (one that is valid and unbiased), and your results will not be useful. You will not know whether the breeze or the Sun made the water evaporate.

As you have been reading, a question may have occurred to you: How is it possible to do a fair test on puddles? How can you be sure that they are the same size? In situations such as this one, scientists often use **models**. A model can be a mental picture, a diagram, a working model, or even a mathematical expression. To make sure your test is fair, you can prepare model puddles that you know are all exactly the same.

You can then place the puddles in controlled conditions, where all variables *except* for sunlight remain constant. For instance, you might construct a cardboard wall around your model puddles to ensure that the wind conditions will be the same for all puddles. You might even carry out your test in a laboratory, using a lamp as a model Sun. **Science Skills Toolkit 7** gives you more information on using models.

Before you begin your investigation, review safety procedures and identify what safety equipment you may need. Refer to page xii in this textbook for more information on safety.

Recording and Organizing Data Another step in performing an investigation is recording and organizing your data. Often, you can record your data in a table like the one shown below. Refer to **Science Skills Toolkit 10** for more information on making tables.

Table 1 Puddle Evaporation Times

Puddle	Evaporation Time (min)
A	37
B	34
C	42
X	100
Y	122
Z	118

Analyzing and Presenting Data After recording your data, the next step is to present your data in a format so that you can analyze it. Often, scientists make a graph, such as the bar graph below. For more information on constructing graphs, refer to **Numeracy Skills Toolkit 4**.

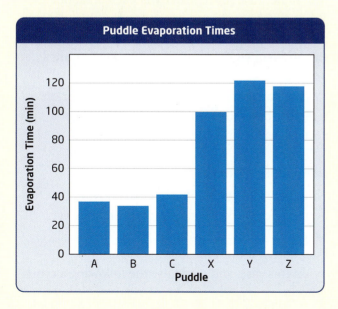

Forming a Conclusion

Many investigations are much more complex than the one described here, and there are many more possibilities for error. That is why it is so important to record careful qualitative and quantitative observations.

After you have completed all your observations, you are ready to analyze your data and draw a **conclusion**. A conclusion is a statement that indicates whether your results support or do not support your hypothesis. First you need to consider whether your predictions were correct. Then you should ask yourself whether you have considered all of the variables. Then you can decide whether your results support your hypothesis. If you had hypothesized that sunlight would have no effect on the evaporation of water, your results would not support your hypothesis. An hypothesis gives you a place to start and helps you design your experiment. If your results do not support your hypothesis, you use what you have learned in the experiment to come up with a new hypothesis to test.

Scientists often set up experiments without knowing what will happen. Sometimes they deliberately set out to show that something will *not* happen in a particular situation.

Eventually, when an hypothesis has been thoroughly tested and nearly all scientists agree that the results support the hypothesis, it becomes a **theory**.

Science Skills Toolkit 3

Technological Problem Solving

Technology is the use of scientific knowledge, as well as everyday experience, to solve practical problems. Have you ever used a pencil to flip something out of a tight spot where your fingers could not reach? Have you ever used a stone to hammer bases or goal posts into the ground? Then you have used technology. You may not know why your pencil works as a lever or the physics behind levers, but your everyday experiences tell you how to use a lever successfully.

A Process for Technological Problem Solving

People turn to technology to solve problems. One problem-solving model is shown below.

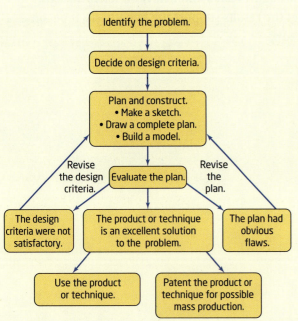

Solving a Technological Problem

Identifying the Problem

When you used that pencil to move the small item you could not reach, you did so because you needed to move that item. In other words, you had identified a problem that needed to be solved. Clearly identifying a problem is a good first step in finding a solution. In the case of the lever, the solution was right before your eyes, but finding a solution is not always quite so simple.

Suppose school is soon to close for a 16-day winter holiday. Your science class has a hamster whose life stages the class observes. Student volunteers will take the hamster home and care for it over the holiday. However, there is a three-day period when no one will be available to feed the hamster. Leaving extra food in the cage is not an option because the hamster will eat it all at once. What devices could you invent to solve this problem?

First, you need to identify the exact nature of the problem you have to solve. You could state it as follows:

The hamster must receive food and water on a regular basis so that it remains healthy over a certain period and does not overeat.

Identifying Criteria

Now, how will you be able to assess how well your device works? You cannot invent a device successfully unless you know what criteria (standards) it must meet.

In this case, you could use the following as your criteria.

1. The device must feed and water the hamster.
2. The hamster must be healthy at the end of the three-day period.
3. The hamster must not appear to be "overstuffed."

How could you come up with such a device? On your own, you might not. If you work with a team, however, each of you will have useful ideas to contribute.

Planning and Constructing

You will probably come up with some good ideas on your own. Like all other scientists, though, you will want to use information and devices that others have developed. Do some research and share your findings with your group. Can you modify someone else's idea? With your group, brainstorm some possible designs. How would the designs work? What materials would they require? How difficult would they be to build? How many parts are there that could stop working during the three-day period? Make a clear, labelled drawing of each design, with an explanation of how it would work.

Examine all of your suggested designs carefully. Which do you think would work best? Why? Be prepared to share your choice and your reasons with your group. Listen carefully to what others have to say. Do you still feel yours is the best choice, or do you want to change your mind? When the group votes on the design that will be built, be prepared to co-operate fully, even if the group's choice is not your choice.

Get your teacher's approval of the drawing of the design your group wants to build. Then gather your materials and build a model of your design. Experiment with your design to answer some questions you might have about it. For example, should the food and water be provided at the same time? Until you try it out, you may be unsure if it is possible (or even a good idea) for your invention to deliver both food and water at the same time. Keep careful records of each of your tests and of any changes you make to your design.

You might find, too, that your invention fails in a particular way. Perhaps it always leaks at a certain point where two parts are joined. Perhaps the food and water are not kept separate. Perhaps you notice a more efficient way to design your device as you watch it operate. Make any adjustments and test them so that your device works in the best and most efficient way possible.

Evaluating

When you are satisfied with your device, you can demonstrate it and observe devices constructed by other groups. Evaluate each design in terms of how well it meets the design criteria. Think about the ideas other groups tried out and why they work better than (or not as well as) yours. What would you do differently if you were to redesign your device?

Science Skills Toolkit 4

Estimating and Measuring

Estimating

How long will it take you to read this page? How heavy is this textbook? You could probably answer these questions by **estimating**—making an informed judgement about a measurement. An estimate gives you an idea of a particular quantity but is not an exact measurement.

For example, if you were responsible for maintaining the grounds in a local park, you might need to know how many insects live in the park. Counting every insect would be very time-consuming. What you can do is count the number of insects in a typical square-metre area.

Then, multiply the number of insects by the number of square metres in the total area you are investigating. This will give you an estimate of the total population of insects in that area.

Number of insects per square meter x total area of site = number of insects on site (estimate)

Measuring Length and Area

You can use a metre stick or a ruler to measure short distances. These tools are usually marked in centimetres and/or millimetres. Use a ruler to measure the length in millimetres between points A and C, C and E, E and B, and A and D below. Convert your measurements to centimetres and then to metres.

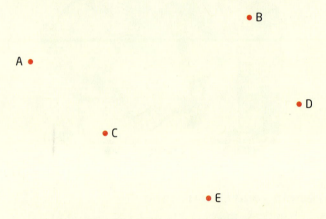

To calculate an area, you can use length measurements. For example, for a square or a rectangle, you can find the area by multiplying the length by the width.

Example 1

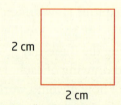

Area of square is 2 cm × 2 cm = 4 cm^2

Example 2

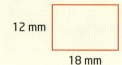

Area of rectangle is 18 mm × 12 mm = 216 mm^2

Make sure you always use the same units—if you mix up centimetres and millimetres, your calculations will be wrong. You will find more information on converting between metric units in **Numeracy Skills Toolkit 3**.

Example 3

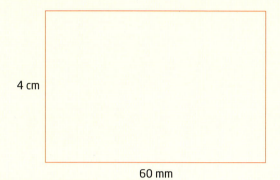

Area of rectangle is 4 cm × 60 mm
= 40 mm × 60 mm
= 2400 mm^2
OR
= 4 cm × 6 cm
= 24 cm^2

Remember to ask yourself if your answer is reasonable (you could make an estimate to consider this). Look at the square in Example 1 on the previous page. It had an area of 4 cm^2. The area calculated for the rectangle in Example 3 on this page was 24 cm^2. Would you estimate that you could fit 6 (24 divided by 4) of those small squares into the large rectangle? This shows you that your answer is reasonable.

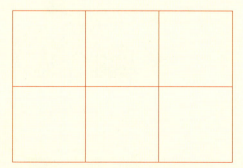

Six of the squares from Example 1 will fit inside the rectangle from Example 3

> **Instant Practice—Estimating and Measuring**
>
> Imagine that all rulers in the school have vanished. The only measurement tool that you now have is a toothpick.
>
> 1. Estimate the length and width of your textbook in toothpick units. Compare your estimates with a classmate's estimates.
>
>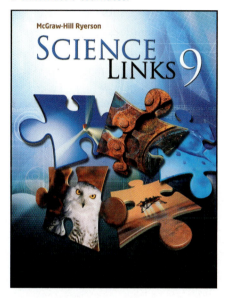
>
> 2. Measure the length and width of your textbook with your toothpick. How close was your estimate to the actual measurement?
>
> 3. If you had a much larger area to measure, such as the floor of your classroom, what could you use instead of toothpicks to measure the area? (Be creative!)
>
> 4. What is your estimate of the number of units you chose (in question 3) for the width of your classroom?

Measuring Volume

The **volume** of an object is the amount of space that the object occupies. There are several ways of measuring volume, depending on the kind of object you want to measure.

As you can see in Diagram A below, the volume of a regularly shaped solid object can be measured directly. You can calculate the volume of a cube by multiplying its sides, as shown on the left in Diagram A. You can calculate the volume of a rectangular solid by multiplying its length × width × height, as shown on the right in Diagram A.

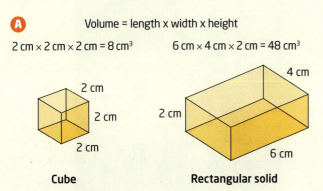

Measuring the volume of a regularly shaped solid

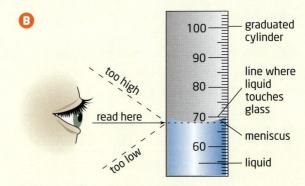

Measuring the volume of a liquid

If all the sides of a solid object are measured in millimetres (mm), the volume will be in cubic millimetres (mm^3). If all the sides are measured in centimetres (cm), the volume will be in cubic centimetres (cm^3). The units for measuring the volume of a solid are called cubic units.

The units used to measure the volume of liquids are called capacity units. The basic unit of volume for liquids is the litre (L). Recall that 1 L = 1000 mL.

Cubic units and capacity units are interchangeable. For example,

$1\ cm^3 = 1\ mL$
$1\ dm^3 = 1\ L$
$1\ m^3 = 1\ kL$

The volume of a liquid can be measured directly, as shown in Diagram B. Make sure you measure to the bottom of the **meniscus**, the slight curve where the liquid touches the sides of the container. To measure accurately, make sure your eye is at the same level as the bottom of the meniscus.

Instant Practice— Measuring Volume

Determine the volume of liquids present in the three graduated cylinders shown here.

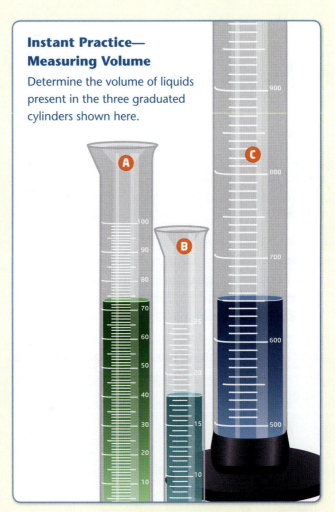

The volume of an irregularly shaped solid object, however, must be measured indirectly, as shown in Diagram C below. This is done by measuring the volume of a liquid it displaces.

When a solid object is placed in liquid, the liquid level will rise. The liquid is displaced, or moved from the place it was originally. The volume of the displaced liquid is equal to the volume of the solid.

1. Record the volume of the liquid.
2. Carefully lower the object into the cylinder containing the liquid. Record the volume again.
3. The volume of the object is equal to the difference between the two volumes of the liquid. The equation below the photographs shows you how to calculate this volume.

Measuring the volume of an irregularly shaped solid

volume of object = volume of water with object
       – original volume of water
    = 85 mL – 60 mL
    = 25 mL

Measuring Mass

Is your backpack heavier than your friend's backpack? You can check by holding a backpack in each hand. The **mass** of an object is the amount of matter in a substance or object. Mass is measured in milligrams, grams, kilograms, and tonnes. You need a balance for measuring mass.

How can you find the mass of a certain quantity of a substance, such as table salt, that you have added to a beaker? First, find the mass of the beaker. Next, pour the salt into the beaker and find the mass of the beaker and salt together. To find the mass of the salt, simply subtract the beaker's mass from the combined mass of the beaker and salt.

If you are using an electronic balance, you will not need to do any calculations to subtract the mass of the beaker. The balance will do the calculation for you. To measure the contents of a beaker, you can place the empty beaker on the balance and hit the "Tare" or "Zero" or "Re-zero" button to reset the balance to zero. Then add the material to be measured into the beaker. The balance subtracts the mass of the beaker before the contents are even added, so it reports only the mass of the contents.

Instant Practice—Measuring Mass

Use the following information to determine the mass of the table salt. The mass of a beaker is 160 g. The mass of the table salt and beaker together is 230 g.

Measuring Angles

You can use a protractor to measure angles. Protractors usually have an inner scale and an outer scale. The scale you use depends on how you place the protractor on an angle (symbol = ∠). Look at the following examples to learn how to use a protractor.

Example 1

What is the measure of ∠XYZ?

Solution

Place the centre of the protractor on point Y. The 0° 180° line should lie along the line YX so that YX crosses 0° on the inner scale. YZ crosses 70° on the inner scale. So ∠XYZ is equal to 70°.

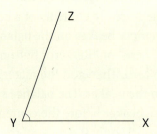

Example 2

Draw ∠ABC = 155°.

Solution

First, draw a straight line, AB. Place the centre of the protractor on B and line up AB with 0° on the outer scale. Mark C at 155° on the outer scale. Join BC. The angle you have drawn, ∠ABC, is equal to 155°.

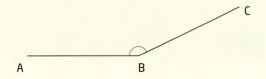

Instant Practice—Measuring Angles

1. State the measure of each of the following angles using the following diagram.
   a. DAF
   b. DAH
   c. IAG
   d. HAF
   e. GAD
   f. DAI
   g. EAG
   h. EAI

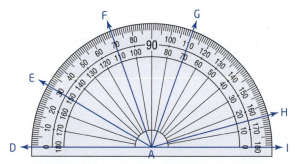

2. Use a protractor to draw angles with the following measurements. Label each angle.
   a. ABC 50°
   b. QRS 85°
   c. XYZ 5°
   d. JKL 45°
   e. HAL 90°

Measuring Temperature

Temperature is a measure of the thermal (heat) energy of the particles of a substance. In the very simplest terms, you can think of temperature as a measure of how hot or how cold something is. The temperature of a material is measured with a thermometer.

For most scientific work, temperature is measured on the Celsius scale. On this scale, the freezing point of water is zero degrees (0°C) and the boiling point of water is 100 degrees (100°C). Between these points, the scale is divided into 100 equal divisions. Each division represents one degree Celsius. On the Celsius scale, average human body temperature is 37°C, and a typical room temperature may be between 20°C and 25°C.

Sometimes scientists use a different unit of temperature called the kelvin (K). Zero on the Kelvin scale (0 K) is the coldest possible temperature. This temperature is also known as absolute zero. It is equivalent to −273°C, which is about 273 degrees below the freezing point of water. Notice that degree symbols are not used with the Kelvin scale.

Most laboratory thermometers are marked only with the Celsius scale. Because the divisions on the two scales are the same size, the Kelvin temperature can be found by adding 273 to the Celsius reading. This means that on the Kelvin scale, water freezes at 273 K and boils at 373 K.

Tips for Using a Thermometer

When using a thermometer to measure the temperature of a substance, here are three important tips to remember.

- Handle the thermometer extremely carefully. It is made of glass and can break easily.
- Do not use the thermometer as a stirring rod.
- Do not let the bulb of the thermometer touch the walls of the container.

Instant Practice—Measuring Temperature

Read the temperature, in °C, from the thermometer in each question. Convert your Celsius reading into Kelvin units.

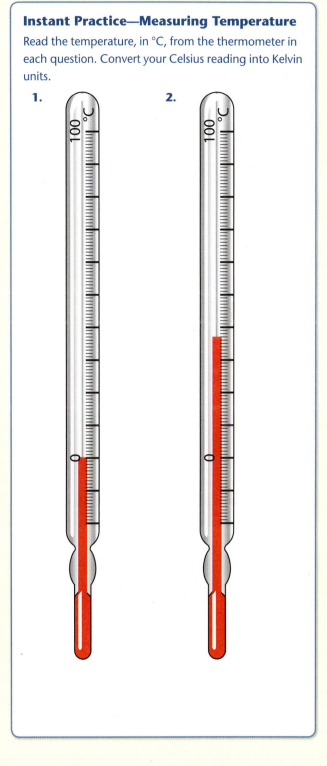

Science Skills Toolkit 5

Precision and Accuracy

No measuring device can give an absolutely exact measure. So how do scientists describe how close an instrument comes to measuring the true result? Quantitative data from any measuring device are uncertain. You can describe this uncertainty in terms of precision and accuracy.

Precision The term **precision** describes both the exactness of a measuring device and the range of values in a set of measurements. The precision of a measuring instrument is usually half the smallest division on its scale.

For example, ruler A below is divided in millimetres, so it is precise to ± 0.05 cm. The length of the object below the ruler would be reported as 8.7 ± 0.05 cm, because it is closer to 8.7 than 8.8 cm. Ruler B below is divided in centimetres, so it is precise to ± 0.5 cm. The length of the object above the ruler would be reported as 9.0 ± 0.5 cm, because it is closer to 9 cm than to 8 cm, and the uncertainty must be included in the measurement.

A precise measuring device will give nearly the same result every time it is used to measure the same object. Consider the following measurements of a 50 g object on a balance. Both give the same average mass, but Scale B is more precise because it has a smaller range of measured values (± 0.3 versus ± 0.5).

Table 1 Measurements of Mass on Two Scales

	Scale A Mass (g)	Scale B Mass (g)
Trial 1	49.9	49.9
Trial 2	49.8	50.2
Trial 3	50.3	49.9
Average	50.0	50.0
Range	± 0.5	± 0.3

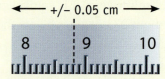

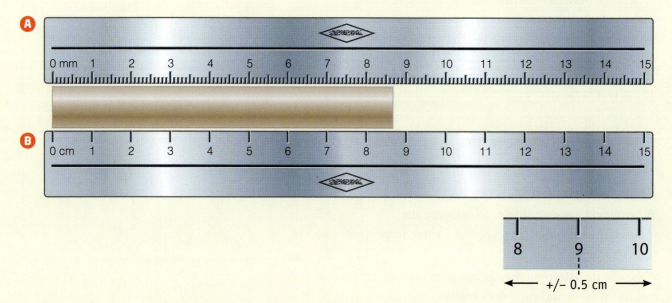

Accuracy How close a measurement or calculation comes to the true value is described as **accuracy**. To improve accuracy, scientific measurements are often repeated and combined mathematically. The average measurements in the table on the previous page are more accurate than any of the individual measurements.

The darts in diagram A below are very precise, but they are not accurate because they did not hit the bull's-eye. The darts in diagram B are neither precise nor accurate. However, the darts in diagram C are both precise and accurate.

A precise but not accurate **B** neither precise nor accurate **C** precise and accurate

> **Instant Practice—Precision and Accuracy**
>
> 1. A student measures the temperature of ice water four times, and each time gets a result of 10.0°C. Is the thermometer precise and accurate? Explain your answer.
>
> 2. Two students collected data on the mass of a substance for an experiment. Each student used a different scale to measure the mass of the substance over three trials. Student A had a range of measurements that was ±0.06 g. Student B had a range of measurements that was ±0.11 g. Which student had the more precise scale?
>
> 3. You want to get an accurate and precise measure of your height. What do you think is the best way to do this? What did you need to consider to make your decision?

Scientific Drawing

Have you ever used a drawing to explain something that was too difficult to explain in words? A clear drawing can often assist or replace words in a scientific explanation. Drawings are especially important when you are trying to explain difficult concepts or describe something that contains a lot of detail. It is important to make scientific drawings clear, neat, and accurate.

Making a Scientific Drawing

Follow these steps to make a good scientific drawing.

1. Use unlined paper and a sharp pencil with an eraser.

2. Give yourself plenty of space on the paper. You need to make sure that your drawing will be large enough to show all necessary details. You also need to allow space for labels. Labels identify parts of the object you are drawing. Place all of your labels to the right of your drawing, unless there are so many labels that your drawing looks cluttered.

3. Carefully study the object that you will be drawing. Make sure you know what you need to include.

4. Draw only what you see, and keep your drawing simple. Do not try to indicate parts of the object that are not visible from the angle of observation. If you think it is important to show another part of the object, do a second drawing, and indicate the angle from which each drawing is viewed. The diagram to the right includes both a front view and a side view of a wheel-and-axle system.

5. Shading or colouring is not usually used in scientific drawings. If you want to indicate a darker area, you can use stippling (a series of dots). You can use double lines to indicate thick parts of the object.

6. If you do use colour, try to be as accurate as you can. Choose colours that are as close as possible to the colours in the object you are observing.

7. Label your drawing carefully and completely, using lower-case (small) letters. Think about what you would need to know if you were looking at the object for the first time. If you are comparing two objects, label each object and use labels to indicate the points of comparison between them. Remember to place all your labels to the right of the drawing, if possible. Use a ruler to draw a horizontal line from the label to the part you are identifying. Make sure that none of your label lines cross.

8. Give your drawing a title. The drawing shown below is from a student's notebook. This student used horizontal label lines for the parts viewed, and a title—all elements of an excellent final drawing.

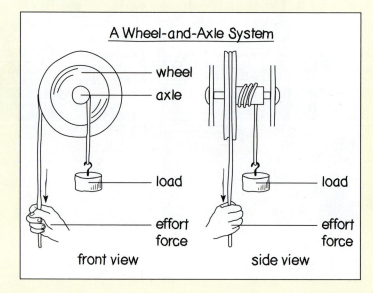

Drawing Circuit Diagrams

When drawing a diagram of an electric circuit, use the following criteria:
- Draw your diagram using a ruler.
- Make all connecting wires and leads straight lines with 90° (right-angle) corners.
- If possible, do not let conductors cross over one another.
- Your finished drawing should be rectangular or square.

An example of a proper circuit diagram is shown below.

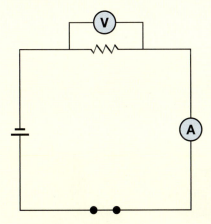

The following symbols are used in circuit diagrams:

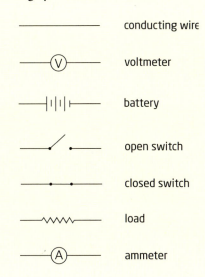

Instant Practice—Scientific Drawings

1. Draw and label a diagram of a pair of scissors. Someone who has never seen a pair of scissors before should be able to use your diagram to understand how scissors work. Remember to use stippling or double lines rather than shading to show darker or thicker parts of the scissors.

2. Briefly describe the circuit shown in the circuit diagram on this page.

3. Draw a circuit diagram with at least four components. Write a brief description of the circuit shown by your diagram.

Science Skills Toolkit 7

Using Models and Analogies in Science

Scientists often use models and analogies to help communicate their ideas to other scientists or to students.

Using Models

When you think of a model, you might think of a toy such as a model airplane. Is a model airplane similar to a scientific model? If building a model airplane helps you learn about flight, then you could say it is a scientific model.

In science, a model is anything that helps you better understand a scientific concept. A model can be a picture, a mental image, a structure, or even a mathematical formula.

Sometimes, you need a model because the objects you are studying are too small to see with the unaided eye. In Unit 2, for example, you will see models used to represent atoms.

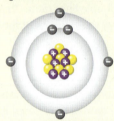

Atoms are so tiny that you cannot see them, even with the strongest of microscopes. A model of an atom can help you to form a mental picture that helps you understand the parts of the atom, even though it doesn't show exactly what an atom looks like.

Sometimes a model is useful because the objects you are studying are extremely large—the planets in our solar system, for example. In other cases, the object may be hidden from view, like the interior of Earth or the inside of a living organism.

A mathematical model can show you how to perform a calculation. If you wanted to explain addition and subtraction to a young child, you might use cookies as a model. By eating a cookie, you could demonstrate subtraction.

Chemical equations are models that are often used in science to help explain how a chemical reaction or series of reactions takes place. In Unit 1, you will see an equation used to represent the process of photosynthesis. Photosynthesis is a complicated process that involves many chemical reactions. An equation helps you to think about the starting materials and end products of the process.

light energy + carbon dioxide + water ⟶ glucose + oxygen

Scientists often use models to test an idea, to find out if an hypothesis is supported, and to plan new experiments in order to learn more about the subject they are studying. Sometimes, scientists discover so much new information that they have to modify their models. Examine the model shown in the photograph below. How can this model help you learn about science?

You can learn about day and night by using a globe and a flashlight to model Earth and the Sun.

> **Instant Practice—Using Models**
> How does using models help each of the following professionals in their work?
> a. architects
> b. aviation engineers
> c. theatre directors
> d. geographers
> e. landscape designers

Using Analogies

An **analogy** is a comparison between two things that have some characteristic in common. Scientists use analogies to help explain difficult concepts. For example, scientists sometimes refer to plants as the lungs of Earth. Recall that plants take in carbon dioxide (CO_2) from the atmosphere to use during photosynthesis. Plants then release the oxygen (O_2) produced by photosynthesis back into the atmosphere.

In a sense, the plants are "breathing" for Earth. When animals breathe, they take oxygen into their lungs and give off carbon dioxide.

Analogies use familiar situations to help explain unfamiliar situations. Picturing an everyday situation, such as the way water moves through a hose, may help you to picture an unfamiliar concept, such as how charge flows through an electric circuit. This is a useful analogy because most people have seen or used a hose, and have an understanding of how water moves through it.

Negative charges are pushed through a circuit in a similar way to how water is pushed through a hose.

Instant Practice—Using Analogies

1. Use an analogy to help explain how organisms in a food web are connected.
2. Look through Unit 4 of this text. Identify two analogies that are used in the unit.

Thinking about photosynthesis in this way may help you to understand the function that plants have in ecosystems. This analogy will only work, though, if you have an idea what lungs do. If you don't know anything about lungs or what they do, an analogy involving lungs won't help you to understand photosynthesis.

Science Skills Toolkit 8

How to Do a Research-Based Project

Imagine if your teacher simply stated that he or she wanted you to complete a research-based project on endangered species.

How do I get started?

This is a really big topic, and it is now your job to decide which smaller part of the topic you will research. One way to approach a research project is to break it up into four stages—exploring, investigating, processing, and creating.

Explore—Pick a Topic and Ask Questions

You need to start by finding out some general things about endangered species. Make a list of questions as you conduct your initial research, such as, What factors cause species to become endangered? Why does it matter? What types of species are endangered? Once you've done some research, you need to focus your topic into a research question.

What is a good research question?

Your research question needs to be specific enough that you can provide a thorough answer within the limits of your project (and in the time you have available). But it shouldn't be so specific that you can answer it in one paragraph!

Suppose, in the course of your research, you decided to learn more about polar bears. A good research question about polar bears would be, Why are polar bears endangered? An even better question could be, What can I do to help prevent polar bear extinction? Both of these questions are deep and can be subdivided into many subtopics.

Investigate—Research Your Topic

When putting together a research project, it is important to find reliable sources to help you answer your question. Before you decide to use a source that you find, you should consider whether it is reliable or whether it shows any bias.

Find Sources of Information There are many sources of information. For example, you can use a print resource, such as an encyclopedia from the reference section of the library.

Another approach is to go on-line and check the Internet. When you use the Internet, be careful about which sites you choose to search for information. You need to be able to determine the validity of a website before you trust the information you find on it.

How can I decide which websites to use?

To do this, check that the author is identified, a recent publication date is given, and the source of facts or quotations is identified. It is also important that the website is published by a well-known company or organization, such as a college, university, or government agency.

> **What if I can't find any sources of information?**

If you are having trouble finding *any* information about your topic, or if the only information you can find is on wiki or personal sites, you may want to consider changing your topic. You may also want to contact an expert on your topic. A credible expert has credentials showing his or her expertise in an area. For example, an expert may be a doctor or have a master's degree. Alternatively, an expert could have many years of experience in a specific career or field of study.

No matter which sources you use, it is your responsibility to be a critical consumer of information and to find trustworthy sources for your research.

You should also ask yourself if the sources you are using are primary or secondary. It is okay to use secondary sources, but you should try to include information from primary sources wherever possible.

> **How do I decide whether a source is reliable and unbiased?**

Two other things to check for in a source are reliability and bias. To check for reliability, try to find the same "fact" in two other sources. But keep in mind that even if you cannot find the same idea somewhere else, the source may still be reliable if it is a research paper or if it was written by an author with strong credentials. To check for bias, look for judgemental statements. Does the author tend to favour one side of an issue more than another? Are all sides of an issue treated equally? A good source shows little bias.

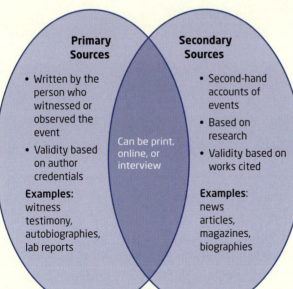

Primary Sources
- Written by the person who witnessed or observed the event
- Validity based on author credentials

Examples: witness testimony, autobiographies, lab reports

Can be print, online, or interview

Secondary Sources
- Second-hand accounts of events
- Based on research
- Validity based on works cited

Examples: news articles, magazines, biographies

Source	Information	Reliability	Bias	Questions I Have
The Canadian Encyclopedia website	Polar bears inhabit ice and coastlines of arctic seas.	• author: Brian Knudsen • secondary source • has links to external sites that are reliable	only lists facts	• Why do they live on ice? • Why don't they move south?
Polar Bears International website	shrinking sea ice habitat	• date at bottom of page 2009 • non-profit organization	designed to save the polar bear	• Why is the ice shrinking?

Record Information

How can I organize my research?

As you find information, jot it down on sticky notes or use a chart similar to the one shown above. Sticky notes are useful because you can move them around, group similar ideas together, and reorganize your ideas easily. Using a different colour for each sub-question is even better! Remember to write the source of your information on each sticky note. In addition to writing down information that you find as you research, you should also write down any questions you think of as you go along.

Process—Ask More Questions and Revise Your Work

Now that you have done some research, what sub-questions have you asked? These are the subtopics of your research. Use the subtopics to find more specific information.

Don't steal ideas!

What if I have too much information—or not enough?

If you find that you have two or three sub-questions that have a lot of research supporting them and a few that do not have much research, do not be afraid to "toss out" some of the less important questions or ideas.

Avoid Plagiarism Copying information word-for-word and then presenting it as though it is your own work is called *plagiarism*. When you refer to your notes to write your project, put the information in your own words. It is also important to give credit to the original source of an idea.

> **Reveal your source!**

Record Source Information Research papers always include a bibliography—a list of relevant information sources the authors consulted while preparing them. Bibliographic entries give the author, title, and facts of publication for each information source. Sometimes, you may want to give the exact source of information within the paper. This is done using footnotes. *Footnotes* identify the exact source (including page number) of quotations and ideas. Ask your teacher how you should prepare your list of works cited and your footnotes.

Create—Present Your Work

Before you choose a format for your final project, consider whether your researched information has answered the question you originally asked.

> **What if my research doesn't answer my question?**

If you have not answered this question, you need to either refine your original question or do some more research! As long as your question still meets the criteria of your original assignment, it is okay to change the question so it focusses on the research you have already done. After all, you don't want your hard work to go to waste!

> **How should I present my work?**

Check the guidelines that your teacher gave you. There may be specific instructions or criteria that will help you decide how to present your work. You also need to consider who the audience is for your project. How you format your final project will be very different if it is meant for a Grade 2 class compared to the president of a company or a government official. You could present your project as a poster, graphic organizer, blog, graphic novel, video, or research paper.

> **Instant Practice**
>
> 1. Describe the steps you should follow in preparing a project on the topic of renewable forms of energy.
>
> 2. The following example is not an effective question on which to base a research project: *How many moons does Jupiter have?* Modify the question to make it an effective research question.
>
> 3. Assume that the target audience for your project is a group of Grade 6 students from a local elementary school. What aspects of your project would you need to modify so that you are reaching the intended audience? What would be the best format to use to present your project to your audience?

Science Skills Toolkit 9

Using Electric Meters

Using Meters to Measure Potential Difference and Current

Types of Meters

The meters you use in your classroom are either analogue meters or digital meters. **Analogue meters** have a needle pointing to a dial. **Digital meters** display measured values directly as numbers, similar to how a digital watch displays the time directly.

The Terminals of a Meter

All meters have two terminals (connecting points) that you connect to the circuit. The negative terminal (−) is black. The positive terminal (+) is red. In order not to damage the meter, you must take care to connect the meter so that its positive (red) terminal is connected to the positive side of the power source. That is, you should be able to trace from the positive (+) terminal on the meter back to the positive terminal at the source. The negative (−) terminal of the meter is always connected to the negative side of the source. The rule is "positive to positive, and negative to negative."

Connecting an Ammeter

An **ammeter** is used to measure the electrical current in a circuit. Current is the amount of charge passing a given point per second. To measure the current at a given location in an electric circuit, the ammeter must be connected so that all the current is allowed to pass through the ammeter. To do this, disconnect one end of a wire to give the same effect as cutting the circuit where you wish to measure the current. Imagine the ammeter and its connecting wires completing the circuit you just disconnected. Make sure the positive terminal on the ammeter traces back to the positive terminal at the source.

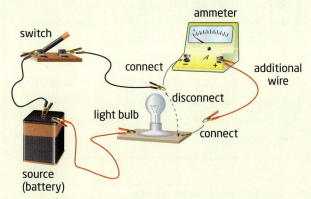

To measure the current through the light bulb, first disconnect the wire connected to the light bulb. Then insert the ammeter into the circuit.

A Analogue meters have a needle pointing to different scales.

B Digital meters display the numerical values directly.

Connecting a Voltmeter

A **voltmeter** is a device used to measure electric potential difference. A potential difference exists between two points in a circuit such as across a battery or a light bulb (load). When connecting a voltmeter to a circuit, you do not need to disconnect or open the circuit. Potential difference is measured between two points in a circuit. Therefore, connect the terminals of the voltmeter to the two connections on the component where you wish to measure the potential difference. Remember the rule "positive to positive and negative to negative," and make sure you can trace the connections on the voltmeter back to the same type of terminal at the source.

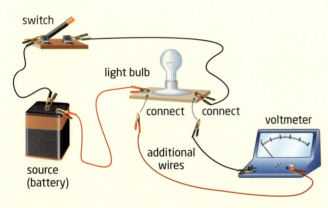

Voltmeters are connected across a component in the circuit.

Connecting a Multimeter

Modern digital meters can also be multimeters. Multimeters can be used to measure potential difference, current, and other electrical properties. When using a multimeter, it is important that you position the dial on the correct setting for your application. As well, the connecting wires must be inserted into the correct meter terminals.

Reading a Meter

A digital meter is easy to read since the measured value is displayed directly as numbers. In order to get the most accurate reading on a digital meter, the meter needs to be set to the appropriate scale.

The dial on a digital meter has several settings. For example, if the dial is set on the 2 V range, the meter will measure potential differences between zero and 2 V. Moving the dial to the 200 V setting will allow the meter to measure between zero and 200 V, but with less accuracy. Therefore, when using meters, you must choose the best setting for your measurement. The best approach is to set the meter on the largest scale to obtain an approximate value. Then lower the scale until you have the highest possible reading without going off scale.

This approach is the same for analogue meters. Some analogue meters have a dial, similar to a digital meter, that is used to change the scale. In other analogue meters, the scale is changed by how the wires are connected to the terminals. Once the scale is selected, you obtain your reading from the most appropriate display on the meter.

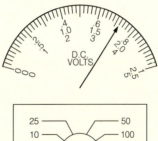

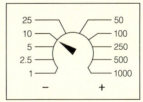

This voltmeter has its dial set at 10 V. To determine the measured potential difference, look for a number at the top of the scale with the same first digit as 10. The top scale has a maximum value of 1, so now the 1 represents 10 V. To read the scale, multiply the number the needle is pointing to by 10. The dial is reporting 7.2 V.

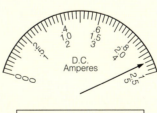

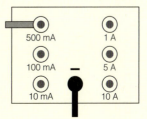

This ammeter has the positive wire connected to the 500 mA scale. The 5 on the bottom scale is the first digit in 500 mA, so the 5 now represents 500 mA. The needle is pointing to 4.7, so the meter is reporting 470 mA of current.

Instant Practice—Using Electric Meters

1. State the colour that is associated with
   a. the positive terminal of a meter
   b. the negative terminal of a meter

2. When you connect a meter to a circuit, to which side of the power source should you always connect the positive terminal of the meter?

3. For which type of meter do you need to disconnect the circuit before connecting the meter to the circuit?

4. A student wishes to use a meter to determine the most accurate measurement possible without damaging the meter. Describe the correct approach for choosing the appropriate scale.

5. Determine the value of current or potential difference indicated by meters A to D shown on this page.

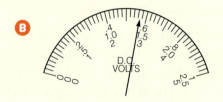

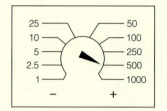

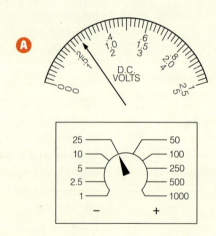

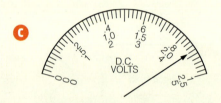

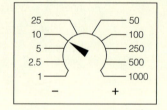

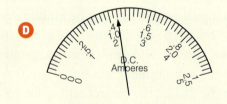

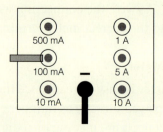

Science Skills Toolkit 10

Creating Data Tables

Scientific investigation is about collecting information to help you answer a question. In many cases, you will develop an hypothesis and collect data to see if your hypothesis is supported. An important part of any successful investigation includes recording and organizing your data. Often, scientists create tables in which to record data.

Planning to Record Your Data Suppose you are doing an investigation on the water quality of a stream that runs near your school. You will take samples of the numbers and types of organisms at two different locations along the stream. You need to decide how to record and organize your data. Begin by making a list of what you need to record. For this experiment, you will need to record the sample site, the pH of the water at each sample site, the types of organisms found at each sample site, and how many of each type of organism you collected.

Creating Your Data Table Your data table must allow you to record your data neatly. To do this you need to create

- headings to show what you are recording
- columns and rows that you will fill with data
- enough cells to record all the data
- a title for the table

In this investigation, you will find multiple organisms at each site, so you must make space for multiple recordings at each site. This means every row representing a sample site will have at least three rows associated with it for the different organisms.

If you think you might need extra space, create a special section. In this investigation, leave space at the bottom of your table, in case you find more than three organisms at a sample site. Remember, if you use the extra rows, make sure you identify which sample site the extra data are from.

Finally, give your table an appropriate title and a number. The title should appear above your table. Your data table might look like the one below.

Table 1 Observations Made at Two Sample Stream Sites

headings show what is being recorded

columns and rows contain data

Sample Site	pH	Type of Organism	Number of Organisms
1		beetle	3
		snail	1
		dragonfly larva	8
2		beetle	6
		dragonfly larva	7

extra rows to collect data in case you need to add observations

Instant Practice—Creating Data Tables

1. You are interested in how weeds grow in a garden. You decide to collect data from your garden every week for a month. You will identify the weeds and count how many there are of each type of weed. Design and draw a data table that you could use to record your data.

2. Many investigations have several different experimental treatments. Copy the following data table into your notebook and fill in the missing title and headings. The investigation tests the effect of increased fertilizer on plant height. There are four plants, and measurements are being taken every two days.

Day 1	Plant 1	5 mL	
	Plant 2	10 mL	10 cm
		15 mL	
		20 mL	

Numeracy Skills Toolkit 1

Scientific Notation

An exponent is the symbol or number expressing the power to which another number or symbol is to be raised, or the number of times the symbol or number (the base) is multiplied by itself. In 10^2, the exponent is 2 and the base is 10. So 10^2 means 10×10.

Table 1 Powers of 10

	Number	Power of 10
Thousands	1000	10^3
Hundreds	100	10^2
Tens	10	10^1
Ones	1	10^0
Tenths	0.1	10^{-1}
Hundredths	0.01	10^{-2}
Thousandths	0.001	10^{-3}

Why use exponents? Consider this. Mercury is about 58 000 000 km, or 58 000 000 000 m, from the Sun. If a zero were accidentally added to this number, the distance would appear to be 10 times larger than it actually is. To avoid mistakes when writing many zeros, scientists express very large and very small numbers in scientific notation.

In scientific notation, a number has the form

$x \times 10^n$ or base $\times 10^{\text{exponent}}$

The base, x, is a number with only one digit to the left of the decimal. The exponent, n, is the number of times 10 is multiplied by itself, and then multiplied by the base.

When you are expressing a decimal number, the exponent will be negative.

$x \times 10^{-n}$

Again, the base, x is a number with only one digit to the left of the decimal. However, 10^{-n}, actually means $\frac{1}{10^n}$. It is the number of times $\frac{1}{10}$ is multiplied by itself, and then multiplied by the base.

Example 1

Mercury is about 58 000 000 000 m from the Sun. Write 58 000 000 000 in scientific notation.

Solution

58 000 000 000. ← The decimal point starts here. Move the decimal point 10 places to the left.
$= 5.8 \times 10\,000\,000\,000$
$= 5.8 \times 10^{10}$

When you move the decimal point to the left, the exponent of 10 is positive. The number of places you move the decimal point is the number in the exponent.

Example 2

The electron in a hydrogen atom is, on average, 0.000 000 000 053 m from the nucleus. Write 0.000 000 000 053 in scientific notation.

Solution

The decimal point starts here. 0.000 000 000 053
Move the decimal point 11 places to the right.
$= 5.3 \times 0.000\,000\,000\,01$
$= 5.3 \times 10^{-11}$

When you move the decimal point to the right, the exponent of 10 is negative. The number of places you move the decimal point is the number in the exponent.

Instant Practice——Scientific Notation

1. Express each of the following in scientific notation.
   a. The approximate number of stars in our galaxy, the Milky Way:
      400 000 000 000 stars
   b. The approximate distance of the Andromeda Galaxy from Earth:
      23 000 000 000 000 000 000 000 m

2. Change the following to standard form (long form of the number).
   a. 9.8×10^5 c. 5.5×10^{-5}
   b. 2.3×10^9 d. 6.5×10^{-10}

Numeracy Skills Toolkit 2

Significant Digits and Rounding

Significant Digits

Significant digits represent the amount of uncertainty in a measurement. The significant digits in a measurement include all the certain digits plus the first uncertain digit. In the example below, the length of the rod is between 5.2 cm and 5.3 cm. Suppose we estimate the length to be 5.23 cm. The first two digits (5 and 2) are certain (we can see those marks), but the last digit (0.03) was estimated, so it is uncertain. The measurement 5.23 cm has three significant digits.

Use these rules to determine the number of significant digits (s.d.) in a measurement.

1. All non-zero digits (1–9) are considered significant.

   Examples:
   - 123 m (3 s.d.); 23.56 km (4 s.d.)

2. Zeros between non-zero digits are also significant.

   Examples:
   - 1207 m (4 s.d.); 120.5 km/h (4 s.d.)

3. Any zero that follows a non-zero digit *and* is to the right of the decimal point is significant.

   Examples:
   - 12.50 m^2 (4 s.d.); 60.00 km (4 s.d.)

4. Zeros used to indicate the position of the decimal are *not* significant. These zeros are sometimes called spacers.

   Examples:
   - 500 km (1 s.d.); 0.325 m (3 s.d.); 0.000 34 km (2 s.d.)

5. All counting numbers have an infinite (never-ending) number of significant digits.

   Examples:
   - 6 apples (infinite s.d.); 125 people (infinite s.d.)

Using Significant Digits in Mathematical Operations

When you use measured values in calculations, the calculated answer cannot be more certain than the measurements on which it is based. The answer on your calculator may have to be rounded to the correct number of significant digits.

Rules for Rounding

1. When the first digit to be dropped is less than 5, the digit before it is not changed.

   Example: 6.723 m rounded to two significant digits is 6.7 m.

2. When the first digit to be dropped is 5 or greater, increase the digit before it by one.

   Example: 7.237 m rounded to three significant digits is 7.24 m. The digit after the 3 is greater than 5, so the 3 is increased by one.

Adding or Subtracting Measurements

Perform the mathematical operation, and then round off the answer to the value having the fewest *decimal places*.

Example: x = 2.3 cm + 6.47 cm + 13.689 cm
  = 22.459 cm
  = 22.5 cm

Since 2.3 cm has only one decimal place, the answer can have only one decimal place.

Multiplying or Dividing Measurements

Perform the mathematical operation, and then round off the answer to the least number of *significant digits* of the data values.

Example: x = (2.342 m)(0.063 m)(306 m)
  = 45.149 076 m^3
  = 45 m^3

Since 0.063 m has only two significant digits, the answer must have two significant digits.

Numeracy Skills Toolkit 3

The Metric System

Throughout history, people have developed systems of numbering and measurement. When different groups of people began to communicate with each other, they discovered that their systems and units of measurement were different. Some groups of people created their own unique systems of measurement.

Today, scientists around the world use the metric system of numbers and units. The metric system is the official system of measurement in Canada.

The Metric System

The metric system is based on multiples of 10. For example, the basic unit of length is the metre. All larger units of length are expressed in units based on metres multiplied by 10, 100, 1000, or more. Smaller units of length are expressed in units based on metres divided by 10, 100, 1000, or more.

Each multiple of 10 has its own prefix (a syllable joined to the beginning of a word). For example, *kilo-* means multiplied by 1000. Thus, one kilometre is 1000 metres.

$1 \text{ km} = 1000 \text{ m}$

The prefix *milli-* means divided by 1000. So one millimetre is one thousandth of a metre.

$1 \text{ mm} = \frac{1}{1000} \text{ m}$

In the metric system, the same prefixes are used for nearly all types of measurements, such as mass, weight, area, and energy. A table of the most common metric prefixes is given at the top of the next column.

Table 1 Commonly Used Metric Prefixes

Prefix	Symbol	Relationship to the Base Unit
giga-	G	$10^9 = 1\ 000\ 000\ 000$
mega-	M	$10^6 = 1\ 000\ 000$
kilo-	k	$10^3 = 1000$
hecto-	h	$10^2 = 100$
deca-	da	$10^1 = 10$
— (Base Unit)	—	$10^0 = 1$
deci-	d	$10^{-1} = 0.1$
centi-	c	$10^{-2} = 0.01$
milli-	m	$10^{-3} = 0.001$
micro-	μ	$10^{-6} = 0.000\ 001$
nano-	n	$10^{-9} = 0.000\ 000\ 001$

Some Common Base Units:
length — m (metre)
mass — g (gram)
liquid volume — L (litre)
electrical energy usage — W (watt)

Instant Practice—Using Metric Measurements

1. What unit would you use to measure a distance one million (1 000 000 or 10^6) times the size of a metre?

2. What unit would you use to measure a volume that is one-thousandth (0.001 or 10^{-3} times) the size of a litre?

3. A hummingbird has a mass of 3.5 g. Express its mass in mg.

4. For an experiment, you need to measure 350 mL of dilute acetic acid. Express the volume in L.

5. A bald eagle has a wingspan up to 2.3 m. Express the length in cm.

6. A student added 0.0025 L of food colouring to water. Express the volume in mL.

The Easy Way to Do Metric Conversions

When you want to find the relationship between different metric units, you can use **Table 1** on page 368 to help you. You can quickly see how many factors of 10 separate the units if you count the number of rows in the table you need to jump to move from one unit to another.

Example 1

Find the relationship between centimetres and kilometers.

Solution

How many rows in the table do you jump to get from cm to km? Five

That means $1 \text{ km} = 10^5 \text{ cm}$ OR $1 \text{ cm} = 10^{-5} \text{ cm}$. There are 10^5 or 100 000 cm in 1 km.

The relationship between centimetres and kilometres is $\frac{1 \text{ cm}}{10^{-5} \text{ km}}$ or $\frac{10^5 \text{ cm}}{1 \text{ km}}$.

Example 2

There are 135 mg of sodium in a granola bar. Express this mass in kilograms.

Solution

There are six rows between mg and kg in the table, so $1 \text{ mg} = 10^{-6} \text{ kg}$ OR $1 \text{ kg} = 10^6 \text{ mg}$.

$135 \text{ mg} = \underline{\qquad} \text{ kg}$

$10^6 \text{ mg} = 1 \text{ kg}$

$(135 \text{ mg}) \left(\frac{1 \text{ kg}}{10^6 \text{ mg}} \right) = \frac{135 \text{ mg} \times 1 \text{ kg}}{10^6 \text{ mg}}$

$135 \text{ mg} = 0.000\ 135 \text{ kg}$ or $1.35 \times 10^{-4} \text{ kg}$

There are 1.35×10^{-4} kg of sodium in the granola bar.

You can actually use the metric prefixes table to help you convert units in another way that you might find easier. You can convert easily between metric units by moving the number's decimal to the left or to the right. The table of metric prefixes tells you how many places to move the decimal.

Converting to Larger Units

If you are converting from a small unit to a larger unit, you will move the decimal to the left and fill in the empty spots with zeros. How many places do you move the decimal?

Count the number of rows you need to jump to get from one unit to the other.

Example 3

Convert 5 cm into a measurement in metres.

Solution

How many rows do you jump to get from cm to m (the base unit) in the chart? Two.
Now move the decimal place in 5 two spaces to the left. Wait a minute! Where is the decimal in 5? Whenever a number doesn't seem to have a decimal, the decimal immediately follows the number.

$5 = 5.$

So if you move the decimal two places to the left, you get.

.5.

Now if you fill in the empty space with a zero, you get .05

So 5 cm = 0.05 m.

Converting to Smaller Units

If you are converting from a large unit to a smaller unit, you will move the decimal to the right and fill in the empty spots with zeros. Again, count the number of rows that you jump between units to decide how many places to move the decimal.

Example 4 Convert 5 km to a measurement in centimetres.

How many rows do you jump to get from km to cm? Five.

Now move the decimal five places to the right.

5.

If you fill in the empty spaces with zeros, you get 500 000.

So 5 km = 500 000 cm.

Numeracy Skills Toolkit 4

Organizing and Communicating Scientific Results with Graphs

In your investigations, you will collect information, often in numerical form. To analyze and report the information, you will need a clear, concise way to organize and communicate the data.

A graph is a visual way to present data. A graph can help you to see patterns and relationships among the data. The type of graph you choose depends on the type of data you have and how you want to present them. You can use line graphs, bar graphs, and pie graphs (pie charts).

The instructions given here describe how to make graphs using paper and pencil. Computer software provides another way to generate graphs. Whether you make them on paper or on the computer, however, the graphs you make should have the features described in the following pages.

Drawing a Line Graph

A line graph is used to show the relationship between two variables. The following example will demonstrate how to draw a line graph from a data table.

Example

Suppose you have conducted a survey to find out how many students in your school are recycling drink containers. Out of 65 students that you surveyed, 28 are recycling cans and bottles. To find out if more recycling bins would encourage students to recycle cans and bottles, you add one recycling bin per week at different locations around the school. In follow-up surveys, you obtain the data shown in **Table 1**. Compare the steps in the procedure with the graph on the next page to learn how to make a line graph to display your findings.

Table 1 Students Using Recycling Bins

Number of Recycling Bins	Number of Students Using Recycling Bins
1	28
2	36
3	48
4	60

Procedure

1. With a ruler, draw an x-axis and a y-axis on a piece of graph paper. (The horizontal line is the x-axis, and the vertical line is the y-axis.)

2. The independent (manipulated) variable is usually shown on the x-axis, while the dependent (responding) variable is shown on the y-axis. To label the axes, write "Number of Recycling Bins" along the x-axis and "Number of Students Using Recycling Bins" along the y-axis.

3. Now you have to decide what scale to use. You are working with two numbers (number of students and number of bins). You need to show how many students use the existing bin and how many would recycle if there were a second, a third, and a fourth bin. The scale on the x-axis will go from 0 to 4. There are 65 students, so you might want to use intervals of 5 for the y-axis. That means that every space on your y-axis represents 5 students. Use a tick mark at major intervals on your scale, as shown in the graph on the next page.

4. You want to make sure you will be able to read your graph when it is complete, so make sure your intervals on the x-axis are large enough.

5. To plot your graph, gently move a pencil up the y-axis until you reach a point just below 30 (you are representing 28 students). Now move along the line on the graph paper until you reach the vertical line that represents the first recycling bin. Place a dot at this point (1 bin, 28 students). Repeat this process for all of the data.

6. If it is appropriate, draw a line that connects all of the points on your graph. A graph showing yearly data that rises and falls without a predictable pattern might have a jagged line connecting all of the points. However, this is not always appropriate. Scientific investigations often involve quantities that change smoothly. In addition, experimental data points usually have some error. On a graph, this means that you should draw a smooth curve (or straight line) that has the general shape outlined by the points. This is called a **line of best fit**. If the points are almost in a straight line, draw a straight line as close to most of the points as possible. There should be about as many points above the line as there are below the line. If the data points do not appear to follow a straight line, then draw a smooth curve that comes as close to the points as possible. Think of the dots on your graph as clues about where the perfect smooth curve (or straight line) should go. A line of best fit shows the trend of the data. It can be extended beyond the first and last points to indicate what might happen.

7. Give your graph a title. Based on these data, what is the relationship between the number of students using recycling bins and the number of recycling bins?

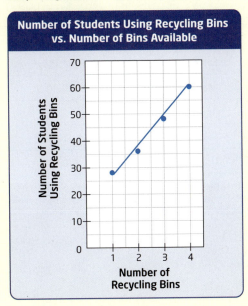

Line Graphs of Potential Difference, Current, and Resistance

While most line graphs have the independent variable on the x-axis and the dependent variable on the y-axis, this is usually not the case for graphs showing the relationship between potential difference and current, potential difference and resistance, or current and resistance.

- In a graph of potential difference and current, the potential difference is shown on the y-axis and current is shown on the x-axis, even when potential difference is the independent variable. This is because there is a special relationship between potential difference and current that can be shown by graphing them this way.

- In a graph of potential difference and resistance, potential difference is usually shown on the y-axis and resistance is shown on the x-axis.

- In a graph of current and resistance, current is usually shown on the y-axis and resistance is shown on the x-axis.

> **Instant Practice——Line Graph**
>
> The level of ozone in Earth's upper atmosphere is measured in Dobson units (DU). Using the information in the table below, create a line graph showing what happened to the amount of ozone over Antarctica during a period of 35 years.
>
> **Table 2 Ozone Levels in Earth's Upper Atmosphere**
>
> | Year | Total Ozone (DU) |
> |------|------------------|
> | 1965 | 280 |
> | 1970 | 280 |
> | 1975 | 275 |
> | 1980 | 225 |
> | 1985 | 200 |
> | 1990 | 160 |
> | 1995 | 110 |
> | 2000 | 105 |

Constructing a Bar Graph

Bar graphs help you to compare a numerical quantity with some other category at a glance. The second category may or may not be a numerical quantity. It could be places, items, organisms, or groups, for example.

Example

To learn how to make a bar graph to display the data in **Table 3** below, examine the graph in the column as you read the steps that follow.

Table 3 Area Covered by Principal Ontario Lakes

Lake	Area (km^2)
Big Trout Lake	661
Lac Seul	1657
Lake Abitibi	931
Lake Nipigon	4848
Lake Nipissing	832
Lake of the Woods	3150
Lake Simcoe	744
Lake St. Clair	490
Rainy Lake	741

Procedure

1. Draw your x-axis and y-axis on a sheet of graph paper. Label the x-axis "Ontario Lakes" and the y-axis "Area (km^2)."

2. Look at the data carefully in order to select an appropriate scale. Write the scale of your y-axis.

3. Using Big Trout Lake and 661 as the first pair of data, move along the x-axis the width of your first bar, then go up the y-axis to 661. Use a pencil and ruler to draw in the first bar lightly. Repeat this process for the other pairs of data.

4. When you have drawn all of the bars, add labels on the x-axis to identify the bars. Alternatively, use colour to distinguish among them.

5. If you are using colour to distinguish among the bars, you will need to make a legend or key to explain the meaning of the colours. Write a title for your graph.

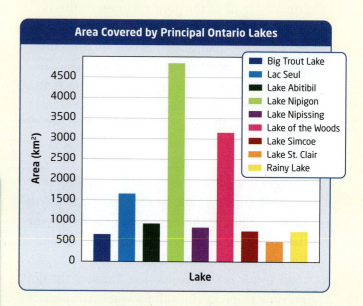

Instant Practice—Bar Graph

Make a vertical bar graph using **Table 5**, which shows each planet's gravitational force in relation to Earth's gravity.

Table 4 Gravitational Pull of Planets

Planet	Gravitational Pull (g)
Mercury	0.40
Venus	0.90
Earth	1.00
Mars	0.40
Jupiter	2.50
Saturn	1.10
Uranus	0.90
Neptune	1.10

Constructing a Pie Graph

A pie graph (sometimes called a pie chart) uses a circle divided into sections (like pieces of pie) to show the data. Each section represents a percentage of the whole. All sections together represent all (100 percent) of the data.

Example

To learn how to make a pie graph from the data in **Table 5**, study the corresponding pie graph on the right as you read the following steps.

Table 5 Birds Breeding in Canada

Type of Bird	Number of Species	Percent of Total	Degrees in Section
Ducks	36	9.0	32
Birds of prey	19	4.8	17
Shorebirds	71	17.7	64
Owls	14	3.5	13
Perching birds	180	45.0	162
Other	80	20.0	72

Procedure

1. Use a mathematical compass to make a large circle on a piece of paper. Make a dot in the centre of the circle.

2. Determine the percent of the total number of species that each type of bird represents by using the following formula.

   $$\text{Percent of total} = \frac{\text{Number of species within the type}}{\text{Total number of species}} \times 100\%$$

   For example, the percent of all species of birds that are ducks is

   $$\text{Percent that are ducks} = \frac{36 \text{ species of ducks}}{400 \text{ species}} \times 100\% = 9.0\%$$

3. To determine the number of degrees in the section that represents each type of bird, use the following formula.

   $$\text{Degrees in "piece of pie"} = \frac{\text{Percent for a type of bird}}{100\%} \times 360°$$

   Round your answer to the nearest whole number. For example, the section for ducks is

   $$\text{Degrees for ducks} = \frac{9.0\%}{100\%} \times 360° = 32.4° \text{ or } 32°$$

4. Draw a straight line from the centre to the edge of the circle. Use your protractor to measure 32° from this line. Make a mark, then use your mark to draw a second line 32° from the first line.

5. Repeat steps 2 to 4 for the remaining types of birds.

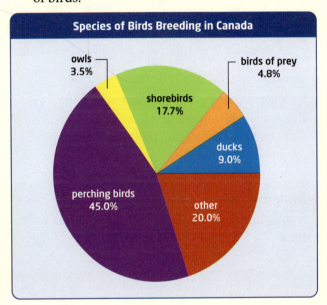

Species of Birds Breeding in Canada

Instant Practice—Pie Graph

Use the following data on total energy (oil, gas, electricity, etc.) consumption for 2004 to develop a pie graph to visualize energy consumption in the world.

Table 6 World Energy Consumption in 2004

Area in the World	Consumption (quadrillion btu)
North America	120.62
Central and South America	22.54
Europe	85.65
Eurasia	45.18
Middle East	21.14
Africa	13.71
Eastern Asia and Oceania	137.61

Choosing the Right Graph for the Job

It is important to choose the appropriate type of graph to organize and communicate your data. Some guidelines are given below.

Line Graphs

Line graphs are useful for

- making comparisons between a large number of categories or across a range of values for the variable that is being tested. For example, the graph below shows the annual energy usage per person from 1980 to 2006. Time is the variable being tested or considered.

- showing general trends in the relationship between variables. Does an increase in the manipulated (independent) variable cause an increase or a decrease in the responding (dependent) variable?

- finding the mathematical relationship between two variables. Rates and ratios can be calculated from a line showing how a variable changes over time.

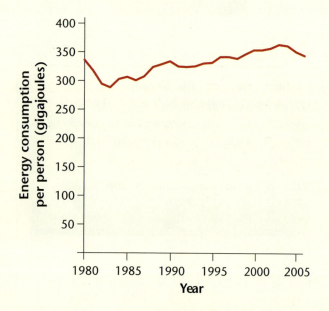

Bar Graphs

Bar graphs are useful for

- comparing a responding (dependent) variable between two distinct types of things, such as plant cells and animal cells, or between competing things, such as brands of a product. For example, this graph compares the distance travelled on a single charge by two different electric cars.

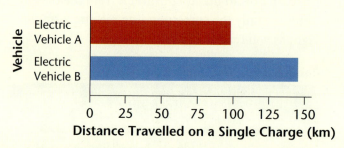

- comparing a responding (dependent) variable among categories within a group, such as provinces in Canada, months in a year, or planets in the Solar System.

- reporting the results of surveys. For example, you might want to show how many people said "Yes" and how many said "No" to each question on a survey.

- showing annual changes. For example, you might use a bar graph to show how energy usage had changed from 2004 to 2006. (However, if you were comparing a large number of categories, such as annual energy use from 1980 to 2006, it would be better to use a line graph.)

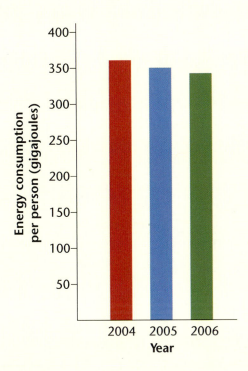

Pie Graphs

Pie graphs are useful for
- quick visual comparisons of proportions between segments of a whole.
- showing, at a glance, the most common category within a fixed set of categories. For example the pie graph on page 373 shows that perching birds are the most common category of the breeding birds sampled.

Limitations of pie graphs include the following:
- They cannot be used to show change over time. They are a snapshot of data collected at one specific time.
- They cannot be used to show complex relationships between variables.
- They must represent categories as percentages of a whole.

- It is difficult to compare similar categories unless the percentages represented by each slice of the pie are clearly labelled, as they are in this example from Unit 2.

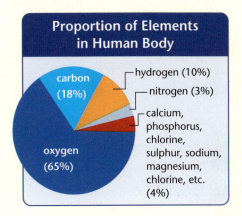

Instant Practice—Choosing Graph Types

What would be the best type(s) of graph to use for each purpose?

1. calculating the rate at which a chemical reaction takes place
2. comparing the gravitational pull of each of the planets in our Solar System
3. showing how production of hydroelectricity has changed from 1960 to today
4. comparing the amount of each chemical element present in a product
5. showing the relationship between world population and degree of global warming.

Numeracy Skills Toolkit 5

The GRASP Problem Solving Method

Solving any problem is easier when you establish a logical, step-by-step procedure. One useful method for solving numerical problems includes five basic steps: **G**iven, **R**equired, **A**nalysis, **S**olution, and **P**araphrase. You can easily remember these steps because the first letter of each word spells the word **GRASP**.

Example of the GRASP Problem Solving Method

Ruby can afford to spend $45.00 this month on electricity. The company that supplies her home with electrical energy charges 10.9¢ per kWh. Based on her budget, how many kWh of electrical energy can she use in a month?

Given—Organize the given data.
budget = $45.00
cost of electrical energy = 10.9¢/kWh

Required—Identify what information the problem requires you to find.
amount of electrical energy that can be used (kWh)

Analysis—Decide how to solve the problem.
1. Convert cents into dollars. (The units given for Ruby's budget—dollars—do not match the units given for the cost of 1 kWh of electrical energy—cents. Both units need to be the same.)
2. Calculate the number of kWh Ruby can afford to use.
   total cost = amount of energy used × cost per unit of energy

Solution—Solve the problem.
1. Convert units
   $1.00 = 100¢
   10.9¢ =
   $(10.9¢) \times \left(\dfrac{\$1.00}{100¢}\right) = \$0.109$

2. Use the total cost equation.
   total cost = (amount of energy used)(cost per unit of energy)

   $\dfrac{\text{total cost}}{\text{cost per unit of energy}} = \dfrac{(\text{amount of energy used})\cancel{(\text{cost per unit of energy})}}{\cancel{\text{cost per unit of energy}}}$

   amount of energy used = $\dfrac{\text{total cost}}{\text{cost per unit of energy}}$

   amount of energy used = $\dfrac{\$45.00}{\$0.109}$

   = 413 kWh

Paraphrase—Restate the solution and check your answer.

Restate Ruby has a budget of $45.00 and electrical energy costs 10.9¢, so she can afford to use 413 kWh of electrical energy this month.

Check Multiply the cost of electrical energy by the answer, and you should get $45.00. Round off the numbers to do a quick estimate. If you multiply $0.11 by 400 kWh, you get $44.00, so you know that your answer is reasonable.

> **Instant Practice—Using GRASP**
>
> The company that supplies Ruby's electrical energy raises the price to 11.1¢/kWh. Use the GRASP method to calculate how much Ruby's monthly energy bill will be if she uses 375 kWh.

Literacy Skills Toolkit 1

Preparing for Reading

Previewing Text Features

Before you begin reading a textbook, become familiar with the book's overall structure and features. This will help you understand where information can be found and how it will be presented. If you look at the Table of Contents on page v, you will see that this textbook is divided into four *units*. Each unit is divided into *topics*.

The Unit Opener

Each unit begins with a unit opener. It includes several features that will help to get you thinking about the unit. Examine the sample unit opener that is reproduced below.

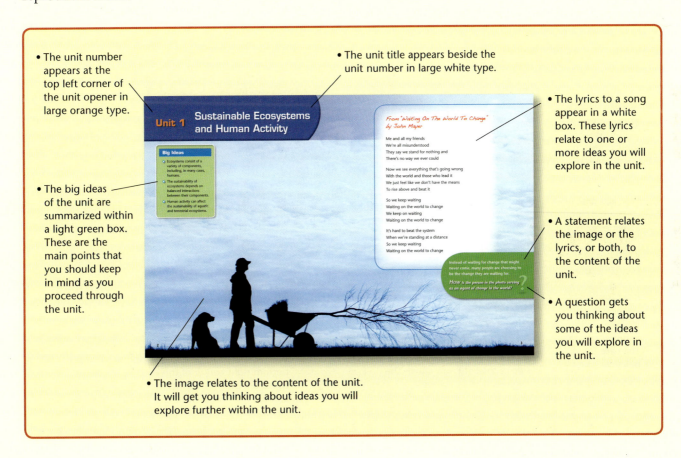

- The unit number appears at the top left corner of the unit opener in large orange type.
- The big ideas of the unit are summarized within a light green box. These are the main points that you should keep in mind as you proceed through the unit.
- The unit title appears beside the unit number in large white type.
- The lyrics to a song appear in a white box. These lyrics relate to one or more ideas you will explore in the unit.
- A statement relates the image or the lyrics, or both, to the content of the unit.
- A question gets you thinking about some of the ideas you will explore in the unit.
- The image relates to the content of the unit. It will get you thinking about ideas you will explore further within the unit.

Instant Practice

1. From what unit is the above sample taken? What does this unit opener tell you about the unit?
2. What is the purpose of each of the features in the unit opener?
3. Find the unit opener for Unit 2. What are the big ideas in that unit?

Preparing for Reading

Previewing Text Features

The Topic Opener

Each unit is broken into *topics*. Each topic explores a question related to the unit. The topic opener contains information about what concepts you will explore, what skills you will develop, and what vocabulary you will learn throughout the topic. It also includes a general introduction to the topic and an activity to get you started. A sample topic opener is shown below.

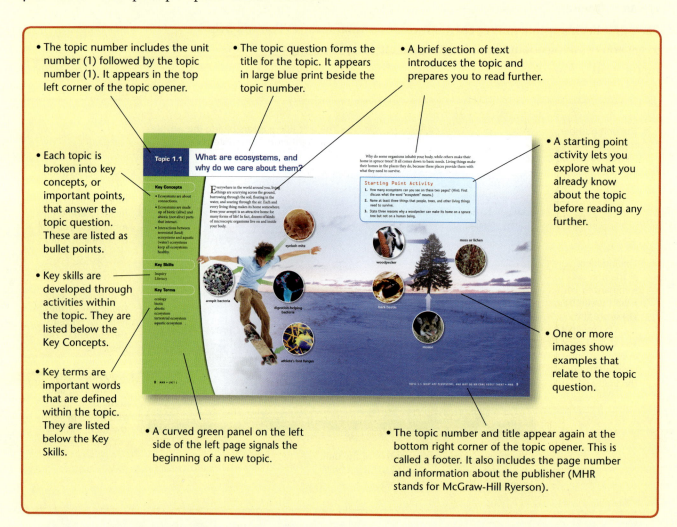

- The topic number includes the unit number (1) followed by the topic number (1). It appears in the top left corner of the topic opener.
- The topic question forms the title for the topic. It appears in large blue print beside the topic number.
- A brief section of text introduces the topic and prepares you to read further.
- Each topic is broken into key concepts, or important points, that answer the topic question. These are listed as bullet points.
- Key skills are developed through activities within the topic. They are listed below the Key Concepts.
- Key terms are important words that are defined within the topic. They are listed below the Key Skills.
- A curved green panel on the left side of the left page signals the beginning of a new topic.
- A starting point activity lets you explore what you already know about the topic before reading any further.
- One or more images show examples that relate to the topic question.
- The topic number and title appear again at the bottom right corner of the topic opener. This is called a footer. It also includes the page number and information about the publisher (MHR stands for McGraw-Hill Ryerson).

Instant Practice

1. Describe two pieces of information that you can find in a topic opener.
2. Find the topic question for another topic in Unit 1. How does the image(s) in the topic opener relate to the question?

Preparing for Reading

Previewing Text Features

The Main Text

Most of the information related to the key concepts of a unit appears in the main text pages of each topic. This is where you will explore the topic questions in detail, find definitions for important vocabulary, and develop skills through activities related to the topic. Each key concept within a topic takes up one or two pages. These pages contain many features designed to help you find your way while reading.

Examine the pages below. They include several features that will help you understand the content.

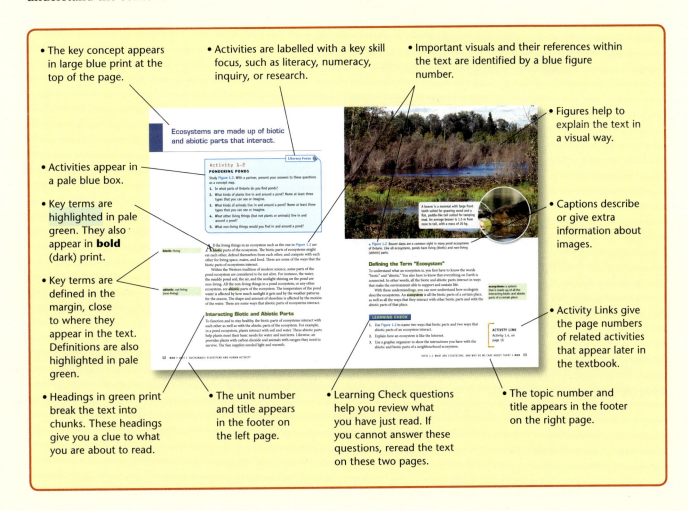

- The key concept appears in large blue print at the top of the page.
- Activities are labelled with a key skill focus, such as literacy, numeracy, inquiry, or research.
- Important visuals and their references within the text are identified by a blue figure number.
- Figures help to explain the text in a visual way.
- Activities appear in a pale blue box.
- Key terms are highlighted in pale green. They also appear in **bold** (dark) print.
- Key terms are defined in the margin, close to where they appear in the text. Definitions are also highlighted in pale green.
- Captions describe or give extra information about images.
- Activity Links give the page numbers of related activities that appear later in the textbook.
- Headings in green print break the text into chunks. These headings give you a clue to what you are about to read.
- The unit number and title appears in the footer on the left page.
- Learning Check questions help you review what you have just read. If you cannot answer these questions, reread the text on these two pages.
- The topic number and title appears in the footer on the right page.

Instant Practice

1. Describe two ways to identify the key terms in a section.
2. Describe two ways to learn more about a visual in this textbook.

LITERACY SKILLS TOOLKIT 1 • MHR 379

Preparing for Reading

Making Connections to Images

Images help to explain or expand on information in the text. Making connections to images help you understand their purpose and meaning. When you look at an image such as the one below (from page 175), start by reading the caption. The caption tells you what is in the image. It may also provide some interesting details.

◀ **Figure 3.6** The Andromeda galaxy is the nearest large galaxy to our own and the most distant object that is still visible to the unaided eye. It is 2.3 million light-years away.

Some people read a textbook by reading the text first and then looking at the images. Other people "read" the images first. If you start by looking at the image, use the figure number to find more information. The figure number in the caption has a matching reference in the text, and both are in bright blue print. The text will tell you more about the figure and how it relates to a key concept.

As you examine an image, think about the answers to these questions:

1. What personal connections can I make to the image, based on what I already know?
2. What do the text and the caption tell me about the image?
3. What else might be in the scene that the image does *not* show?
4. What questions do I have about the image that the text and caption do not answer?

In some places in the textbook, an image may appear without a figure number or a caption. Sometimes the caption is left out because the image is part of an activity. In other places, such as the topic opener, the main text tells you more about the image. Sometimes, the caption is left out on purpose, so that you will ask yourself questions about the image:

1. What does the image show?
2. Why did the author include it?
3. How does it relate to the text?

Literacy Skills Toolkit 2

Reading Effectively

Identifying the Main Idea and Details

The *main* idea of a text is the *most important* idea. Here are some strategies for identifying the main idea of a topic or paragraph.

- Pay attention to titles, headings, and subheadings. Note how print size and colour help you identify each of these.
- Look at the images on the page to get a general idea of the content.
- Note any terms that appear in **bold** or *italic* print. Bold print is used to identify key terms. Italics are used to add emphasis to other important words.

Details in the text *support and explain* the main idea. Details might be facts or examples. These phrases are clues that details will follow:

- For instance
- For example
- ...such as

Instant Practice

1. Examine the pages below and identify clues that hint at the main idea.

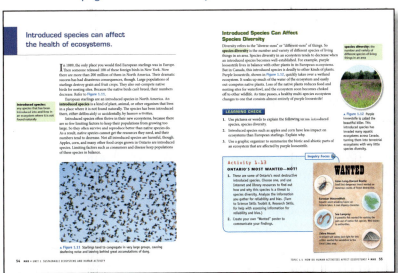

2. Here is a section of text from the pages shown above. Can you identify any details in this text that might support the main idea?

Introduced Species Can Affect Species Diversity

Diversity refers to the "diverse-ness" or "different-ness" of things. So **species diversity** is the number and variety of different species of living things in an area. Species diversity in an ecosystem tends to decrease when an introduced species becomes well-established. For example, purple loosestrife lives in balance with other plants in its European ecosystems. But in Canada, this introduced species is deadly to other kinds of plants. Purple loosestrife, shown in Figure 1.12, quickly takes over a wetland ecosystem. It soaks up much of the water of the ecosystem and easily out-competes native plants. Loss of the native plants reduces food and nesting sites for waterfowl, and the ecosystem soon becomes choked off to other wildlife. As time passes, a healthy multi-species ecosystem changes to one that consists almost entirely of purple loosestrife!

Reading Effectively

Making Connections to What You Already Know

You may already know some facts about the concept you are studying. You may have gathered knowledge from reading other texts, from the news, or from your own experiences. This knowledge can help you understand new information.

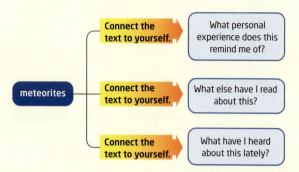

In Topic 3.3, you will learn about meteorites. You can use a **concept map** like the one above to organize what you already know about meteorites.

Making Inferences (Reading Between the Lines)

Often, a text does not contain *all* the details related to a particular topic. Some details or connections between ideas may be hinted at rather than stated clearly. The writer relies on you to make inferences, or "read between the lines." This involves combining information in the text with what you already know. It also involves thinking about how two pieces of information are related.

Read the following text:

By cutting down a forest near a stream, tree roots that once trapped soil wither and die. Soil and nutrients wash into the stream, and this can harm or kill fish and other living things.

Ask yourself questions about the text and organize your thoughts in a graphic organizer like the one below.

What information does the text provide? The text says…	→ Cutting down a forest near a stream can kill or harm fish.
What does this tell you about forests and nearby streams? The text says…/I think…	→ Forests and nearby streams are connected.
What do you already know about forests? I say…	→ People sometimes cut down trees in forests.
What connection can you make between what you already know and what you just read? And so…	→ Cutting down trees could harm or kill fish in nearby streams.

Reading Effectively

Skim, Scan, or Study

Not all parts of a textbook should be read at the same speed. In general, how fast you should read a chunk of text depends on your purpose for reading. **Table 1** shows three reading speeds, each suiting a different purpose for reading.

Sometimes the features of the text can help you decide how fast you should read. For example, if you see a page that contains several bold, highlighted key terms, you should read the text slowly and carefully. Text in a topic opener can be read more quickly, since it is only an introduction to the topic. It will not likely explore key concepts in detail.

**Table 1
Purposes of Reading Speeds**

Purpose	Reading Approach (Skim, Scan, or Study)
Preview text to get a general sense of what it contains.	Read quickly (skim).
Locate specific information.	Read somewhat quickly (scan).
Learn a new concept.	Read slowly (study).

Instant Practice

1. Look through Topic 2.4 with a partner. Identify two sections of text that should be read slowly and carefully. Then identify two sections of text that could be skimmed.
2. Can you think of a reason why you would need to scan a section of text?

Asking Questions

As you are reading, stop every now and then to ask yourself questions starting with *who, what, where, why, when,* and *how*.

Read the following paragraph from page 62:

> It began with a single city, Sydney, Australia, in 2007. As 2 million people and 2000 businesses shut down their lights for one hour, they sent a message to the world, to all who might hear. "We are concerned about climate change, and we hope our simple act inspires you to reduce activities that release excess carbon and other substances that contribute to it."

Who sent a message to the world?
What message did they send?
Where did their message begin?
Why did they send this message?
When did this happen?
How did they send a message to the world?

If you can't answer these questions about the text you've just read, you might need to go back and read more carefully.

You can also use these questions to predict what you will read next. Then continue reading to see if your questions are answered by the text. If they are not, write them down. You can discuss them with a partner, ask your teacher, or do some research to find out more.

Instant Practice

Use the coloured question words to ask questions that are not answered in the paragraph on the left. Then go to page 62 to see if your questions are answered in the next paragraph.

Reading Effectively

Checking Your Understanding

When you are reading text that contains new ideas and new key terms, stop after each chunk of text to make sure that you understand what you have just read. You can use the steps in the following flowchart to do this.

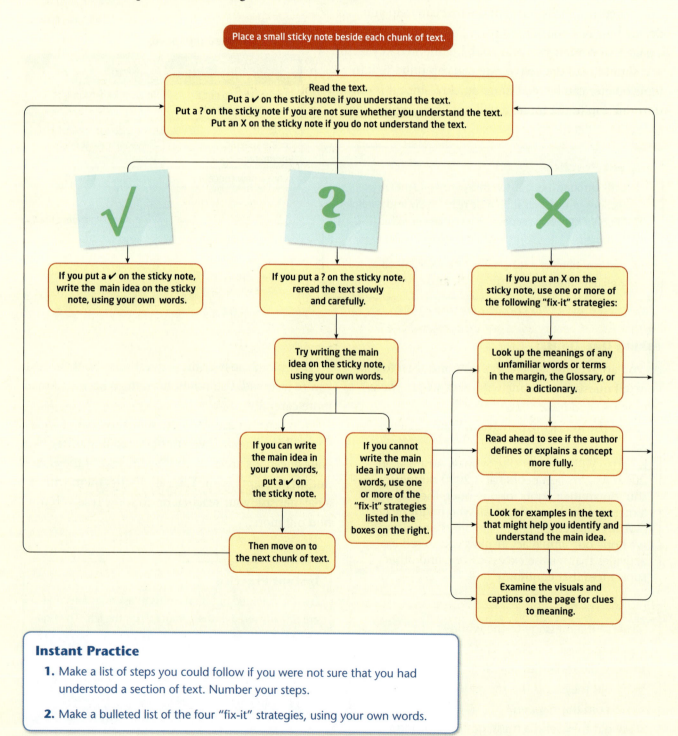

> ### Instant Practice
> 1. Make a list of steps you could follow if you were not sure that you had understood a section of text. Number your steps.
> 2. Make a bulleted list of the four "fix-it" strategies, using your own words.

Literacy Skills Toolkit 3

Reading Graphic Text

Reading Diagrams

A diagram is a simplified drawing that uses symbols to represent objects, directions, and relationships. Reading the labels of a diagram can help you understand these symbols.

To read a diagram

1. **Read the title or caption to understand the main idea of the diagram.**

   For example, the caption of **Figure 3.10** tells you that the movement of the Moon around the Earth results in different *phases* of the Moon. (You will learn about this in Topic 3.2.)

2. **Consider how each part illustrates the main idea.**

   The diagram shows that the moon has a different appearance when it has a different position in relation to the Earth and the Sun.

3. **Look closely at the labels and reread the caption, if you need to, to understand the details of the diagram.**

   The labels and the illustration show how the Moon appears in different phases.

4. **Find the reference to the diagram in the text to find out additional information, and to understand how the diagram relates to the main idea in the text.**

In the main text, near the figure reference, you find the following information: "The lit-up side of the Moon is always fully lit up, but we can't always see the whole lit-up side. Instead, we see changes in the amount of lit-up surface during a month: the phases of the Moon."

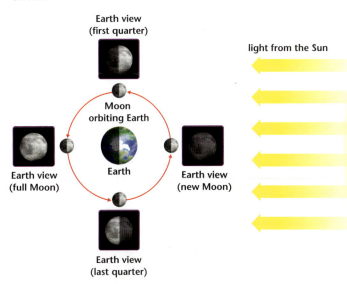

▲ **Figure 3.10** The phases of the Moon are caused by the amount of lit-up surface that we can see from Earth as the Moon orbits our planet.

Instant Practice

Examine **Figure 1.4A** on page 22. Follow the steps above to read the diagram.

1. Explain the main idea of the diagram.
2. How did the caption, labels, and information in the main text help you to understand the diagram?

Reading Graphic Text

Reading Tables

A table contains cells that are organized in rows and columns. Each cell contains data. Each column and row has a heading to help you understand the information in each cell.

To read a table or to find patterns in a table

1. **Read the title of the table.**

   Based on the title, you should be able to predict what kind of information you will see in the cells.

2. **Read the column and row headings carefully.**

3. **Move your eyes left and right across the rows, and up and down along the columns.**

4. **When you look at a cell, look again at the headings.**

   What is the heading of the column containing this cell? What is the heading of this row?

5. **Look for units.** If measurements are included in the table, the column headings should tell you what units are being used to report the measurements.

6. **Look for patterns as you move left to right across a row, or top to bottom down a column.**

   If the column contains numbers, do the numbers increase steadily as you move down the column? Do they decrease steadily?

7. **Look for breaks in patterns.**

   Is there one cell that doesn't fit the pattern in the rest of its column? Think about why this might be the case.

8. **Look for relationships between columns or rows.**

   Do the numbers in one column increase as the numbers in another column decrease? Do numbers increase from top to bottom in every column? What does this tell you?

In **Table 1** below, the number 500 is found in the column labelled "Number of Zebra Mussels (per m^2)" and in the row labelled "1992." This number tells you that 500 zebra mussels per m^2 were found in Lake Ontario in 1992.

The year increases as you look down the first column. The number of zebra mussels per m^2 also increases as you look down the second column. This tells you that the number of zebra mussels in Lake Ontario increased over time.

Table 1
Zebra Mussels in Lake Ontario

Year	Number of Zebra Mussels (per m^2)
1990	0
1991	230
1992	500

Reading Graphic Text

Reading Graphs

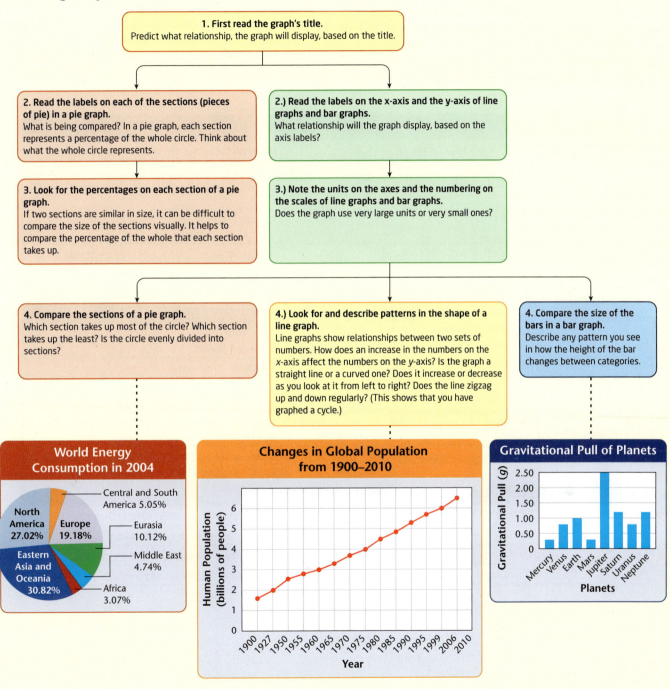

Literacy Skills Toolkit 4

Word Study

Common Base Words, Prefixes, and Suffixes in Science

Understanding how words are put together can help you figure out their meanings. The list below includes some common *base words* that are used in science. Also listed are some common *prefixes* and *suffixes*, which change the meaning of a base word when they are combined with the base word.

Base Word	Definition	Example
conduct	To lead or act as a channel for	A **conductor** allows electrons to move easily between atoms.
electr(o)	Having to do with electricity	An **electroscope** is a device for detecting an electric charge.
phot(o)	Having to do with light	A **photometer** measures the amount of light that is emitted from a source.
resist	To hold off; to prevent or oppose	A **resistor** decreases the electric current that is flowing through a component.
sustain	To keep going; to maintain	**Unsustainable** means not able to keep going.

Prefix	Definition	Example
bio-	Having to do with life	**Biocontrol** is the use of living things to control unwanted species.
dis-	Not; the opposite of; having an absence of	A **disinfectant** helps to remove and prevent infection.
infra-	Below; beneath	**Infrared** light has a a lower frequency than red light.
semi-	Half or partial	A **semiconductor** allows electrons to move fairly well between atoms.
non-	Not; having an absence of	A **non-metal** is an element that does *not* have the properties of a metal.

Suffix	Definition	Example
-al	Relating to	**Environmental** means relating to the environment.
-ic	Relating to; characterized by	**Atomic** means relating to an atom.
-ity	The state or quality of	**Reactivity** is the quality of being reactive.
-ion	The action or process of	**Pollution** is the process of polluting.

Instant Practice

1. Use the table to predict the meaning of conductivity.
2. Think of a word that ends in one of the suffixes listed above. (You can browse through this textbook or a dictionary to find a word, if you wish.) Explain the meaning of your word. Compare your word and definition with words and definitions that your classmates suggest.

Word Study

Word Family Webs and Word Maps

Science textbooks include many words that you may not have seen before. On the last page, you learned that looking at base words, prefixes, and suffixes can help you to understand the meanings of unfamiliar words.

Word Family Webs

Words that share the same base, prefix, or suffix are related. They make up a word family. A *word family web* can help you see the connections between words in a word family. Then you can figure out unfamiliar words in the family.

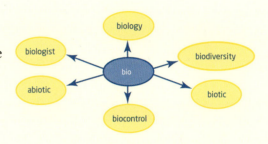

The web to the right shows words that all have the prefix *bio*, from the Greek word meaning life. *Biology*, for example, means the study of life. If *ogist* means someone who studies, what does *biologist* mean?

Instant Practice

The prefix *astro* means star. Use this knowledge of the prefix to predict the meanings of unfamiliar words in the web shown. Check your predictions by locating these words in the text. Use the glossary and the index to help you.

Word Maps

A *word map* helps you organize your thoughts about a word. Look at the word map for the word *insulator*. The map contains the definition of insulator, but it also includes other information that explains the *concept* of an insulator. You might want to add a section to your word map—*non-examples*. What materials are NOT insulators? If you are able to think of non-examples, you know you have a good understanding of the concept.

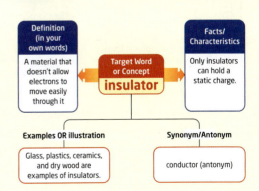

Instant Practice

1. Create a word map for the word *ecosystem*.
2. Exchange word maps with a partner. Are your maps the same?

Literacy Skills Toolkit 5

Organizing Your Learning: Using Graphic Organizers

When deciding which type of graphic organizer to use, consider your purpose: to brainstorm, to show relationships among ideas, to summarize a section of text, to record research notes, or to review what you have learned before writing a test. Twelve different graphic organizers are shown here. A chart at the end of this toolkit summarizes the function of each organizer to help you decide on the best one for the information you are working with.

T-Chart

A *t-chart* is a simple two-column chart that can be used to compare or show a relationship between two things.

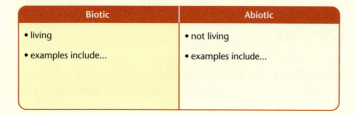

PMI Chart

A *PMI chart* has three columns. PMI stands for "Plus," "Minus," and "Interesting." A PMI chart can be used to state the good and bad points about an issue. The third column in the PMI chart is used to list interesting information related to the issue. PMI charts help you to organize your thinking after reading about a topic that is up for debate or that can have positive or negative effects.

K-W-L Chart

A *K-W-L chart* is used to record what you already know about a topic, what you want to learn about the topic, and what you know after you have read about the topic. It can help you to plan your research about a subject or check your learning after reading about a subject.

What You Know	What You Want to Know	What You Learned
• There are many stars in the sky. • The Sun is a star. • The Sun is the centre of our solar system.	• How many stars are there? • How far away is the nearest star other than our Sun? • Is it part of a solar system too?	• There are so many stars that scientists find it difficult to even estimate how many there are. They estimate that there are 100 000 000 000 (10^{11}) stars just in our galaxy, and possibly 10^{22} to 10^{24} in the whole universe. • The nearest star, Proxima Centauri, is more than four light years away. • Proxima Centauri is part of the Alpha Centauri solar system.

Concept Map

A *concept map* uses shapes and lines to show how ideas are related. Each idea, or concept, is written inside a circle, a square, a rectangle, or another shape. Words that explain how the concepts are related are written on the lines that connect the shapes.

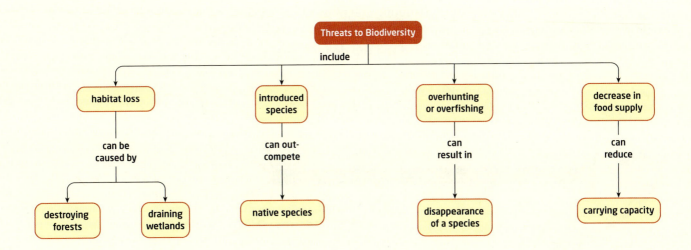

Spider Map

A *spider map* shows a main idea and several ideas related to the main idea. It does not show the relationships among the ideas. A spider map is useful when you are brainstorming or taking notes.

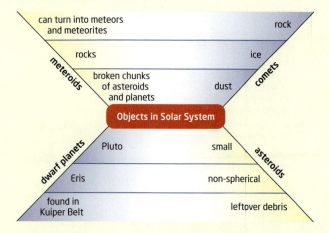

Fishbone Diagram

A *fishbone diagram* looks similar to a spider map, but it organizes information differently. A main topic, situation, or idea is placed in the middle of the diagram. This is the "backbone" of the "fish". The "bones" (lines) that shoot out from the backbone might be used to list reasons that the main situation exists, issues that affect the main idea, or arguments that support the main idea. Finally, supporting details shoot outward from these issues. Fishbone diagrams are useful for planning and organizing a research project. You can clearly see when you don't have enough details to support an issue. Then you can do more research.

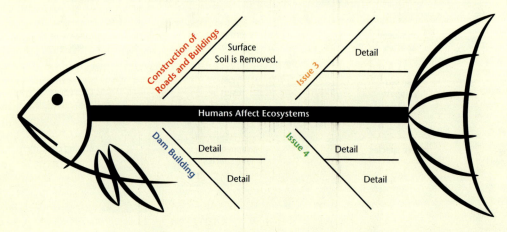

Main Idea Web

A *main idea web* shows a main idea and several supporting details. The main idea is written in the centre of the web, and each detail is written at the end of a line going from the centre.

Flowchart

A *flowchart* shows a sequence of events or the steps in a process. A flowchart starts with the first event or step. An arrow leads to the next event or step, and so on, until the final outcome. All the events or steps are shown in the order in which they occur.

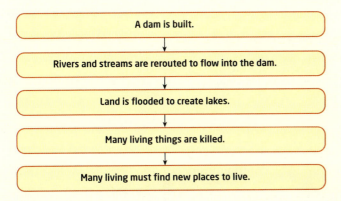

Cycle Chart

A *cycle chart* is a flowchart that has no clear beginning or end. All the events are shown in the order in which they occur, as indicated by arrows, but there is no first or last event. Instead, the events occur again and again in a continuous cycle.

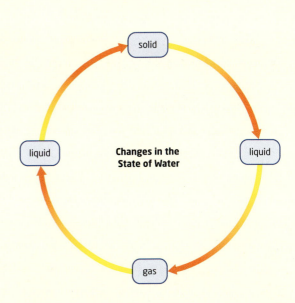

Venn Diagram

A *Venn diagram* uses overlapping shapes to compare concepts (show similarities and differences).

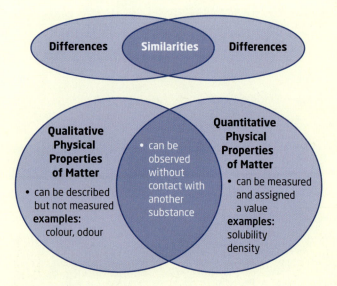

Double Bubble Organizer

Like a Venn diagram, a *double bubble organizer* is used to concepts (show similarities and differences). It separates the details that two concepts share and the details that they do not share.

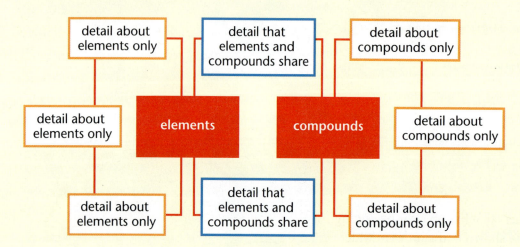

Cause-and-Effect Map

The first *cause-and-effect map* below shows one cause that results in several effects. The second map shows one effect that has several causes.

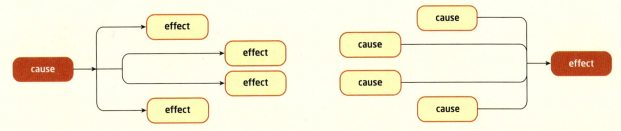

Which Organizer Should I Choose?

When you are trying to decide how to organize information, you can use the following chart to help you.

What are you trying to do with your graphic organizer?	t-Chart	PMI-Chart	K-W-L Chart	Concept Map	Spider Map	Fishbone Diagram	Main Idea Web	Flowchart	Cycle Chart	Venn Diagram	Double Bubble Organizer	Cause-and-Effect Map	Word Map (see page 389)
Brainstorm				X	X		X						
Show relationships among ideas or words				X		X	X						X
Check your understanding			X			X		X	X	X	X	X	X
Compare (show similarities and differences)	X									X	X		
Examine the pros and cons of an issue	X	X											
Examine the causes and/or effects of an action or issue				X		X		X				X	
Take notes			X	X	X	X	X	X					
Plan your research			X			X		X					
Show a process or series of events								X	X				
Show a continuous series of events									X				

Instant Practice

1. Create a Venn diagram that compares two of your favourite science topics.
2. Draw a spider map that reflects what you know about electricity.

Appendix

Properties of Common Substances

KEY TO SYMBOLS:
Common names of substances are enclosed in parentheses.
(*) water solution of a pure substance   (e) element   (c) compound

Name	Formula	Common Use or Important Feature
acetic acid (vinegar) (c)	CH_3COOH	used in the manufacture of cellulose ethanoate; vinegar is a 5 to 7 percent solution in water
aluminum (e)	Al	used in aircraft, cooking utensils, and electrical apparatus
ammonia (c)	NH_3	used as refrigerant and in manufacture of resins, explosives, and fertilizers
argon (e)	Ar	used in electric lights
beryllium (e)	Be	used for corrosion-resistant alloys
boron (e)	B	used for hardening steel and for producing enamels and glasses
bromine (e)	Br_2	used to make certain pain-relieving drugs; liquid causes severe chemical burns; vapour is harmful to lungs
calcium (e)	Ca	very abundant; essential to life
calcium carbonate (limestone) (c)	$CaCO_3$	main ingredient in chalk and marble
calcium hydroxide (slaked lime) (c)	$Ca(OH)_2$	aqueous solution used to test for CO_2
carbon (diamond) (e)	C	very hard; used for drilling through rock
carbon (graphite) (e)	C	very soft; used in lubricants, pencil leads, and electrical apparatus
carbon dioxide (c)	CO_2	does not support combustion and is denser than air; used in fire extinguishers and as a refrigerant at $-78.5°C$
chlorine (e)	Cl_2	poisonous; used to kill harmful organisms in water
copper (e)	Cu	soft metal; good conductor of heat
ethanol (ethyl alcohol) (c)	C_2H_5OH	derived from fermentation of sugar; used as solvent or fuel; found in wine
fluorine (e)	F_2	similar to chlorine
gold (e)	Au	very soft metal; highly resistant to tarnishing
glucose (c)	$C_6H_{12}O_6$	simple sugar; human body converts most sugars and starches to glucose
hydrochloric acid (*)	HCl	corrosive acid; properties vary according to concentration
hydrogen (e)	H_2	highly flammable; liquid form used as rocket fuel
hydrogen peroxide (c)	H_2O_2	thick and syrupy when pure; an antiseptic
iodine (e)	I_2	crystals sublime readily to form poisonous violet vapour
iron (e)	Fe	rusts readily; soft when pure

DEFINITIONS:
deliquescent: able to absorb water from the air to form a concentrated solution
sublime: to form a vapour directly from a solid

Appearance (at room temperature: 20°C)	Melting Point (°C)	Boiling Point (°C)	Density (g/cm^3 or g/mL)
colourless liquid with pungent smell	16.6	118.1	—
silver-white metal	659.7	2519	2.7
very soluble gas with pungent smell	−77.8	−33.4	less dense than air
inert gas	−189	−185	denser than air
hard, white metal	1280	2471	1.85
brown, amphorous powder or yellow crystals	2075	4000	2.37(brown), 2.34 (yellow)
red-brown liquid	−7.2	58.8	3.12
soft, white metal that tarnishes easily	845	1484	1.55
white solid	decomposes at 900°C	—	2.93
white solid	decomposes at 522°C	—	2.24
colourless, solid crystals	3500	3930	3.51
grey-black solid	4492	4492	2.25
colourless gas with a faint tingling smell and taste	—	—	—
green gas	−101.6	−34.6	denser than air
shiny, reddish solid	1084	2562	8.95
colourless liquid	−114.5	78.4	0.789
greenish yellow gas	−270	−188	—
shiny, yellow solid	1063	2856	19.3
white solid	146	decomposes before it boils	1.54
colourless liquid	varies	varies	varies
colourless gas	−259	−253	much less dense than air
colourless liquid	−0.4	150.2	1.45
violet-black, solid crystals	114	184	4.95
shiny, silver solid	1535	2861	7.86

Appendix

Properties of Common Substances

KEY TO SYMBOLS:
Common names of substances are enclosed in parentheses.
(*) water solution of a pure substance   (e) element   (c) compound

Name	Formula	Common Use or Important Feature
lead (e)	Pb	soft metal; forms poisonous compounds
lithium (e)	Li	used in alloys; its salts have various medical uses
magnesium (e)	Mg	used in photography; compounds used in medicine; essential to life
mercury (e)	Hg	only liquid metal; forms poisonous compounds
methane (c)	CH_4	main constituent in natural gas
neon (e)	Ne	discharge of electricity at low pressures through neon produces an intense orange-red glow
nickel (e)	Ni	used for nickel plating and coinage, in alloys, and as a catalyst
nitrogen (e)	N_2	will not burn or support burning; makes up 80 percent of air
oxygen (e)	O_2	must be present for burning to take place; makes up 20 percent of air
platinum (e)	Pt	used in jewellery; alloyed with cobalt, used in pacemakers
polyethylene (polythene) (c)	$(C_2H_4)n$	polymer of ethylene; used as insulating material; flexible and chemically resistant
potassium (e)	K	essential to all life; found in all living matter; salts used in fertilizers
propane (c)	C_3H_8	flammable; used as fuel
silver (e)	Ag	soft metal; best-known conductor of electricity
sodium (e)	Na	used in preparation of organic compounds, as coolant, and in some types of nuclear reactors
sodium chloride (table salt) (c)	NaCl	used to season or preserve foods
sucrose (sugar) (c)	$C_{12}H_{22}O_{11}$	made from sugar cane or sugar beets
sulfur (brimstone) (e)	S_8	used to make dyes, pesticides, and other chemicals
tin (e)	Sn	soft metal; rust resistant
titanium (e)	Ti	alloys are widely used in the aerospace industry
water (c)	H_2O	good solvent for non-greasy matter
zinc (e)	Zn	used in alloys such as brass and galvanized iron

DEFINITIONS:
deliquescent: able to absorb water from the air to form a concentrated solution
sublime: to form a vapour directly from a solid

Appearance (at room temperature: 20°C)	Melting Point (°C)	Boiling Point (°C)	Density (g/cm³ or g/mL)
shiny, blue-white solid	327.4	1750	11.34
silver-white metal (least dense solid known)	179	1340	0.534
light, silvery-white metal that tarnishes easily in air	651	1107	1.74
shiny, silvery liquid	−38.9	356.6	13.6
odourless, flammable gas formed from decaying organic matter	−182.5	−161.5	—
colourless, odourless gas	−248	−246	—
silvery-white, magnetic metal that resists corrosion	1455	2913	8.90
colourless gas	−209.9	−195.8	slightly less dense than air
colourless gas	−218	−183	slightly denser than air
silver-white solid	1769	3824	21.41
tough, waxy, thermoplastic material	—	—	—
silvery-white, soft, highly reactive, alkali metal	63.5	759	0.86
colourless gas	—	−42.17	—
shiny, white solid	961	2162	10.5
soft, silvery-white metal; very reactive	97.5	892	0.971
white, crystalline solid	801	1465	2.16
white solid	170	decomposes at 186°C	1.59
yellow solid	112.8	444.6	2.07
shiny, slightly yellow solid	231.9	2602	7.31
lustrous white solid	1666	3287	4.5
colourless liquid	0	100	1.00
hard, bluish-white metal	419	907	7.14

Glossary

How to Use This Glossary

This Glossary provides the definitions of the key terms that are shown in boldface type in the text. Definitions for other important terms are included as well. The Glossary entries also show the numbers of the topics where you can find the boldface words.

1.1 = Unit 1, Topic 1
U2GR = Unit 2 Get Ready
SST1 = Science Skills Toolkit 1
NST1 = Numeracy Skills Toolkit 1
LST1 = Literacy Skills Toolkit 1
App = Appendix

A pronunciation guide, using the key below, appears in square brackets after selected words.

a = mask, back
ae = same, day
ah = car, farther
aw = dawn, hot
e = met, less
ee = leaf, clean
i = simple, this
ih = idea, life
oh = home, loan
oo = food, boot
u = wonder, Sun
uh = taken, travel
uhr = insert, turn

A

abiotic [ae-bih-AW-tik] not living (non-living); for example, rocks or sunlight (1.1)

accuracy how close a measurement or calculation comes to the true value (SST5)

ammeter a device that measures the current in an electrical circuit; measured in the unit amperes (A) (4.4, SST9) ▼

analogue meter a device that displays measurements, such as current or potential difference, with a needle pointing to a dial (SST9) ▼

ampere (A) [AM-peer] the unit of measure that describes the amount of current flowing through a wire in an electrical circuit; ampere is abbreviated as amps (4.4)

analogy a comparison between two things that have some feature in common; uses a familiar situation to help explain an unfamiliar situation; for example, the Internet is sometimes called an information superhighway because it allows large amounts of information to travel back and forth very efficiently; this is an analogy that helps you to understand the Internet (SST7)

aquatic ecosystem an ecosystem that is based mostly or totally in water; for example, the beaver pond shown here is an aquatic ecosystem; it includes the beaver, the plant life in and around the water, and other living parts, as well as non-living parts such as the water itself and the dead trees that the beavers have used to make their dam. (1.1) ▼

asteroid a rocky object in space that orbits the Sun and is found in the area between the orbits of Mars and Jupiter (3.3)

asteroid belt the region of the solar system where most asteroids are located; contains rocky chunks of various sizes that orbit the Sun between the orbits of Mars and Jupiter (3.3) ▼

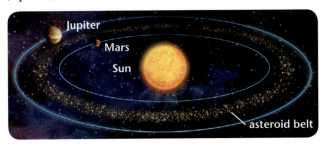

astronomer [uh-STRON-uh-mer] a scientist who studies astronomy (3.1)

astronomical unit [as-truh-NOM-i-kuhl YOO-nit] a measurement equal to the distance between Earth and the Sun, about 150 million km; abbreviation is A.U. (3.1)

atmosphere the layer of gases above Earth's surface (3.2) ▼

atom the smallest unit of an element that displays the same properties as the element, for example, hydrogen (2.4) ▼

atomic number [uh-TOM-ik NUM-ber] the number of protons in the nucleus of an atom; for example, the atomic number for helium is 2 (2.4)

atomic structure the arrangement of the parts of an atom; for example, the atomic number for helium is 2 (2.4)

aurora light shows in Earth's upper atmosphere; created by solar wind; the Northern Lights are an aurora often seen in Canada (3.2) ▼

B

battery a source of electrical energy; for example, a 9 V battery (4.4) ▼

bias a point of view that influences a decision and prevents a fair and balanced judgement (SST1, SST8)

bibliography a list of information sources you have used to help you create a research-based project (SST8)

biocontrol the use of living things to control the introduced species in an ecosystem; for example, this European beetle has been used to help control the growth of purple loosestrife (a species that is harmful to aquatic ecosystems)(1.6) ▼

biodiversity all the different species that live in an ecosystem, as well as all the different ecosystems within and beyond that ecosystem (1.6)

biomass energy energy produced from living or recently living things; for example burning wood produces biomass energy (4.1)

biotic [bih-AW-tik] living; for example, insects or plants (1.1)

bog a type of wetland in which the water is acidic and low in nutrients (1.6) ▼

boiling point a constant for known substances, it is the temperature at which a liquid turns into a gas and its abbreviation is B.P.; the boiling point of water is 100°C (App)

C

calendar a way of showing days, organized into a schedule of larger units of time such as weeks, months, seasons, or years; usually a table or a chart; our calendar is linked to the appearance of the Moon in the night sky (3.2)

carrying capacity the largest population that an ecosystem can support (1.4) ▼

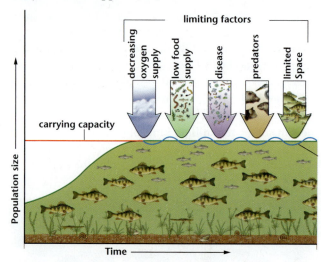

celestial object [suh-LES-chuhl AWB-jekt] any object that exists in space, such as a planet, a star, or the Moon; a star is a celestial object (3.1)

cellular respiration a process in the cells of most living things that converts the energy stored in chemical compounds into usable energy; the equation below shows the materials needed for and the materials produced by this process (1.2) ▼

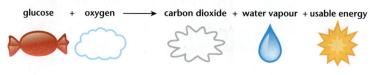

402 MHR • GLOSSARY

charging by contact causing a neutral object to become charged by touching it with a charged object; for example, getting a shock when you touch a charged doorknob ▼

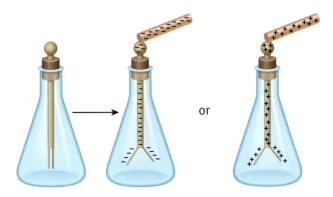

charging by induction causing a neutral object to become charged by bringing a charged object near to, but not touching, the object; for example, having your hair stand up when you bring a charged comb near to, but not touching it (4.3) ▼

chemical formula a short form used to represent a molecule; uses letters and numbers; only pure substances have chemical formulas; H_2O is the chemical formula for water (2.5)

chemical property the ability of a substance to change or react, and to form new substances when interacting with other substances, for example, zinc reacts with hydrochloric acid to produce hydrogen gas (2.2)

chemical reaction any change that occurs when substances interact to produce new substances with new properties; changes in the chemical and physical properties of the pure substances let you know that a chemical reaction has occurred; when vinegar and baking soda are combined, they react to produce frothy bubbles, as shown (2.6) ▼

chemical symbol a short form used to represent the name of an element; C is the chemical symbol for carbon (2.5)

circuit breaker a safety device that keeps a circuit from carrying too much current and starting a fire; a circuit breaker is placed in series with circuits that lead to outlets and appliances; each of the large black switches in the photo below is a circuit breaker and each controls a separate circuit within a building's electrical system (4.6) ▼

GLOSSARY • MHR 403

circuit diagram a diagram that uses standard symbols to represent the parts in an electric circuit and how they are connected; the circuit below includes a source and three loads (U4GR, 4.4, SST6) ▼

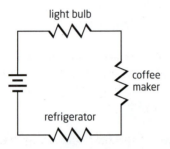

combustibility the ability of a substance to catch fire and burn in air (2.2)

comet chunks of loosely held rock and ice that come from the outer parts of the solar system; Halley's Comet is a well known comet that passes by the Earth about every 75 or 76 years. (3.3) ▼

competition when two or more members of a population compete for the same resource in the same location at the same time; for example, wolves and cougars are in competition for rabbits, which they eat for food (1.4)

compound a pure substance that can be broken down into smaller parts using chemical properties; water and carbon dioxide are both examples of compounds; all parts of a compound will contain identical molecules; in distilled water, as shown below, all parts will contain identical H_2O molecules (2.3) ▼

conclusion a statement that indicates whether your results support or do not support your hypothesis (SST2)

conductivity describes how easily a substance lets heat or electricity move through it (2.2, 4.2)

conductor a material that lets heat and electricity move through it easily; conductors are usually metals (2.3, 4.2)

constellation a group of stars that seem to form a shape or pattern in the night sky; the Little Dipper is an example of a constellation (3.1) ▼

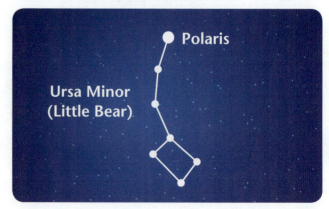

consumer any living thing that gets the energy it needs by eating producers or other consumers; dogs and cats are both consumers (1.2)

control a test that an experimenter carries out with no variable; this test can be compared to a test in which an independent variable is manipulated by the experimenter, to see whether the independent variable does indeed cause a change in the dependent variable (SST2)

current a flow of electrical charges; an electric current carries energy from a source (such as a battery) to an electrical device (such as a flashlight), along wires; electric current is measured in amperes; this measurement represents how many units of charge pass a point every second, as shown here (4.4) ▼

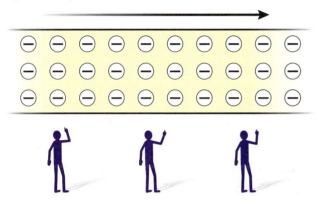

cycle a pattern of change that is continuous, or a process that repeats itself forever; for example, in the water cycle, shown here, water moves continuously through ecosystems (1.3) ▼

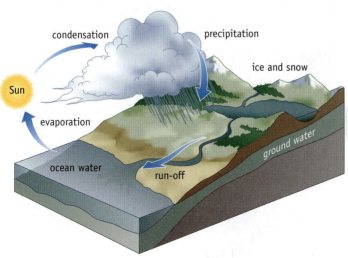

D

decomposer an organism that obtains energy by consuming dead plant and animal matter; mushrooms are one type of decomposer, which is why they are often found growing on decaying logs (1.3)

decomposition a kind of reactivity that can break down a substance into its parts; for example, water can be broken down into hydrogen and oxygen by decomposition (2.2)

deforestation the practice of clearing forests for logging or other human uses, and never replanting them (1.5) ▼

density describes how compact a substance is, and is calculated by dividing mass by volume, or D = m/v (2.2)

dependent variable the factor that is observed; the experimenter in a test looks for changes in the dependent variable in response to the independent variable (the one controlled by the experimenter); also called the responding variable (SST2)

digital meter a device that displays measured values, such as current or potential difference, directly as numbers on a small screen, similar to how a digital watch displays the time directly (SST9) ▶

discharged state of an object when it has lost its excess charge; for example, a doorknob with excess negative charge becomes discharged when it passes the excess charge to the hand of someone who touches the knob (4.3)

ductile easily stretched into a wire or hammered thin (2.3)

ductility ability to be stretched into a wire without snapping; copper has high ductility, so it can be stretched into wire (2.3) ▼

dwarf planet a rocky object whose gravitational pull isn't strong enough to keep other rocks out of its orbit; most are found in orbit beyond Neptune; Pluto is a dwarf planet (3.3)

E

eclipse the phenomenon in which one celestial object moves directly in front of another celestial object, as viewed from Earth; eclipses seen from Earth can be solar (when the Moon moves between the Earth and the Sun) or lunar (when the Earth moves between the Sun and the Moon) (3.2)

ecology the scientific study of the connections between everything on Earth, both biotic and abiotic (1.1)

ecosystem a system that is made up of all the interacting biotic and abiotic parts of a certain place (1.1)

electrical circuit the connection of a source, a load, and a conductor that allows a current to flow; an electrical circuit is always a closed path (4.4) ▼

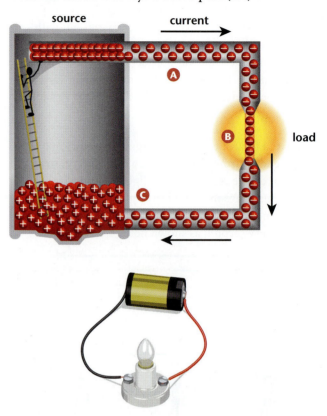

electrical current *see current*

electrical energy a form of energy given off when charged particles interact; like all forms of energy it can be changed into other forms to power electrical devices (4.1)

electrically neutral an object that has an equal number of positive charges and negative charges; this object is not electrically charged; for example, both the balloon and the person's hair in this diagram are electrically neutral because they both have equal numbers of positive and negative charges (4.2) ▼

electron a particle of an atom that surrounds the nucleus; has a charge of negative one (2.4)

electroscope a device for testing an object to find out if it is charged (4.3) ▼

element a pure substance that cannot be broken down into simpler parts by chemical methods; all the elements are listed on the Periodic Table; all of the particles in an element are identical, as shown below in the element aluminum (2.3) ▼

EnerGuide label a label that shows about how much energy an appliance uses in one year of normal use; this label is put on all new appliances sold in Canada (4.7) ▼

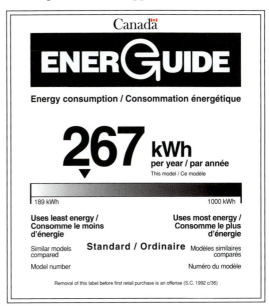

energy level the cloud-like region around the nucleus of an atom (2.4) ▼

Adding a specific amount of energy causes an electron to move to a higher energy level.

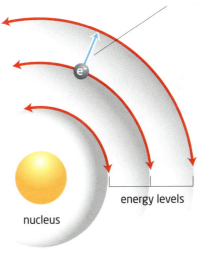

GLOSSARY • MHR 407

ENERGY STAR® label a label on an electrical appliance that means that the appliance or equipment meets or exceeds the government standards for electrical efficiency (4.7) ▼

environmental farm plan a volunteer-membership program in which farmers examine and make plans to reduce environmental impacts of farms (1.6)

equilibrium a stable, balanced, or unchanged system (1.6)

estimating making an informed judgement about a measurement; an estimate gives you an idea of a particular quantity but is not an exact measurement (SST4)

evaporation the change of state from a liquid to a gas; evaporation of the water in a puddle can cause the puddle to dry up (U2GR, SST2)

extinction the death of all of the individuals of a species; for example, dinosaurs became extinct millions of years ago (SST8)

F

fair test a test with only one independent (manipulated) variable; a that is valid and unbiased (SST2)

family a vertical column of elements in the periodic table also known as a group; Group 17 is shown here (5.3) ▼

food chain a model that describes how the energy that is stored in food is transferred from one living thing to another (1.2) ▼

floating algae (producer)
mosquito larva (consumer)
minnow (consumer)
perch (consumer)

food web a model that describes how energy in an ecosystem is transferred through two or more food chains (1.2) ▼

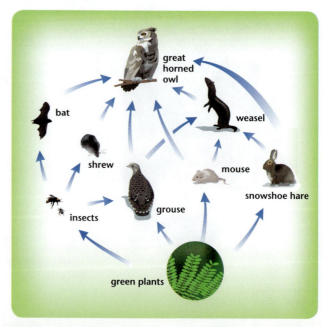

fuse a safety device that is found in older buildings and some appliances; like a circuit breaker, it keeps a circuit from carrying too much current, and is placed in series with other circuits that lead to appliances and outlets (4.6) ▼

G

galaxy a collection of many billions of stars, planets, gas, and dust, held together by gravity, as shown below; Earth is in the Milky Way galaxy (3.1) ▼

graphic organizer a chart or a combination of words and shapes that helps you to record and organize information; often used to show relationships between different ideas or pieces of information; graphic organizers include flowcharts, Venn diagrams, spider maps, and many more; a main idea web is shown here (LST5) ▼

gravitational pull the force of attraction that any two masses have for each other; the Sun's gravitational pull keeps Earth orbiting the Sun, and the gravity of the Earth pulls objects towards the center of the planet (3.1)

grounding connecting a conductor to Earth's surface so that charges can flow safely to the ground; a metal rod is used to ground the metal parts of large tank trucks while they are refueling, as shown (4.3) ▼

H

habitat area where an organism lives (1.6, SST8)

host the living thing on which a parasite feeds (1.4)

hypothesis a statement about an idea that you can test, based on your observations (SST2)

I

independent variable the factor that an experimenter changes in a test (also called the manipulated variable) (SST2)

inner planets the four planets in our solar system that are closest to the Sun: Mercury, Venus, Earth, and Mars; they are rocky, have cratered surfaces, are relatively small, and have no or few moons (3.3) ▼

GLOSSARY • MHR **409**

insulator material that does not allow heat and electricity to move through it easily; insulators are often made of plastic or rubber (2.6)

introduced species any species that has been introduced into and lives in an ecosystem where it is not found naturally; for example, purple loosestrife has invaded Ontario's wetlands, forcing out other plants that used to live there (1.5) ▼

issue a topic that can be seen from more than one point of view (SST1)

K

kilowatt (kW) a unit of electrical energy usage; 1 kW = 1000 W (4.7)

kilowatt hour (kWh) a measure of the number of watts of energy used in one hour (4.7)

L

lander a spacecraft designed to land on a celestial object; for example, the *Phoenix Mars Lander* has done some exploration of the surface of Mars (3.4) ▼

lightning rod a metal sphere or point that is attached to the highest part of a building and connected to the ground; protects buildings and people by allowing charges from lightning to travel safely from the air to the ground (4.3) ▼

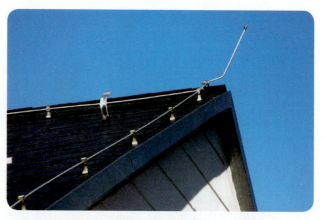

light-year a measurement equal to the distance that light travels in one year, about 9.5×10^{12} km; abbreviation is *ly* (3.1)

limiting factor any resource that limits the size to which a population can grow; food is a limiting factor to the growth of a population (1.4)

literacy the ability to identify, understand, analyze, create, and communicate using printed and written materials in various situations; for example, having the skills needed to view a video or read a newspaper article or web page, question the text, apply it to what you are learning in science, and express your opinion about it. (LST1-5)

load the part of an electrical circuit that requires electricity to work; an oven, a light bulb, and a computer are all examples of loads (4.4) ▼

lunar eclipse an event where Earth moves directly between the Sun and the Moon so that Earth casts its shadow on the Moon (3.2) ▼

lustre the ability to reflect light, or shine; silver has a high lustre and so is popular for jewellery; sodium metal is also lustrous, as shown here (2.2) ▼

M

magnetosphere the area of space that contains a planet's magnetic field; the magnetosphere offers Earth some protection from the damaging effects of the solar wind (3.2)

malleability ability of a substance to be bent or shaped by hammering without breaking; the malleability of steel makes it possible to make a sheet of steel into a car door panel or a trash can (2.3) ▼

manipulated variable the factor that an experimenter changes in a test (also called the independent variable) (SST2)

mass the amount of matter in a substance or object; mass is measured in milligrams, grams, kilograms, and tonnes (U2GR, 2.1, SST4)

matter anything that has mass and occupies space (volume) (2.1)

meniscus the slight curve where a liquid touches the sides of its container; to measure the volume of a liquid accurately, make sure your eye is at the same level as the bottom of the meniscus (SST4) ▼

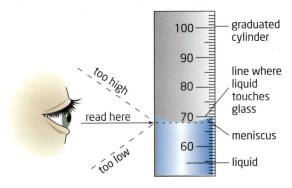

metal typically, an element that is solid at room temperature, shiny, malleable, ductile, and a good conductor; for example, aluminum, iron, calcium (2.3)

metalloid an element that shares some properties with metals and some properties with non-metals (2.4)

meteor [MEE-tee-uhr] a meteoroid that hits Earth's atmosphere and burns up, making a streak of light across the sky, as in the Leonid meteor shower, shown (3.3) ▼

GLOSSARY • MHR **411**

meteorite [MEE-tee-uhr-iht] a meteoroid that survives Earth's atmosphere and lands on Earth's surface; the crater shown below was formed by a meteorite (3.3) ▼

meteoroid [MEE-tee-uhr-oid] in space, a chunk of rock, metal, or both, shed from asteroids or comets; the streak of light in this image is a meteoroid passing through the Milky Way (3.3) ▼

Milky Way the galaxy that includes the solar system; looks like a hazy white band in the night sky (3.1) ▼

mixture matter that can be separated into parts using differences in physical properties; saltwater is a mixture of salt and water (2.3)

model anything that helps you understand a concept or process; can be a mental picture, a diagram, a structure, a working device, a chemical equation, or even a mathematical expression; can also be used to simplify a situation to allow you to test an hypothesis under conditions in which only one independent variable is changing (SST2, SST7)

molecule [MAWL-uh-kyul] a type of particle that is made up of two or more atoms bonded together; carbon monoxide is an example of a molecule (2.5) ▼

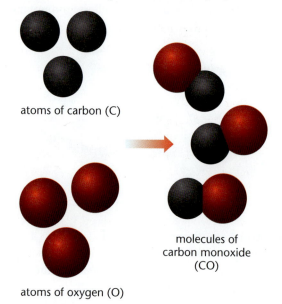

atoms of carbon (C)

atoms of oxygen (O)

molecules of carbon monoxide (CO)

multimeter a device that measures several different electrical quantities, including potential difference, current, and resistance (SST9) ▼

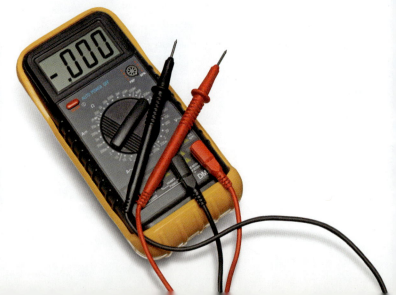

N

negative charges the type of electrical charges that can be rubbed off a material; negative charges are associated with electrons (4.2)

neutron located in the nucleus of an atom, has no charge (2.4)

non-metal typically, an element that is solid at room temperature, dull, brittle, not ductile, and a poor conductor (2.3)

non-renewable energy source a source of energy, such as fossil fuels and uranium, that cannot be replaced or restocked in a human lifetime (4.1)

nuclear fusion [NOO-klee-er FYUSH-uhn] the process of energy production in which hydrogen atoms collide violently to form a helium atom, releasing great amounts of energy; happens naturally in our Sun; nuclear fusion is occurring continuously in the Sun and creating huge amounts of heat and light; nuclear fusion reactions involving atoms of other elements, such as uranium, are used to produce nuclear energy, which then can be converted into electrical energy (3.2, 4.1)

nucleus in chemistry, the charged centre of an atom; contains the atom's protons and neutrons (2.4)

numeracy the ability to identify, understand, analyze, solve problems, apply solutions, and communicate using numbers, mathematical language, symbols, equations, and graphs; for example, having the skills needed to take information from a graph and use formulas, addition, subtraction, multiplication, and/or division to solve a problem and clearly express your answer (NST1-5)

nutrient a chemical that a living thing needs to live and grow; plants absorb nutrients from the soil in the form of minerals, such as potassium and phosphorus (1.3)

nutrient cycle the pattern of continual use and reuse of a nutrient; the carbon cycle is a nutrient cycle; a general nutrient cycle is shown below (1.3) ▼

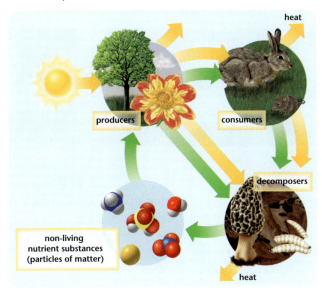

O

ohm (Ω) [OHM] the unit used to measure resistance in an electric circuit (4.4)

orbit a circular path, caused by gravitational pull, in which one object travels around another; for example, Earth travels in an orbit around the Sun (3.1) ▼

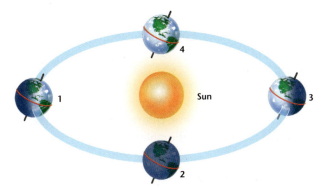

outer planets the four planets of our solar system that are farthest from the Sun: Jupiter, Saturn, Uranus, and Neptune; they are gassy with no solid surface, relatively large, and have many moons (3.3) ▼

P

parallel circuit a circuit that has two or more paths for the current to follow; the current in each branch in a parallel circuit is less than the current through the source (4.5) ▼

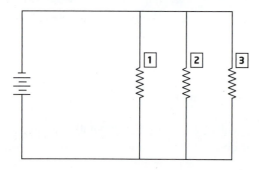

parasite a living thing that lives on or inside another living thing and uses it or its tissue for food or shelter; a tapeworm is a parasite (1.4)

period a horizontal row of elements in the periodic table; for example, sodium, magnesium, aluminum, silicon, phosphorus, sulfur, chlorine, and argon make up period 3 of the periodic table (2.4)

periodic table a system for organizing the elements into columns and rows, so that elements with similar properties are in the same column (2.4) ▼

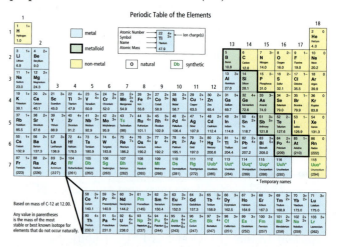

phantom load the electricity that is used by an appliance or device when it is turned off or in stand-by mode; some people are measuring the phantom load in their homes to find out how much energy they are wasting (4.7)

phases of the Moon the monthly changes in the amount of the *lit-up* side of the Moon we can see; the Moon's light is actually reflected sunlight (3.2) ▼

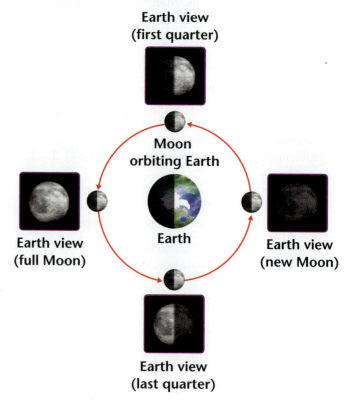

414 MHR • GLOSSARY

photosphere the surface layer of the Sun (3.2)

photosynthesis a process in the cells of plants, algae, and some bacteria that converts light energy from the Sun into stored chemical energy that can be used by other living things; the equation for this process, shown below, summarizes the materials necessary for and the materials produced by photosynthesis (1.2) ▼

physical property the way matter looks, feels, smells or tastes; one physical property of gold is its lustre (2.2)

plagiarism copying information word-for-word and then presenting it as though it is your own work (SST8)

planet an object in the sky that orbits one or more stars (and is not a star itself), is spherical, and does not share its orbit with another object (U3GR, 3.1, 3.3) ▼

population all the individuals of a species that live in a certain place at a certain time (1.4)

positive charges the type of electrical charges that are left behind when negative charges are rubbed off a material; positive charges are associated with protons (4.2)

potential difference used to describe the amount of energy a source can provide; the potential difference across a source is the difference between the energy of a unit of charge entering one end of a source, and the energy of a unit of charge leaving the other end of the source; potential difference is measured in volts, so its measurement is called voltage (4.4)

precipitate a solid substance that can form when certain dissolved substances are mixed together (2.2) ▼

precision describes both the exactness of a measuring device and the range of values in a set of measurements; the precision of a measuring instrument is usually half the smallest division on its scale (SST5)

predation a relationship between two different species in which one species feeds on another (1.4)

predator an organism that hunts, kills, and eats other organisms; for example, a lynx is a predator that eats hares (1.4) ▼

prediction a forecast about what you expect to observe; a prediction will help you decide if your hypothesis is correct (SST2)

prefix a part added to the beginning of a word that changes the meaning of a base word; for example, in the word *undo*, *un-* is a prefix (meaning *not*) that changes the meaning of the base word, *do*, to mean the opposite or reverse action (LST4)

prey an organism that is eaten as food by a predator; for example a snowshoe hare is prey to a lynx (1.4) ▼

producer any living thing that gets the energy it needs by making its own food; for example, grass is a producer that makes its food by photosynthesis (1.2)

proton [PROH-tawn] a particle of an atom that is inside the nucleus; has a positive charge (2.4)

pure substance a substance made up of only one kind of matter; for instance, copper, distilled water, and aluminum are all pure substances (2.3) ▼

copper particles

distilled water particles

aluminum particles

Q

qualitative observation an observation that can be described, but not measured using numbers; for example, "The ruler is brown," or "The ruler is long." (SST2)

quantitative observation an observation that can be measured and assigned a numerical value; for example, "The ruler is 30 cm long." (SST2)

R

radioactive the property of some elements to give off rays of energy as the element breaks down; these rays are given off as the result of a nuclear reaction, and can be harmful to living things (3.5)

reactivity the degree to which a substance can change (2.2)

renewable energy source a source of energy, such as water, that can be replaced or restocked in a human lifetime (4.1)

resistance a measure of how much a load pushes against a current in an electrical circuit; measured in the unit ohms (4.4)

responding variable the factor that is observed; the experimenter in a test looks for changes in the responding variable in response to the manipulated variable (the one controlled by the experimenter); also called the dependent variable (SST2)

robotics using machines to replace human actions; robotics are very important to the space program; an illustration of the *Mobile Base System* from the International Space Station is shown (3.4) ▼

rotation the turning of an object around an imaginary axis running through it; Earth's rotation around its axis takes 24 h (3.1, 3.2, 3.3) ▶

416 MHR • GLOSSARY

S

satellite human-made object or vehicle that orbits Earth, the Moon, or other celestial bodies; also, a celestial body that orbits another body of larger size; for example, the Moon is Earth's natural satellite; Canada's *NEOSSat,* shown in the illustration below, is used to monitor asteroids (3.2, 3.4) ▼

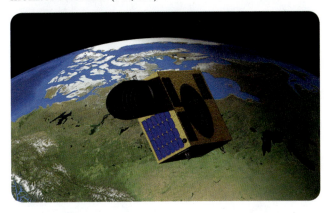

series circuit an electrical circuit that has only one path for the current to follow; in a series circuit, the current is the same at every point in the circuit (4.5) ▼

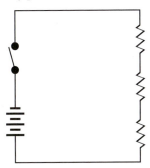

significant digits all the certain digits plus the first uncertain digit in a measurement; represent the amount of uncertainty in a measurement (NST1)

smart growth expansion of human communities by concentrating growth in the centre of a city, rather than in less populated areas; homes and businesses are found within the same areas, while parks and other natural areas are preserved (1.6)

smart meter a meter that measures total electrical energy used every hour and sends this information to the utility company automatically (4.7) ▼

solar eclipse an event where the Moon moves directly between the Sun and Earth so that the Moon casts its shadow on part of Earth (3.2) ▼

solar energy energy that is directly converted from the energy of the Sun into electricity (4.1)

solar system the system of planets, including Earth, moons, and other objects that orbit the Sun; in the illustration shown here, the eight planets are drawn to scale, but the distance between the planets is not drawn to scale (3.1) ▼

solar wind a stream of fast-moving charged particles that the Sun sends into the solar system; solar wind is deadly to living things; Earth's magnetic field protects us from this extreme energy (3.2)

solubility describes how much of a substance dissolves in another substance (2.2)

source a material whose energy is used to create electricity; moving water, fossil fuels (coal, oil, natural gas), uranium, wind, and the Sun are all sources of electrical energy; also, the device that supplies electrical energy to operate any electrical device; for example, a battery or an outlet (4.1, 4.4) ▼

species diversity the number and variety of species of living things in an area (1.5)

star a huge celestial body made of superheated gases (3.1) ▼

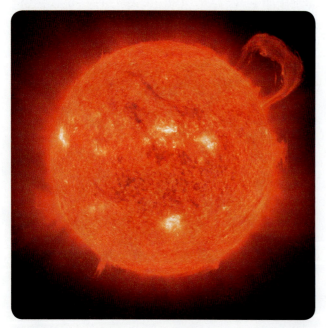

static electricity the result when positive and negative charges are separated; creates an excess of charges that stay where they are (4.2) ▼

sunspot an area of strong magnetic fields on the surface layer of the sun (photosphere); as shown, sunspots move as the Sun rotates (3.2) ▼

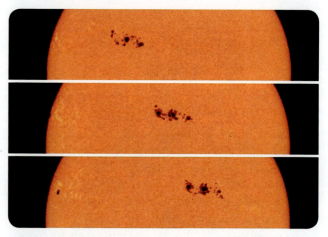

sustain to continue; to support (1.1)

sustainability maintaining an ecosystem so that present populations can get the resources they need without risking the ability of future generations to get the resources that they need (1.6)

sustainable describes practices that do not lead to long-term exhaustion of a resource (1.6, 4.7)

sustainable ecosystem an ecosystem that can withstand pressure and support a variety of organisms now and for the future (1.6)

suffix an ending that changes the meaning of a base word; for example, in the word *fearless*, *-less* is a suffix (meaning *without*) that changes the meaning of the base word *fear*, to give the meaning *without fear* (LST4)

switch a control device connected to an electrical circuit that completes or breaks the circuit; opening the switch stops the current and turns the item off (4.4) ▼

switch

T

technology the use of scientific knowledge, as well as everyday experience, to perform tasks or solve practical problems; for example, a computer is a form of technology that allows you to write reports, surf the Internet, play games, and perform many other tasks (SST3)

temperature a measure of the thermal (heat) energy of the particles of a substance; a measure of how hot or how cold something is (SST4)

terrestrial ecosystem an ecosystem that is based mostly or totally on land; forests are examples of terrestrial ecosystems (1.1) ▼

texture describes how the surface of a substance feels (its roughness, softness, or smoothness) (2.2)

theory a statement that explains why or how something happens; eventually, when an hypothesis has been thoroughly tested and nearly all scientists agree that the results support the hypothesis, it becomes a theory (SST2)

tidal energy energy captured from the movement of waves; for example the tidal energy from waves on the Atlantic Ocean is captured and converted to electrical energy at a generating station near the Bay of Fundy in Nova Scotia, Canada (4.1) ▼

U

universe all the celestial objects we see, and that can be seen, in the sky (3.1)

unsustainable a pattern of activity in an ecosystem that leads to the ecosystem not working well; in general terms, a pattern of activity or use that will lead to a shortage of resources in the future; for example, current use of fossil fuels will lead to these resources not being available to provide energy for future generations of people (1.6, 4.7)

urban sprawl the growth of relatively low-density development on the edges of urban areas; the spread of housing and businesses into less populated areas, taking over areas that were once home to wildlife (1.6)

V

variable any possible factor that could affect a test (SST2)

volt the unit to measure potential difference; the potential difference of a AA battery is 1.5 volts; the potential difference of a typical electrical outlet in your home is 120 volts (4.4)

voltmeter a device that measures volts (4.3, SST9) ▼

volume the amount of space that a substance or object occupies (U2GR, 2.1, SST4)

W

watershed any area of land (natural, human-made, or both) that drains into a body of water, as shown below; for example, water in the Hudson Bay watershed drains into the Hudson Bay (1.5) ▼

watt a unit of electrical energy usage; 1 kilowatt = 1000 W (4.7)

wind farm many large wind turbines placed in one location to capture wind energy; wind farms capture wind energy and convert it into electrical energy (4.1) ▼

Index

A

abiotic, 12
  ecosystems, 44–45
  human activities, effect of, 50–53
accuracy, 353
acetic acid, 140
acetylene, 131
acid rain, 142
Activity
  ammeter, 285
  aquatic ecosystems, 16
  Arctic ecosystem, 67
  atoms, building, 123
  battery size, 277
  building a periodic table, 128
  building atoms, 123
  building constellations, 177
  building molecules, 135
  Canadian astronauts/space explorers, 210, 213
  celestial orbits, 172
  changes in ecosystems over time, 60
  charge, 267
  charging an electroscope, 270
  charging by contact, 270
  charging by induction, 270
  chemical-free, 96
  chemicals, pros and cons, 98
  chemical properties, 109
  chemical reactions, 143
  circuit, 285
  circuit breaker, 309
  classifying elements, 116
  classifying galaxies, 176
  colony on another planet, 194
  conductor, 259
  connections, 11
  constellations, building, 177
  cycles, 33
  delivering electrical energy, 312
  discharge, 271
  ebonite, 260
  eclipses, modelling, 191
  ecosystem balance, 34
  ecosystem interactions, 14
  ecosystems, local, 16
  electroscope, 266
  elements and compounds, 138
  energy and food chain, 25
  energy source, 246
  food chain and energy, 25
  food webs, 67
  forest ecosystem, 67
  galaxies, classifying, 176
  grounding an electroscope, 271
  human activities, effect of, 50–53
  insulator, 258
  interactions and nutrient cycles, 31
  introduced species, 55
  limiting factors for ecosystems, 45
  mapping the solar system, 206
  modelling eclipses, 191
  molecules, building, 135
  Moon facts, 189
  Moon, distance, 205
  near earth objects, 204
  non-renewable energy sources, 249
  off-world Earths, 225
  patterns in the periodic table, 125
  physical properties, 107, 109
  physical properties of metals, 118
  physical properties of non-metals, 118
  pith ball, 265
  ponds, 12
  power, 312
  product labels, dangerous and hazardous, 145
  reading EnerGuide labels, 319
  recycling on Mars, 38
  renewable energy sources, 249
  responsibilities, 64
  retort stand, 265
  safety, 101
  solar system, mapping, 206
  static electricity, 260
  Sun facts, 185
  terrestrial ecosystems, 16
  time-of-use pricing, 317
  town council meeting, 73
  travelling bombs, 224
  types of cleaning products, 147
  units, 175
  voltmeter, 285
  watershed mind map, 57
aerospace mechanical technician, 219
Aldrin, Buzz, 227
algae growth, 46–47
algea, 22
aluminum, 94, 114
ammeter, 282, 285, 286–287, 355, 362
ammonia, 98, 142
ampere, 278, 281
analogies, 357
analogue meters, 362
Andromeda Galaxy, 175
angles, 350
antonym, 389
aquatic ecosystems, 14
  abiotic and biotic interactions, 30
  characteristics, 4–5
  watershed, 56
arborist, 79
area, 346–347
Armstrong, Neil, 227
aspen, 45
asteroid belt, 202
asteroids, 202, 204
astronomical unit (AU), 174
atmosphere, 186
  life, 187
atom, 122
atomic number, 122
atomic structure, 122
aurora, 186

B

Babbar, Vishvek, 284
bacteria, 48
baker, 152
baking soda, 140
Balakrishnan, Nishant, 228
bar graph, 372, 374
base words, 388
bat, 23
battery, 355
beaver, 13
bias, 336
  research based project, 359
Big Dipper, 171, 177
biocontrol, 68
biodiversity, 5, 66, 68
Bionic Energy Harvester, 321
biotic, 12
  ecosystems, 44–45
  human activities, effect of, 50–53
blue jay, 22
Bondar, Roberta, 210
Bowman, Patrick, 322
bromine, 133
Bullfrog Power, 250

C

Camelopardalis, 171
Canadarm2, 212
carbon, 130, 131, 132
carbon cycle, 36–37
  cellular respiration, 32
  photosynthesis, 32
carbon dioxide, 140
cardinals, 74
careers, 152–153, 218–219, 324–325
carrying capacity, 42
  ecosystems, 42–43
Case Study Investigation
  energy efficiency, 320–321
  salt, 136–137
  solar storms, 192–193
  songbirds, 74–75
  Ursa Major native tale, 178–179

Cassini-Huygens space probe, 224
Cassiopeia, 171
caterpillar, 45
cause-and-effect map, 394
celestial, 170
celestial sphere, 170
cell, 355
cellular respiration, 20
   carbon cycle, 32
   cycle, 32
   oxygen cycle, 32
   photosynthesis, 21
Celsius, 351
Cepheus, 171
ceramicist, 153
Ceres, 202
charge, 241, 252–260, 264
   contact by, 266
   discharge, 241
   flow of, 241, 274
   friction, by, 254
   induction, by, 241, 267
   like, 256
   movement of, 274
   negative, 241, 254–255, 276
   neutral, 254, 257
   opposite, 241, 256
   positive, 254–255, 276
   static electricity, 262
charging by contact, 266, 270
charging by induction, 267, 270
chemical formulas, 133-134
chemical properties, 108, 109, 110
   periodic table, 125
chemical reactions, 140–141, 142
   energy, 142
chemical symbols, 91, 132–133
chemical technician, 153
chemicals and energy, 97
chlorine, 115, 133, 144
circuit, 274, 280, 285
   connection, 241
   diagrams, 281
   load, 241
   source, 241
   symbols, 355
   symbols for diagrams, 281
circuit breakers, 240, 308
circuit diagrams, drawing, 355
circuit panel, 307
cleaning products, types of, 146, 147
combustibility, 108
comets, 203, 204
competition, 44
   populations, 45
compound, 90, 114
concept web, 391
conclusion, 343

conductivity, 106, 117, 258
conductor, 258
constellation, 171
construction electrician, 325
construction millwright, 325
consumers, 22
control, 341
   variables, 342
copper, 114, 132, 258
Corelli, Dayna, 58
curatorial assistant, 219
current, 278–279, 282, 283, 288–289, 290
   parallel circuit, 295
   series circuit, 294
current electricity, 241, 274, 276
cycle
   cellular respiration, 32
   nitrogen, 35
   photosynthesis, 32
cycle chart, 393

D

data, 342
   analyzing, 343
   organizing, 342
   presenting, 343
   recording, 342
   tables, 365
decomposer, 30
decomposition, 108
density, 106
dental assistant, 153
dependent variables, 341
Dextre (space robot), 212, 216
digital meters, 362
discharge, 264, 268, 271
Donelan, Max, 321
double bubble organizer, 394
Draco, 171
draftperson, 219
ductility, 117
Duimering, Adrienne, 150
dwarf planets, 202

E

Earth, 164, 174, 198, 199
Earth hour, 62
ebonite, 260
ebonite rod, 265
eclipses, 164, 190–191
eco-adventure guide, 79
ecology, 11
   careers, 78–79
ecosystems, 13
   abiotic, 4, 12, 44–45
   biotic, 4, 12, 44–45

   carrying capacity, 42–43
   connections, 10–11
   cycles, 28
   described, 8
   energy, 18–19
   growth, 5
   growth limits, 40
   health of, 54–55
   human activities, 34–35, 50–53
   introduced species, 54
   limiting factors, 42–43
   matter, 28–29
   nitrogen cycle, 35
   nutrient cycles, 34–35
   population, 42–43
   resource availability, 42–43
   size of, 11
   species diversity, 55
   watershed, 56
electric circuit, 280
   safety, 304–305, 310–311
electric motor system technician, 325
electrical energy, 278
electrically neutral, 254
electricity, 238–239, 244
   charges. See charges
   circuits. See circuits, parallel circuits, series circuits
   current. See current electricity
   demand, 314
   energy, 238, 240, 244, 246–247
   generation of, 244–250
   static. See static electricity
electrons, 122, 254
electroscope, 266
elements, 88–89, 90, 114
   number of, 130
EnerGuide label, 318
energy
   chemical reactions, 142
   chemicals, 97
   conservation, 314, 318–319
   ecosystems, 18
   electricity, 238, 240, 244, 246–247
   flows. See food chains; food webs
   food chain. See food chains; food webs
   food webs. See food chains; food webs
   non-renewable, 240, 244, 248
   photosynthesis, 20–21
   renewable, 240, 244, 248
   Sun, 184–185
   transfer between ecosystems, 4
   use of, 274
energy levels, 122
ENERGY STAR® label, 318

entertainment industry power technician, 325
environmental enforcement officer, 79
equilibrium, 66
Eris, 202
Essibrah, Florell, 152
estimation, 346, 367
European starlings, 54
Eves, Chandler, 78
experiments, 340–341
exponents, 366

F

fair test, 342
family, 124
fertilizer, 142
fishbone diagram, 392
flow of charges, 241
flowchart, 393
fluorine, 133
food chains, 4, 22, 24
    energy, 4, 24
food webs, 4, 22
    energy, 4
forest technologist, 79
fossil fuels, 248
fuses, 240, 309

G

Gagarin, Yuri, 227
galaxy, 173
    distances, 175
Gandhi, Mohandas, 136
Garneau, Marc, 210
Garney, David, 211
gas, 115, 131
gas giant planets, 200
gasoline, 131
generator, 246
geographic information systems specialist, 79
glass, 94, 258
global population, 40–41
gold, 120–121, 258
graphic organizers, 390–395
graphic text, reading, 385–387
graphs, 370–375
GRASP method of problem solving, 376
grass, 22
grasshopper, 22
gravitational pull, 172
gravity, 184
great horned owl, 23
green electricity, 250
ground fault interrupter (GFI), 311
ground(ing), 240, 269

safety, 310–311
grouse, 23

H

Hadfield, Chris, 210
hairstylist, 153
Hando, Tiffany, 324
Haumea, 202
horticultural technician, 79
Household Hazardous Products Symbol (HHPS), 144, 145
human activities in ecosystems, 34–35
hydroelectric energy, 247
hydrogen, 131, 133
hydrogen peroxide, 134
hypothesis, 340

I

independent variables, 341
induction, charging by, 241
information in research based project, 359
information technology hardware technician, 325
inner planets, 198
insulator, 258
International Space Station, 212
introduced species, 5, 54
Investigation
    ammeter, 286–287
    charging materials, 262
    chemical properties, 110
    current, 288–289, 290
    Dextre (space robot), 216
    ecosystem and energy flow, 26
    forest ecosystem, 26
    gas, identifying an unknown, 148–149
    green electricity, 250
    human activity in local ecosystem, 59
    lightning rods, 272
    limiting factors on growth, 46–47
    local environmental project, 76
    parallel circuit, 300–301
    physical properties, 110
    potential difference, 290
    resistance, 288, 289
    series circuit, 298–299
    star-finder wheel, 180
    voltmeter, 286–287
investigation, performing, 341
iodine, 133
iron, 132, 258
issues, 335
    action, 337
    alternatives, 336

    analyzing, 335
    bias, 336
    decision, 337
    identifying, 335, 336
    internet, 335–336
    research, 335–336

J

Jupiter, 174, 200, 201

K

Kelvin, 351
kilowatt, 316
kilowatt-hour, 316
kinetic energy, 246
Kolodziejczak, Jason, 218
KWL chart, 391

L

landscape technician, 79
Large Magellanic Cloud, 176
law of electrical charge, 256
lead, 132
light year, 175
lightning, 264, 269
limiting factors, 42, 43–45
    ecosystems, 42–43
line graph, 370, 374
literacy skills, 377
litres, 348
Little Dipper, 171
load, 274, 278, 280
    parallel circuit, 297
    phantom, 319
    series circuit, 296
lunar eclipses, 191
lustre, 106, 117
lye, 134
lynx, 44

M

machine-tool builder, 219
MacLean, Steve, 210
magnetosphere, 186
main idea web, 391
maintenance electrician, 325
Makemake, 202
malleability, 117
manipulated variable. *See* independent variable
Mars, 198, 199
mass, 349
matter, 88–89, 97
    ecosystems, 28–29
    universe, 97
measurement, 346

angles, 350
area, 346–347
mass, 349
temperature, 351
volume, 348
mechanical energy, 278
Mediouni, Sarah, 150
megawatts, 312
Meilhausen, Shelby, 228
meniscus, 348
mercury, 132
Mercury, 174, 198, 199
metals, 90, 117
meteorite, 203
meteoroids, 203, 204
meters, 241, 274
electrical usage, 316
reading, 362–363
methane, 98
metric system, 368–369
conversions, 369
micro electronics manufacturer, 325
microwaves, 184
Milky Way, 173
millilitres, 348
minnow, 22
Mobile Base Station, 213
models, 342, 356
molecules, 133-135
Moon, 164, 182–183, 188–189
phases of, 188
mosquito, 22
MOST (space telescope), 211
mouse, 23
MSDS, 100
multimeter, 363

N

natural gas, 131
near Earth objects, 204
negative charge, 254
NEOSSat (near earth object surveillance satellite), 211
Neptune, 174, 200, 201
network cabling specialist, 325
neutron, 122
night sky, 168–169
nitrogen, 132, 133
nitrogen cycle in ecosystems, 35
non-metals, 90, 117
non-renewable energy source, 248
Northern Lights, 186
nuclear-powered planetary probes, 224
nuclear reactor, 247
nucleus (of atom), 122
nutrient, 30

nutrient cycles, 30
ecosystems, 34–35

O

observations, 339
off-peak use, 317
ohm, 279, 281
orbit (of planets), 172
outer planets, 200
oxygen, 133
oxygen cycle
cellular respiration, 32
photosynthesis, 32

P

parallel circuit, 240, 292–294, 295, 296–297, 300–301
current, 295
load, 297
potential difference, 297
practical wiring, 306–307
parasites, 44
Parker, Rebekah, 58
Payette, Julie, 210
perch, 22
period, 124
periodic table, 91, 120, 124, 125-128
chemical properties, 125
patterns, 125
physical properties, 125
pictorial version, 126–127
simplified version, 124
PET plastics, 94
phantom load, 319
Phantom Torso, 214–215
Phoenix Mars Lander, 208–209, 211
phosphorus, 133
photosynthesis, 4, 20, 187
carbon cycle, 32
cellular respiration, 21
cycle, 32
energy, 20–21
oxygen cycle, 32
physical properties, 90, 104–105, 106, 107
common substances, 396–397
periodic table, 125
pie graph, 373, 375
Pietrzakowski, Katie, 322
pith ball, 265
plagiarism, 360
plastics, 94, 99, 131, 258
Pluto, 202
PMI chart, 390
Polaris, 171
pollution, 5
populations, 42

abiotic factors, 44–45
biotic factors, 44–45
competition, 45
ecosystems, 42–43
positive charge, 254
potassium, 132
potential difference, 240, 274, 276, 282, 283, 290
parallel circuit, 297
series circuit, 296
power, 312
powerline technician, 325
precipitate, 108
precision, 352
precision metal fabricator, 219
predators, 44
prediction, 340
prefixes, 388
prey, 44
producers, 22
Project CHIRP, 75
proton, 122, 254
pure substances, 90, 114
purple loosestrife, 55
PVA (polyvinyl alcohol) plastics, 99

Q

qualitative observation, 339, 340
quantitative observation, 339, 340

R

radar technician, 219
radiowaves, 184
reading effectively, 381–384
inferences, 382
scan, 383
skim, 383
study, 383
understanding, 384
renewable energy source, 248
research based project, 358–359
bias, 359
experts, 359
information, 359, 360
internet, 358
plagiarism, 360
presentation, 361
primary source, 359
reliability, 359
revision, 360
secondary source, 359
topic, 358
resistance, 241, 274, 279, 281, 288–289
resistor, 355
responding variable. *See* dependent variable

retort stand, 265
robotics technician, 219
rounding, 367

S

safety, 100–101
    electric circuit, 304–305, 310–311
    grounding, 310–311
salt, 115, 134
Saturn, 200, 201
science inquiry, 102
scientific drawing, 354–355
scientific inquiry, 339
scientific notation, 366
series circuit, 240, 292–293, 294, 295–297, 298–299
    current, 294
    load, 296
    potential difference, 296
shrew, 23
Siddiqui, Ghufran, 284
significant digits, 367
silver, 120, 132, 258
Smith, Peter H., 208
snowshoe hare, 23, 45
sodium, 115, 132
sodium acetate, 140
sodium bicarbonate, 140
solar eclipses, 190
solar storms, 192–193
solar system, 164, 172
    mapping, 206
    space exploration, 196–197
solubility, 106
songbirds, 74–75
source, 276, 280
space exploration, 162–163
    benefits from, 165, 220–228
    Canada's role, 165, 208–213
    human effects, 226–227
    solar system, 196–197
    technologies from, 222–223
space shuttle, 10
species diversity, 55
spectrum, 184
spider map, 392
star, 172
star-finder wheel, 180
static electricity, 238, 252, 260
    charge, 262
Su, Yvonne, 72
subatomic particles, 122
substances
    classification, 112
    dangerous, 144
Sudbury, 196–197
suffixes, 388

sugar, 114
sulfur, 133
Sun, 164, 172, 182–183
    characteristics, 184–185
    energy, 184–185
sustainability, 64
sustainable ecosystems, 2
    human activity, 2
    maintaining, 68–69
    promoting, 62
switch, 281

T

t-chart, 390
tare, 349
technological problem solving, 344–345
technology, 344
temperature, 351
terraforming, 225
terrestrial ecosystems, 14
    abiotic and biotic interactions, 30
    characteristics, 4–5
Tesla coils, 302
Tesla, Nicola, 302
texture, 106
thermoelectric energy, 247
thermometer, 351
Thirsk, Robert, 210
ticks, 44
time-of-use prices, 317
tin, 132
tool designer, 219
topic opener, 378
Tremblay, Isabelle, 208
Tryggvason, Bjarni, 210
turbine, 246

U

ultraviolet rays, 184, 187
unit opener, 377
universe, 172
    matter, 97
uranium, 248
Uranus, 200, 201
urban sprawl, 69
Ursa Major (Great Bear), 171
Ursa Minor (Little Bear), 171

V

variables, 341
    control, 342
Venn diagram, 393
Venus, 174, 198, 199
veterinary technician, 153
vinegar, 140

visible light, 184
volt, 281
voltmeter, 282, 285, 286–287, 355, 363
volume, 348
Voyager 1 & 2, 227

W

water, 114, 134
water quality technician, 79
watershed, 56
watt, 316
weasel, 23
welder, 153
white-tailed deer, 44
Whiteway, Jim, 211
WHMIS, 100, 101
wildlife technologist, 79
Williams, Dave, 210
word family webs, 389
word maps, 389
word study, 388–389

X

X rays, 184, 187

Y

yellow warbler, 45

Credits

Photo Credits

p2-3, 4-5 Kari Marttila/Alamy; p4 top James Woodson/Digital Vision/Getty Images; N. Lightfoot/IVY IMAGES; Visuals Unlimited/Corbis; p5 top David Coder/iStock; bottom Joseph Sohm/Visions of America/Corbis; p8-9 Russell Kord/Alamy; p8 centre James Woodson/Digital Vision/Getty Images, insets from top Andrew Syred/Photo Researchers, Inc.; David M. Phillips/Photo Researchers, Inc.; SPL/Photo Researchers, Inc.; Biophoto Associates/Photo Researchers, Inc.; p9 top left Steve & Dave Maslowski/Photo Researchers, Inc.; Edward H. Holsten/USD Forest Services; Bill Ivy; right D. Roitner/IVY IMAGES; p10 NASA; p13 Leo Wehrstedt, inset Jason Kasumovic/iStock; p15 top left Jim Zipp/Photo Researchers, Inc.; Ronald Wittek/age footstock/MaXx Images; right Bill Ivy; p17 Riccardo Savy/Getty Images; p18-19 WEJ Scenics/Alamy, inset David Coder/iStock Photo; p30 Natalie Fobes/Science Faction/Getty Images; p34 Jon Patton/iStock; p38 Photo Courtesy of Canadian Space Agency; p40 Oasis/Photos 12/Alamy; p41 COC/J Merrithew/THE CANADIAN PRESS; Image Source/Getty Images; p50-51 W. Sproul/IVY IMAGES; p53 top Ashley Cooper/Corbis; Russ Heinl/All Canada Photos; Bill Ivy; p54 Richard R. Hansen/Photo Researchers, Inc.; p55 Visuals Unlimited/Corbis; p56 Bill Ivy; p58 Photos courtesy of Dayna Corelli and Rebekah Parker; p60 Bill Aron/PhotoEdit; 62-63 Konstantin Dikovsky/iStock; p62 David Tanaka; p64-65 Yves Marcoux/First Light/Getty Images; p65 Wayne Levin/Photo Resource/Alamy; p68-69 Noah Poritz/Photo Researchers, Inc.; p68 top Photo courtesy of Suzanne Lafrance, Ontario Ministry of Natural Resources; N. Lightfoot/IVY IMAGES; Noah Poritz/Photo Researchers, Inc.; p69 Bill Ivy; p70-71 Kari Marttila/Alamy; p70 top Comstock/Corbis; T. Meyers/IVY IMAGES; p71 Sarah Ivy/IVY IMAGES; p72 Photos courtesy of Yvonne Su and Chaminade College; p74 top Steve Byland/iStock; Bill Ivy; p75 T. Parent/ResourceEye Services Inc.; p77 Leo Wehrstedt; p78 Photo courtesy of Chandler Eves; p79 left O. Bierwagon/IVY IMAGES; T. Willis/iStock; Bill Ivy; p80 top James Woodson/Digital Vision/Getty Images; David Coder/iStock Photo; p81 middle Visuals Unlimited/Corbis; N. Lightfoot/IVY IMAGES; p82 Sebastien Cote/iStock; p83 Joseph Sohm/Visions of America/Corbis; Roel Smart/iStock; p88-91 © 2009 periodictable.com; p90 top Bill Ivy; Ray Ellis/Photo Researchers, Inc.; David Tanaka; p91 top Skip ODonnell/iStock; S. McCutcheon/IVY IMAGES; *The Raven and the First Men*, 1990, 22k gold, lost wax technique, textured 7cm x 6.9cm x 5.5cm, Bill Reid Foundation Collection BRFC#25,Gift of Dr. Martine Reid with assistance from the Government of Canada and the BRF Trustees, Photo Credit: Kenji Nagai; Mark Joseph/Getty Images; p91 bottom Mark Joseph/Getty Images; p94-95 Tim Graham/Getty Images; p94 Bill Ivy; p95 top row Phil Augustavo/iStock; Grant Dougall/iStock; Michael-John Wolfe/iStock; Lenscraft Imaging/iStock; bottom Pattie Calfy/iStock; Cordelia Molloy/Photo Researchers Inc.; Kshishtof/iStock; p96 Ian Chrysler; K. Bruce Lane Photography; p97 Julie Deshaies/iStock; p99 Helene Rogers/Alamy; p104-105 Bill Ivy; p104 top Skip ODonnell/iStock; Andrea Skjold/iStock; p105 Falk Kienas/iStock; David Tanaka; bottom David Tanaka; p106 2nd from top McPhoto/KPA/maXx Images; Stephanie Tomlinson/iStock; Alexey Kuznetsov/iStock; Tim Pieloth/iStock; p108 top Charles D. Winters/Photo Researchers, Inc.; David Tanaka; David Tanaka; Richard Megna/Fundamental Photographs NYC.; Charles D. Winters/Photo Researchers, Inc.; p111 Eddie Gerald/Alamy; p112-113 S. McCutcheon/IVY IMAGES; p113 Guy Croft SciTech/Alamy; p114 Charles V. Angelo/Photo Researchers Inc.; Maximilian Stock Ltd.; G.P. Bowater/Alamy; p115 Martyn F. Chillmaid/Photo Researchers Inc.; Charles D. Winters/Photo Researchers, Inc.; The McGraw-Hill Companies, Inc./Stephen Frisch, photographer; p116 Richard Megna/Fundamental Photographs NYC; p117 The McGraw-Hill Companies, Inc./Stephen Frisch, photographer; p120 *Grizzly Bear Bracelet*, 1975, silver, engraved, 2.2 cm x 5.5, Bill Reid Foundation Collection BRFC#8, Gift of Dr. Martine Reid, Photo Credit: Kenji Nagai; p121 top Tom Hanley/Alamy; *The Raven and the First Men*, 1990, 22k gold, lost wax technique, textured 7cm x 6.9cm x 5.5cm, Bill Reid Foundation Collection BRFC#25, Gift of Dr. Martine Reid with assistance from the Government of Canada and the BRF Trustees, Photo Credit: Kenji Nagai; p126-127 (c) 2009 periodictable.com; p129 E.R. Degginger/Photo Researchers, Inc.; p130-131 Pali Rao/iStock; p130 top Chris Knapton/Photo Researchers, Inc.; David Tanaka; p131 top left David Tanaka; Mark Evans/iStock; middle left Fred Chartrand/THE CANADIAN PRESS; Jason Verschoor/iStock; Khanh Trang/iStock; bottom row Diane Diederich/iStock; Izabela Habur/iStock; Jorgen Udvang/iStock; p134 David Tanaka; p136-137 Dirk Wiersma/Photo Researchers, Inc.; p136 top Central Press/Hulton Archive/Getty Images; p137 top Diane Diederich/iStock; Sawayasu Tsuji/iStock; p138 Richard Megna/Fundamental Photographs NYC; Photo courtesy of Canadian Space Agency; Phil Degginger/Alamy; p139 Charles D. Winters/Photo Researchers, Inc.; p140-141 Ray Ellis/Photo Researchers, Inc.; p141 David Tanaka; p142 left MistikaS/iStock; Grant Heilman Photography; p143 top David Tanaka; NASA; left David Tanaka; Christina Ivy/IVY IMAGES; left Matt Meadows/Peter Arnold/Alamy; Chris Rose/Alamy; p148-149 Dave Starrett; p150 Photos courtesy of Adrienne Duimering and Sarah Mediouni; p151 Dave Starrett; p152 Photos courtesy of Kirk Kolas; p153 left erel photography/iStock; jeffrey hochstrasser/iStock; Christian Carroll/iStock; p154 top Bill Ivy; Skip ODonnell/iStock; S. McCutcheon/IVY IMAGES; p155 *The Raven and the First Men*, 1990, 22k gold, lost wax technique, textured 7cm x 6.9cm x 5.5cm, Bill Reid Foundation Collection BRFC#25, Gift of Dr. Martine Reid with assistance from the Government of Canada and the BRF Trustees, Photo Credit: Kenji Nagai; middle David Tanaka; Ray Ellis/Photo Researchers, Inc.; p156 Bill Ivy; p157 Mark Joseph/Getty Images; p158 Charles D. Winters/Photo Researchers, Inc.; p162-165 The Picture Desk; p164-165 NASA; p165 lower NASA/Photo Researchers Inc.; p166-167 NASA; p169 NASA/JPL-Caltech; p170 David Tanaka; p173 SOHO/EIT; p175 NASA; p176 top left NOAO/AURA/NSF, NOAO/AURA/NSF/02326, Mark Westmoquette, University College London, Jay Gallagher, University of Wisconsin- Madison, Linda Smith University College London WIYN/NSF/NASA/ESA/04872, NOAN/AURA/NSF/02329, C. Smith, S. Points, the MCELS Team and NOAO/AURA/NSF/LMC, Bill Schoening/NOAO/AURA/NSF/02676, NOAO/AURA/NSF/02328, N.A. Sharp/NOAO/AURA/NSF; p180 David Tanaka; p182-183 Mike Agliolo / Photo Researchers, Inc.; p184-185 NASA/Photo Researchers Inc.; p186 Chuck Graves of Majic Country; pp 188-189 NASA; p190 left Jerry Schad/Photo

Researchers, Inc.; John R. Foster/Photo Researchers, Inc.; p191 NASA; p192 top Getty Images; NASA; p193 Cary Anderson/Getty Images; pp197-202 NASA; p203 left Jerry Lodriguss/Photo Researchers, Inc.; Used by permission of Keon Miskotte/NASA; p204 E. De Jong and S. Suzuki, JPL, NASA; p208-209 NASA; p208 NASA/UA/Isabelle Tremblay, Canadian Space Agency; p209-210 Canadian Space Agency; p211 top NASA; NASA; AGI Inc./Dynacon Inc./University of Calgary; p212-213 NASA; p212 MD Robotics; p213 NASA; p218 Photo courtesy of Jason Kolodziejczak; David Furst/AFP/Getty Images; p219 NASA; p220-221 David Tanaka; p221 Piotr Rydzkowski/iStock; NASA; p224 NASA; p225 Detley van Ravenswaay/Photo Researchers, Inc.; p226 top Dr. Michael A. Rappenglück, M. A., Gilching, Germany; Sisse Brimberg/National Geographic/Getty Images; p227 NASA; p228 Photo courtesy of Shelby Mielhausen. Copyright Ethan Meleg, André Van Vugt, courtesy of Youth Science Canada. Used by permission of Nishant Balakrishnan; pp230-231 NASA; p232 NASA; p233 NASA; p235 Asbjorn Aakjaer/iStock; pp 238-241 Mark Burnham/First Light; p240 top Masterfile; PKruger/iStock; David Tanaka; p241 top David Parsons/iStock; David Tanaka; Paul Edmondson/Getty Images; p243 Rene Mansi/iStock; pp244-245 Lester Lefkowitz/CORBIS, inset Daniel Timiraos/iStock; p248 Rolf Hicker Photography/Alamy; Jarek Szymanski/iStockphoto; p250 Liz Richardson; Masterfile; Reimar/Alamy; p252 Bill Ivy; p257 Bill Ivy; Theresa Sakno; p258 David Tanaka; pp264-265 David Parsons/iStock; p269 top Masterfile; Mediacolor's/Alamy; pp270-271 David Parsons/iStock; pp274-275 Planetary Visions Ltd/Photo Researchers, Inc; pp276, 277, 279, David Tanaka; p284 Photos courtesy of Vishvek Babbar and Ghufran Siddiqui; p302 Bettman/CORBIS; p308 top Tom Kelley/Getty Images; David Tanaka; pp309, 310, 311 David Tanaka; p312 Bill Ivy; p315 top left row Jamie Grill/Blend Images/Corbis, Rich Legg/iStock; nathan winter/iStock; middle row left Zave Smith/UpperCut Images/Alamy; Tomaz Levstek/iStock; Nick M. Do/iStock; Bottom row left Webphotographeer/iStock; David Tanaka; p316 left Bill Ivy; Pkruger/iStock; LiJA/iStock; Geoffrey Hammond/iStock; David Falconer/Corbis; p317 Bill Ivy; p318, 319 David Tanaka; p320-321 eric lefrancois/iStock; p321 Courtesy of Max Donelan; p322 Photos courtesy of Katie Pietrzakowski and Patrick Bowman; p324 Photo courtesy of Tiffany Hando; Lester Lefkowitz/Getty Images; p325 left Tom Paiva/Taxi/ Getty Images; Sandi Krasowski/THE CANADIAN PRESS; Mychele Daniau/AFP/Getty; p326 top Masterfile; David Parsons/iStock; David Tanaka; p327 David Tanaka; Pkruger/iStock; p329 Paul Edmondson/Getty Images; Jim West/Alamy. p3 From "Waiting on the World to Change" by John Mayer. Hal Leonard Corporation. Permission Pending. p65 Source: Kanatiio (Gabriel, Allen), "A Thanksgiving Address" in Report of the Royal Commission of Aboriginal Peoples. V. 1. Looking Forward, Looking Back (Ottawa: Supply and Services, 1996) p.xxi. Reproduced with the permission of the Minister of Public Works and Government Services, 2009, and courtesy of the Privy Council Office. p89 "The Elements" by Tom Lehrer, used by permission of the author. p163 "Moonshot" by Buffy Sainte Marie. Universal Music Publishing. Permission Pending. p 239 "Electricity" by OMD, Hal Leonard. Permission Pending.

Front Matter credits.
iv-v Yves Marcoux/First Light/Getty Images; v James Woodson/Digital Vision/Getty Images.; N. Lightfoot/IVY IMAGES; Steve Byland/iStock; vi – vii Pali Rao/iStock; vi illus., Tom Hanley/Alamy; Ray Ellis/Photo Researchers, Inc; vii Jerry Lodriguss/Photo Researchers, Inc.; NASA; NASA; viii-ix eric lefrancois/iStock; viii Masterfile; David Parsons/iStock; Bettman/CORBIS; illus; xii Michael Thompson/iStock; xiii Stockbyte/iStock; John Clines/iStock; xiv SW Productions/Getty Images;

Back Matter Credits
p353 © 2007 Getty Images, Inc.; p358 (c) Eyewire/Getty Images; p359 © Photodisc/PunchStock; p360 (c) Stockbyte/Getty Images; p361 Buccina Studios/Getty Images p377 (reproduction of p2-3 Kari Marttila/ Alamy); p378 (reproduction of p8-9 Russell Kord/Alamy; p8 centre James Woodson/Digital Vision/Getty Images, insets from top Andrew Syred/Photo Researchers, Inc.; David M. Phillips/Photo Researchers, Inc.; SPL/Photo Researchers, Inc.; Biophoto Associates/Photo Researchers, Inc.; p9 top left Steve & Dave Maslowski/Photo Researchers, Inc.; Edward H. Holsten/USD Forest Services; Bill Ivy; right D. Roitner/IVY IMAGES); p379 (reproduction of p12-13 Leo Wehrstedt, inset Jason Kasumovic/iStock); p380 NASA; p381 (reproduction of p54-55 Richard R. Hansen/Photo Researchers, Inc., Visuals Unlimited/Corbis

Back Matter Illustration Credits
p335 bottom left The McGraw-Hill Companies

GLOSSARY
p400 top left sciencephotos/Alamy; bottom right Leo Wehrstedt, inset Jason Kasumovic/iStock; p401 middle left © Tom Van Sant/CORBIS; top right Cary Anderson/ Getty Images; p402 top left Noah Poritz/Photo Researchers, Inc.; bottom left Photo courtesy of Suzanne Lafrance, Ontario Ministry of Natural Resources; p403 top right David Tanaka; bottom right David Tanaka; p 404 bottom left Jerry Lodriguss/Photo Researchers, Inc.; top right Maximilian Stock Ltd.; top right W.Sproul/IVY IMAGES; bottom right McGraw-Hill Ryerson; p406 left Charles V. Angelo/Photo Researchers Inc.; p407 top left ©Visuals Unlimited/Corbis; bottom left G.P. Bowater/Alamy; top right David Tanaka; p408 left David Tanaka; p409 top left David Tanaka; middle left NASA; top right Masterfile; bottom right NASA; p410 top left Visuals Unlimited/CORBIS; bottom left NASA/JPL/UA/Lockheed Martin; top right Mediacolor's/Alamy; p411 middle left (c) Andrew Lambert Photograph/Photo Researchers, Inc.; bottom left Ryan McVay/Getty Images; bottom right B.A.E. Inc./Alamy; p412 top left StockTrek/Getty Images, inset © Dr. Parvinder Sethi; middle left StockTrek/Getty Images; bottom left Baback Tafreshi / Photo Researchers, Inc; bottom right JUPITERIMAGES / Polka Dot / Alamy; p415 bottom left NASA; top right Richard Megna/Fundatmental Photographs NYC; p416 bottom left Charles V. Angelo/Photo Researchers Inc., Maximilian Stock Ltd, G.P. Bowater/Alamy; top right NASA; p417 top left AGI Inc./Dynacon Inc./University of Calgary; top right Bill Ivy; p418 bottom left NASA/Photo Researchers, Inc., top right Bill Ivy; bottom right SOHO/ESA/NASA; p419 top right Stephen Saks Photograph/Alamy; p420 middle Bill Ivy; bottom Todd Arbini Photography/iStockphoto

Glossary Illustration Credits

p400 bottom left McGraw-Hill Ryerson; p401 top left Argosy Publishing; bottom right Neil Stewart; p410 bottom right Ralph Voltz; p419 bottom right David Wysotski; p420 top Ralph Voltz

Illustration Credits

p. 11 Ralph Voltz; p. 14 Phil Wilson; p. 20 Deborah Crowle; p. 21 Deborah Crowle; p. 22, 84 Phil Wilson; p. 25 Neil Stewart; p. 36 Emmanuel Ceriser; p. 37 Emmanuel Ceriser; p. 48 Eric Kim; p. 55 Phil Wilson; p. 59 Rob Schuster; p. 78 Joe LeMonnier; p. 81 Eric Kim; p. 102 Eric Kim; p. 147 Steve Attoe; p. 171 Argosy Publishing; p. 173 Argosy Publishing; p. 174 Argosy Publishing; p. 178-179 Rob Schuster; p. 187, 195 Ralph Voltz; p. 197 Joe LeMonnier; p. 197 Rob Schuster; p. 198, 200 Argosy Publishing; p. 214 Emmanuel Ceriser; p. 215 Emmanuel Ceriser; p. 222–223 Eric Kim; p. 253 Ralph Voltz; p. 255 Ralph Voltz; p. 256 Ralph Voltz; p. 257 Neil Stewart; p. 262 Ralph Voltz; p. 263 Neil Stewart; p. 264 Ralph Voltz; p. 266 Ken Batelman; p. 267 Ken Batelman; p. 268 Joe LeMonnier; p. 268 Ralph Voltz; p. 272 Deborah Crowle; p. 273 Ralph Voltz; p. 277 Neil Stewart; p. 278 Neil Stewart; p. 279 Deborah Crowle; p. 280 Deborah Crowle; p. 292 Neil Stewart; p. 293 Neil Stewart; p. 294 Neil Stewart; p. 295 Neil Stewart; p. 304 Ralph Voltz; p. 305 Ralph Voltz; p. 306 Ken Batelman; p. 307 Deborah Crowle; p. 314 Joe LeMonnier; p. 331 Ken Batelman; p. 332 Ken Batelman

Periodic Table of the Elements

Legend	
■ metal	
■ metalloid	
■ non-metal	

Atomic Number → 22, 4+/3+ ← Ion charge(s)
Symbol → **Ti**
Name → Titanium
Atomic Mass → 47.9

O natural
Db synthetic

	1	2	3	4	5	6	7	8	9	10	11	12	13	14	15	16	17	18
1	1 1+ **H** Hydrogen 1.0																	2 0 **He** Helium 4.0
2	3 1+ **Li** Lithium 6.9	4 2+ **Be** Beryllium 9.0											5 3+ **B** Boron 10.8	6 **C** Carbon 12.0	7 3− **N** Nitrogen 14.0	8 2− **O** Oxygen 16.0	9 1− **F** Fluorine 19.0	10 0 **Ne** Neon 20.2
3	11 1+ **Na** Sodium 23.0	12 2+ **Mg** Magnesium 24.3											13 3+ **Al** Aluminum 27.0	14 4+ **Si** Silicon 28.1	15 3−/5+ **P** Phosphorus 31.0	16 2− **S** Sulfur 32.1	17 1− **Cl** Chlorine 35.5	18 0 **Ar** Argon 39.9
4	19 1+ **K** Potassium 39.1	20 2+ **Ca** Calcium 40.1	21 3+ **Sc** Scandium 45.0	22 4+/3+ **Ti** Titanium 47.9	23 5+/4+ **V** Vanadium 50.9	24 3+/2+ **Cr** Chromium 52.0	25 2+/3+/4+ **Mn** Manganese 54.9	26 3+/2+ **Fe** Iron 55.8	27 2+/3+ **Co** Cobalt 58.9	28 2+/3+ **Ni** Nickel 58.7	29 2+/1+ **Cu** Copper 63.5	30 2+ **Zn** Zinc 65.4	31 3+ **Ga** Gallium 69.7	32 4+ **Ge** Germanium 72.6	33 3+/5+ **As** Arsenic 74.9	34 2− **Se** Selenium 79.0	35 1− **Br** Bromine 79.9	36 0 **Kr** Krypton 83.8
5	37 1+ **Rb** Rubidium 85.5	38 2+ **Sr** Strontium 87.6	39 3+ **Y** Yttrium 88.9	40 4+ **Zr** Zirconium 91.2	41 3+/5+ **Nb** Niobium 92.9	42 2+/3+ **Mo** Molybdenum 95.9	43 7+ **Tc** Technetium (98)	44 4+/3+ **Ru** Ruthenium 101.1	45 3+ **Rh** Rhodium 102.9	46 2+/4+ **Pd** Palladium 106.4	47 1+ **Ag** Silver 107.9	48 2+ **Cd** Cadmium 112.4	49 3+ **In** Indium 114.8	50 4+/2+ **Sn** Tin 118.7	51 3+/5+ **Sb** Antimony 121.8	52 2− **Te** Tellurium 127.6	53 1− **I** Iodine 126.9	54 0 **Xe** Xenon 131.3
6	55 1+ **Cs** Cesium 132.9	56 2+ **Ba** Barium 137.3	57 3+ **La** Lanthanum 138.9	72 4+ **Hf** Hafnium 178.5	73 5+ **Ta** Tantalum 180.9	74 6+ **W** Tungsten 183.8	75 4+/7+ **Re** Rhenium 186.2	76 4+/7+ **Os** Osmium 190.2	77 3+/4+ **Ir** Iridium 192.2	78 4+/2+ **Pt** Platinum 195.1	79 3+/1+ **Au** Gold 197.0	80 2+/1+ **Hg** Mercury 200.6	81 1+/3+ **Tl** Thallium 204.4	82 2+/4+ **Pb** Lead 207.2	83 3+/5+ **Bi** Bismuth 209.0	84 2+/4+ **Po** Polonium (209)	85 1− **At** Astatine (210)	86 0 **Rn** Radon (222)
7	87 1+ **Fr** Francium (223)	88 2+ **Ra** Radium (226)	89 3+ **Ac** Actinium (227)	104 **Rf** Rutherfordium (261)	105 **Db** Dubnium (262)	106 **Sg** Seaborgium (263)	107 **Bh** Bohrium (262)	108 **Hs** Hassium (265)	109 **Mt** Meitnerium (266)	110 **Ds** Darmstadtium (281)	111 **Rg** Roentgenium (272)	112 **Uub*** Ununbium (285)	113 **Uut*** Ununtrium (284)	114 **Uuq*** Ununquadium (289)	115 **Uup*** Ununpentium (288)	116 **Uuh*** Ununhexium (292)		118 **Uuo*** Ununoctium (294)

* Temporary names

58 3+/4+ **Ce** Cerium 140.1	59 3+/4+ **Pr** Praseodymium 140.9	60 3+ **Nd** Neodymium 144.2	61 3+ **Pm** Promethium (145)	62 3+ **Sm** Samarium 150.4	63 3+/2+ **Eu** Europium 152.0	64 3+ **Gd** Gadolinium 157.3	65 3+/4+ **Tb** Terbium 158.9	66 3+ **Dy** Dysprosium 162.5	67 3+ **Ho** Holmium 164.9	68 3+ **Er** Erbium 167.3	69 3+/2+ **Tm** Thulium 168.9	70 3+/2+ **Yb** Ytterbium 173.0	71 3+ **Lu** Lutetium 175.0
90 4+ **Th** Thorium 232.0	91 5+/4+ **Pa** Protactinium 231.0	92 6+/4+/5+ **U** Uranium 238.0	93 5+/3+/4+/6+ **Np** Neptunium (237)	94 4+/6+/3+/5+ **Pu** Plutonium (244)	95 3+/4+/5+/6+ **Am** Americium (243)	96 3+ **Cm** Curium (247)	97 3+ **Bk** Berkelium (247)	98 3+ **Cf** Californium (251)	99 3+ **Es** Einsteinium (252)	100 3+ **Fm** Fermium (257)	101 2+/3+ **Md** Mendelevium (258)	102 2+/3+ **No** Nobelium (259)	103 3+ **Lr** Lawrencium (262)

Based on mass of C-12 at 12.00.

Any value in parentheses is the mass of the most stable or best known isotope for elements that do not occur naturally.